Molecular Methods of Plant Analysis

Editors:
J.F. Jackson (Managing Editor)
H.F. Linskens
R.B. Inman

Volume 21

Springer

Berlin
Heidelberg
New York
Barcelona
Hong Kong
London
Milan
Paris
Tokyo

Volumes Already Published in this Series (formerly "Modern Methods of Plant Analysis"):

Analysis of Taste and Aroma

Edited by
J.F. Jackson and H.F. Linskens

With 72 Figures

Springer

Prof. JOHN F. JACKSON
Dept. of Horticulture,
Viticulture and Oenology
Waite Campus
SA 5064 Glen Osmond
Australia

Prof. HANS F. LINSKENS
Goldberglein 7
91056 Erlangen
Germany

Prof. ROSS B. INMAN
Dept. of Molecular Virology and Biochemistry
University of Wisconsin
Madison, Wisconsin
USA

ISSN 0937-8340
ISBN 3-540-41753-2 Springer-Verlag Berlin Heidelberg New York

Library of Congress Cataloging-in-Publication Data

Analysis of taste and aroma / edited by J.F. Jackson and H.F. Linskens.
p. cm. – (Molecular methods of plant analysis ; v. 21)
Includes bibliographical references.
ISBN 3540417532 (alk. paper)
1. Taste – Molecular aspects. 2. Smell – Molecular aspects. 3. Plants – Analysis. 4. Botanical chemistry. I. Jackson, J.F. (John F.), 1935– II. Linskens, H.F. (Hans F.), 1921– III. Series.
QK865 .M57 vol. 21
[QP456]
571.2′028 s – dc21
[572′.362] 2001042002

Springer-Verlag Berlin Heidelberg New York
a member of BertelsmannSpringer Science + Business Media GmbH

http://www.springer.de

Printed in Germany

Production: PRO EDIT GmbH, Heidelberg, Germany
Cover design: design & production GmbH, Heidelberg, Germany
Typesetting: Best-set Typesetter Ltd., Hong Kong

Printed on acid-free paper SPIN 10795738 11/3130/Di 5 4 3 2 1 0

Preface

Molecular Methods of Plant Analysis

Concept of the Series

The powerful recombinant DNA technology and related developments have had an enormous impact on molecular biology. Any treatment of plant analysis must make use of these new methods. Developments have been so fast and the methods so powerful that the editors of *Modern Methods of Plant Analysis* have now decided to rename the series *Molecular Methods of Plant Analysis*. This will not change the general aims of the series, but best describes the thrust and content of the series as we go forward into the new millennium. This does not mean that all chapters a priori deal only with the methods of molecular biology, but rather that these methods are to be found in many chapters together with the more traditional methods of analysis which have seen recent advances. The numbering of the volumes of the series therefore continues on from 20, which is the most recently published volume under the title *Modern Methods of Plant Analysis*.

As indicated for previous volumes, the methods to be found in *Molecular Methods of Plant Analysis* are described critically, with hints as to their limitations, references to original papers and authors being given, and the chapters written so that there is little need to consult other texts to carry out the methods of analysis described. All authors have been chosen because of their special experience in handling plant material and/or their expertise with the methods described. The volumes of the series published up to now fall into three groups: Volumes 1–5 and Volume 11 dealing with some basic principles of methods, Volumes 6, 7, 8, 10, 14, 16, 18 and 20 being a group determined by the raw plant material being analysed, and a third group comprising Volumes 9, 12, 13, 15, 17 and 19 which are separated from the other volumes in that the class of substances being analysed for is indicated in the volume title. Volume 21 and future volumes of *Molecular Methods of Plant Analysis* will continue in a similar vein but will include more chapters involved with the methods of molecular biology.

Development of the Series

The handbook, *Modern Methods of Plant Analysis*, was first introduced in 1954, and was immediately successful, seven volumes appearing between 1956 and 1964.

This first series was initiated by Michael Tracey of Rothamsted and Karl Paech of Tübingen. The so-called *New Series of Modern Methods of Plant Analysis*, Volumes 1–20, began in 1985 and has been edited by Paech's successor, H.F. Linskens of Nijmegen, The Netherlands, and John F. Jackson of Adelaide, South Australia. These same editors have now teamed up with a third, Ross B. Inman of Madison, Wisconsin, USA, to produce the renamed series *Molecular Methods of Plant Analysis*. As before, the editors are convinced that there is a real need for a collection of reliable, up-to-date methods of plant analysis covering large areas of applied biology ranging from agricultural and horticultural enterprises to pharmaceutical and technical organizations concerned with material of plant origin.

Future volumes will include such topics as Testing for Genetic Manipulation in Plants, Genetic Transformation of Plants and Various Aspects of Plant Genomics.

Volume 21: ***Taste and Aroma***

Chapters dealing in many cases with analytical procedures involving molecular biology are presented in Volume 21, beginning with an introductory chapter on the molecular biology of human taste and aroma receptors with implications for taste and aroma of plant products.

A subsequent chapter reports the use of DNA microarrays in identifying genes involved in strawberry flavour formation; further chapters deal with taste and flavour of beer, soybean and other plant products, hop aroma extraction and analysis, wine olfactometry evaluation, and analysis of citrus flavours.

The use of antisense genes in depressing certain aromas in fruits is also described, and articles on the use of electroantennography in analysing flower volatiles, analysis of rose flower volatiles, analysis of flavour by GC olfactometry (finger span method and solid phase microextraction method) and methods describing RNA gel blot analysis in determining floral scent gene expression round off this volume.

J.F. JACKSON, Managing Editor, H.F. LINSKENS, R.B. INMAN

Contents

1 Molecular Biology of Taste and Aroma Receptors: Implications for Taste and Aroma of Plant Products

J.F. Jackson

1.1 Introduction

Our knowledge of the "outside" world is to a large extent dependent on our five senses. Taste and aroma are two of these that, initially at least, humans and other animals were dependent on for their very survival. Now of course they also hold the key to our enjoyment of life as well. Recently, several research teams have reported on the molecular biology of receptors for taste (Hoon et al. 1999) and aroma (Malnic et al. 1999), which has potential for explaining some of the known characteristics of these two senses, and for predicting other unsuspected properties.

1.2 Taste Buds and Receptor Cells

The tongue, responsible for our taste sense, has long been known to have "taste buds" that are distributed over the upper surface of the tongue within a variety of projections known as papillae. The five basic taste modalities of sweet, bitter, sour, salty and umami (sodium glutamate) are preferentially detected by taste buds in different parts of the tongue, with some overlap between them. Thus the circumvallate papillae at the back of the tongue contain thousands of taste buds that are sensitive to bitter substances. The folate papillae at the sides of the tongue are sensitive to sour materials, whilst the fundiform papillae at the front of the tongue preferentially detect sweet compounds. Salty substances are detected at the front and sides of the tongue. As far as is known, each taste bud contains 50 to several 100 cells which may be precursor cells, support cells or taste receptor cells (Lindemann 1996). Receptor cells have at their base nerve endings conducting impulses (information) towards the central nervous system.

How do the receptor cells work? Are the receptor cells specific for each taste modality or are all receptor cells sensitive to the various modalities, and the information decoded allowing for tastant variation? It is known that salty and sour tastants affect taste cell function by entry of Na^+ or H^+ ions through membrane

Department of Horticulture, Viticulture and Oenology, University of Adelaide, Waite Campus, Glen Osmond SA 5064, Australia channels on the surface (Heck et al. 1984). The molecular components were not known in these cases, but for sweet, bitter and umami transduction it is thought that G-protein-coupled receptors are involved (Wong et al. 1996).

1.3 Taste Receptors

In order to work out the way in which the various taste modalities operate, the identification of genes encoding taste receptors would undoubtedly be of assistance. The first steps towards achieving this have been made by Hoon et al. (1999). These research workers have used subtractive and differential single-cell screening techniques to isolate two G-protein-coupled receptor cells. They generated a subtracted cDNA library from rat circumvallate papillae (Hoon and Ryba 1997) and constructed a collection of single-cell cDNA libraries using RNA isolated from individual rat taste receptor cells. These libraries were screened for sequences preferentially expressed in taste cells by differential hybridizations with cDNA probes prepared from taste and non-taste tissue. Candidate clones were assayed for taste cell specificity by in situ hybridization to tissue sections of rat tongue.

In this way a novel G-protein-coupled receptor TRI was identified which was expressed in a few of the receptor cells, and a further G-protein-coupled receptor TR2 was identified using PCR and low stringency screens on rat circumvallate, foliate and fungiform cDNAs. Both TR1 and TR2 belong to a new subgroup of G-protein-coupled receptors most closely related to metabotropic glutamate receptors (Nakanishi 1992), a family of putative pheromone receptors (Ryba and Tirindelli 1997) and the Ca^{2+}-sensing receptor of Brown et al. (1993). TRI and TR2 both have a very long N-terminal extracellular domain which is presumably the ligand binding site. They have a series of conserved cysteine residues in the extracellular domain, and seven putative trans-membrane segments, but share only approximately 40% sequence identity.

1.4 Taste Receptor Expression Patterns

The patterns of expression of TRI and TR2 were examined by Hoon et al. (1999) using in situ hybridizations of sections of various taste papillae. They found a correlation between TR1 expression and sweet sensitivity, and TR2 with bitter sensitivity. They also found that while TR1 and TR2 can indeed be expressed in the same taste buds, they are not usually expressed in the same cells. Also, they found a small population of cells which expressed both.

They concluded that a given taste cell may express more than one receptor. This is consistent with other results suggesting that some taste receptor cells may respond to more than one taste modality.

1.5 Conclusions for Taste Modality

1. Two seven-transmembrane domain proteins have been identified as likely mammalian taste receptors. They are G-protein-coupled receptors akin to mammalian candidate sensory receptors.
2. These receptors, TR1 and TR2, are specifically expressed in subsets of taste receptor cells.
3. They show high topographic selectivity among different taste papillae.
4. Taste receptors are localized in the taste pore of taste buds which is the only part exposed to taste stimuli.

The evidence strongly shows a correlation of TR1 with sweet taste and TR2 with bitter taste. Further research with expression systems, genetic studies and in situ physiological studies with cells tagged with fluorescent reporters are needed to resolve unequivocally this correlation.

Several questions are posed by Hoon et al. (1999) as a result of their molecular biological approach. For example, how many taste receptor genes are there; we have two, are there any more? Southern blot analysis did not show any more, but a number of genetic and cross-desensitization studies do suggest different receptors for different bitter compounds (Spielman et al. 1992). It is likely then that there are additional receptors; they may well have little sequence homology with TR1 and TR2 as, for example, exists in *Caenorhabditus elegans* where there are literally hundreds of highly divergent chemosensory receptors (Troemel et al. 1995).

Now that two taste receptor genes have been isolated, it should be possible to identify individual receptor cells and determine their relationship to specific nerve fibres, yielding more about information processing in taste perception as has been done for odour detection.

1.6 Aroma Detection in Mammals

Studies on the rodent olfactory systems have demonstrated that volatile odorants that enter the nose are detected by millions of olfactory sensory neurons (Shepherd 1988; Buck 1996). These neurons transmit signals to the olfactory bulb of the brain, which in turn sends signals to the primary olfactory cortex. After this signals go on further to higher cortical areas or to the limbic system giving rise to conscious perception of aromas and emotional and motivational effects.

The detection of odorants by the neurons in the nose is mediated by approximately 1,000 different G-protein-coupled odorant receptors encoded by a multigene family (Buck and Axel 1991). These odorant receptors share characteristic sequence motifs but vary in sequence, consistent with an ability to recognize diverse ligands. It is also known that each olfactory neuron expresses only one odorant receptor gene. Within the nose, neurons expressing a given odorant receptor are confined to one of four odorant receptor expression zones. Within each zone they are randomly interspersed with neurons expressing other odorant receptors. Axons of neurons expressing the same odorant receptors converge at fixed sites in only a small number of the olfactory bulb's approximately 2,000 glomeruli (Mombaerts et al. 1996).

1.7 Model for the Olfactory System

All of this suggests that olfactory information is first organized into four sets in the nose (the four odorant receptor expression zones mentioned above), and then reorganized in the olfactory bulb into a "sensory map" which is identical for all individuals. Further, each olfactory neuron in the nose and each glomerulus in the bulb receives input from only one odorant receptor type.

It has been observed that single olfactory neurons and single glomeruli in the olfactory bulb are stimulated by several odorants. It therefore seems that each odorant receptor must recognize several odorants. Odorant receptor specificity has been studied by Malnic et al. (1999) using a combination of calcium imaging and single cell reverse transcriptase–polymerase chain reaction. They showed that single odorant receptors do indeed recognize multiple odorants, and further that one odorant is recognized by many odorant receptors. Additionally different odorants are recognized by different combinations of receptors. Their work also showed that slight changes in odorant structure or concentration result in changes in the combination of receptors recognizing the odorant. Malnic et al. (1999) have therefore quite elegantly shown that the olfactory system uses a combinatorial receptor coding scheme to encode aroma identities.

1.8 Conclusions for Aroma Perception in Humans

Since there are approximately 1,000 odorant receptor genes in the genome the combinatorial receptor coding scheme should permit recognition of a vast number of diverse aromas. As pointed out by Malnic et al. (1999) even if each odorant were encoded by only three receptors, the number of odorants that could be discriminated would be nearly 1 billion!

Using a number of aliphatic odorants with different chain length, these workers also showed that there was a very fine discrimination in ligand binding – a single receptor can distinguish between odorants that differ in carbon chain length by only one carbon atom or between odorants that have the same carbon chain but with different functional group. Thus, for example, functional groups such as carboxylic acids and alcohols give compounds very different aromas (e.g. unpleasant or rancid for the former and fresh or pleasant for the latter).

Aliphatic alcohols of increasing chain length were recognized by increasing numbers of receptors (and neurons) in the study by Malnic et al. (1999), while it is known that the detection threshold for these compounds decreases with increasing chain length (Cain 1988). It seems that size or complexity of an aroma code could be an important factor in determining how easily an odorant can be detected.

It is not yet known how signals generated from different receptors in the nose are dealt with in the brain. It is known that signals derived from olfactory neurons in the nose are ultimately sent to a number of different brain areas, in some of which the conscious perception of odours mediates emotional and physiological effects (e.g. for rotten food or for pheromones).

References

Avenet P, Hofmann F, Lindemann B (1988) Transduction taste receptor cells requires cAMP-dependent protein kinase. Nature 331:351–354

Brown EM, Gamba G, Riccardi D, Lomardi M, Butlers R, Kifor O, Sun A, Hediger MA, Lytton J, Herbert SC (1993) Cloning and characterization of an extracellular Ca^{2+}-sensing receptor from bovine parathyroid. Nature 366:575–580

Buck L (1996) Information coding in the vertebrate olfactory system. Annu Rev Neurosci 19: 517–544

Buck L, Axel R (1991) A novel multigene family may encode odorant receptors: a molecular basis for odor recognition. Cell 65:175–187

Cain WS (1988) Olfaction. In: Atkinson RJ, Hermstein RJ, Lindzey G, Luce RD (eds) Steven's handbook of experimental psychology – perception and motivation. Wiley, New York, pp 409–459

Gilbertson TA, Avenet P, Kinnamon SC, Roper SD (1992) Proton currents through amiloride-sensitive Na channels in hamster taste cells. Role in acid transduction. J Gen Physiol 100: 803–824

Heck GL, Mierson S, DeSimone JA (1984) Salt taste transduction occurs through an amiloride-sensitive sodium transport pathway. Science 223:403–405

Heck MA, Adler E, Lindemeier J, Battery JF, Ryba NJP, Zuker CS (1999) Putative mammalian taste receptors: a class of taste-specific GPCRs with distinct topographic selectivity. Cell 96: 541–551

Hoon MA, Ryba NJP (1997) Analysis and comparison of partial sequences of clones from a taste-bud-enriched cDNA library. J Dent Res 76:831–838

Hoon MA, Adler E, Lindemeier J, Battery JF, Ryba NJP, Zuker CS (1999) Putative mammalian taste receptors: a class of taste-specific GPCRs with distinct topographic selectivity. Cell 96: 541–551

Kinnamon SC, Dionne VE, Beam KG (1988) Apical localization of K^{+} channels in taste cells provides the basis for sour taste transduction. Proc Natl Acad Sci USA 85:7023–7027

Lindemann B (1996) Taste reception. Physiol Rev 76:718–760

Malnic B, Hirono J, Sato T, Buck LB (1999) Combinational receptor codes for odors. Cell 96: 713–723

Mombaerts P, Wang F, Dulac C, Chao S, Nemes A, Mendelsohn M, Edmondson J, Axel R (1996) Visualizing an olfactory sensory map. Cell 87:675–686

Nakanishi S (1992) Molecular diversity of glutamate receptors and implications for brain function. Science 258:597–603

Ryba NJP, Tirindelli R (1997) A new multigene family of putative pheromone receptors. Neuron 19:371–379
Shepherd GM (1988) Neurobiology. Oxford University Press, New York
Spielman AI, Huque T, Whitney G, Brand JG (1992) The diversity of bitter taste signal transduction mechanisms. Soc Gen Physiol Ser 47:307–324
Troemel ER, Chou JH, Dwyer ND, Colbert HA, Bargmann CI (1995) Divergent seven transmembrane receptors are candidate chemosensory receptors in *C. elegans*. Cell 83:207–218
Wong GT, Gannon KS, Margolskee RF (1996) Transduction of bitter and sweet taste by gustducin. Nature 381:796–800

2 Use of DNA Microarrays in the Identification of Genes Involved in Strawberry Flavor Formation

A. Aharoni and A.P. O'Connell

2.1 Introduction

Controlling gene expression levels is one of the key regulatory mechanisms used by living cells to sustain and execute their operations. Monitoring gene expression has been for more than a decade an important molecular tool for providing clues to gene function and for new perspectives on spatial and temporal cellular activities. Genome sequencing and the availability of large sets of expressed sequence tags (ESTs) from numerous organisms urged the development of efficient and accurate methods for large-scale and genome-wide analyses of genetic variation and expression patterns. Novel methods based on either gel separation (Bruce et al. 2000), counting of tags and signatures of DNA fragments (Brenner et al. 2000) or on specific hybridizations of nucleic acids to macro – and microarrays (Schena et al. 1995; Lockhart et al. 1996; Desprez et al. 1998) are currently available.

The attractive features of microarray technologies are miniaturization, the high degree of assay sensitivity (fluorescence-based detection) and large number of simultaneous determinations. Two primary, fundamentally different array-based technologies have been developed, both capable of large-scale expression analyses. Fodor and colleagues introduced a method for high-density spatial synthesis of oligonucleotides to produce arrays which contain a few hundred thousand distinct features (Fodor et al. 1991). Oligonucleotide arrays allow most accurate measurements of sequence discrimination and are therefore well suited for DNA variation analyses. Manufacturing such arrays requires prior sequence knowledge and complicated design and production methodologies (Lipshutz et al. 1999). Spotted arrays, often referred to as cDNA microarrays in which pre-synthesized nucleic acids are robotically deposited on a solid surface (such as glass) are more flexible in their design and fabrication (Duggan et al. 1999). However, it is worth mentioning that oligonucleotide arrays can also be manufactured by mechanical microspotting.

The basic microarray assay used in both methods is similar. It includes first the preparation of biological samples to be compared, followed by a biochemical reaction between the samples and the bound material. The next phase includes

Business Unit Cell Cybernetics, Plant Research International, P.O. Box 16, 6700 AA, Wageningen, The Netherlands detection of signal (often fluorescent, using confocal scanning) allowing simultaneous analysis of mixed samples. The last phase of the microarray assay involves data analyses and mining in which quantitative comparison between the samples is carried out. Here we focus primarily on aspects and use of cDNA microarrays although oligonucleotide arrays will be mentioned where appropriate. For the sake of clarity we define the tethered nucleic acid as "probe" and the free nucleic acid as "target".

Presently, the main large-scale applications of microarrays are detection of comparative expression and DNA variation analyses (Lander 1999). Variation in DNA sequence underlies most of the differences we observe within and between species. Locating and identifying these genotypic differences allows linkage of genetic to phenotypic variation. Single nucleotide polymorphisms (SNPs) are the most frequent type of variation in genomes. Microarrays, in particular oligonucleotide arrays, may be used for large-scale SNP detection and discovery. Arrays using targets originating from small sequenced genomes such as yeast may be used for the determination of genome-wide allelic variation (Lander 1999).

Microarray technologies used for expression monitoring of genes in specific tissues and under different circumstances provide important information for the elucidation of gene function. Beyond answering questions on an individual gene basis, gene expression arrays are well suited for the genetic dissection of functional pathways and for studying the co-ordinate action of cellular components in regulating and carrying out cellular activities. In expression monitoring, a single hybridization experiment can provide quantitative results for as many as 18,000 (Alizadeh et al. 2000) and even more genes at once. The flexible nature of the fabrication and hybridization methods of cDNA microarrays allows the application of the technology to non-model organisms. To date, reports on the use of microarrays in plants have been limited. This technology has been primarily applied to the model species *Arabidopsis thaliana* and *Petunia hybrida* (Schena et al. 1995; Lemieux et al. 1998; Ruan et al. 1998; Reymond et al. 2000) and more recently to a non-model and more complex species such as strawberry (Aharoni et al. 2000).

In the first part of this chapter we will describe the basic principles and current global applications of the microarray approach. The second part of the chapter will demonstrate the power of microarrays to link gene to function, and its use for studying pathways central to strawberry fruit flavor development. It will describe the cloning of the *SAAT* gene encoding an ester-forming enzyme in the ripe strawberry fruit. Volatile esters are important compounds providing fruity odors. The identification of other candidate flavor-associated genes through the association of gene expression patterns with existing biochemical knowledge will also be demonstrated.

2.2 The Microarray Method

2.2.1 Principle

Like other hybridization methods the specificity of microarray technology relays on the selective and differential annealing of nucleic acid strands. Earlier methods such as DNA and RNA gel blot analyses use a uniquely labeled nucleic acid molecule in solution for the hybridization with a complex mixture such as total RNA samples attached to a solid support. Information obtained from such experiments relates to the abundance of one single polynucleotide of interest. Array-based methods use the reverse strategy (Fig. 2.1). They use complex mixtures of labeled polynucleotide (such as cDNA derived from mRNA) for the hybridization with large amounts of individual elements (most often unique PCR products), attached to a solid surface. Fluorescent labeling with dyes possessing different excitation and emission characteristics allows simultaneous hybridization of two or more targets (often a reference and a test target). The array is then scanned at two (or more in the case of more dyes) wavelengths, and the ratio of the fluorescence emissions of the two wavelengths represents the differential gene expression. In this manner a single experiment provides quantitative hybridization data for hundreds and thousands of probes.

2.2.2 Microarray Procedure

2.2.2.1 Array Fabrication

The first step of cDNA microarray fabrication requires the selection of "probes" to be spotted as PCR products. The choice of probes is flexible and they could be

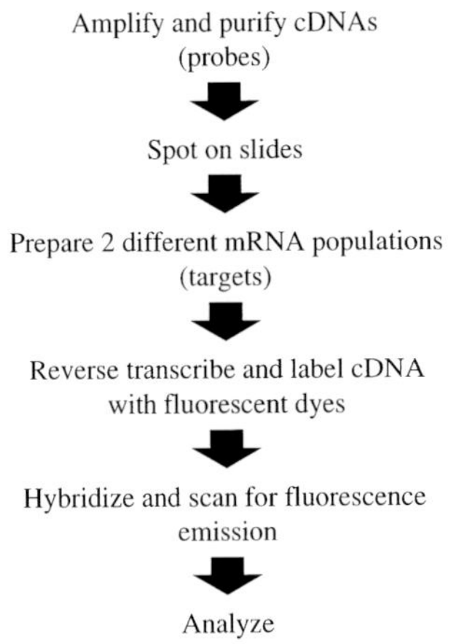

Fig. 2.1. Schematic of the microarray procedure

products of direct amplification of genomic DNA (by the use of gene-specific primers) or amplified inserts from cDNA libraries (for instance ESTs) or any other library of interest. For example, the first microarray allowing genome-wide expression monitoring was generated by amplifying genomic DNA with specific primer pairs designed on 6,200 open reading frames (ORFs) of yeast (DeRisi et al. 1997).

Prior to spotting, probes are often purified from unwanted PCR components and concentrated by precipitation or gel-filtration. Apart from PCR products (ranging in size from approximately 0.2 to 2.5 kb), synthetic oligonucleotides might also be spotted on solid surface (Mir and Southern 1999). Each array element is generated by the deposition of a few nanoliter of a PCR product using a motion control system that spots or prints a sample of each probe onto multiple surfaces (often 50–100 microscope slides) in a serial operation (Fig. 2.2A). The actual printing technologies fall into two distinct categories: contact printing (mechanical deposition) and non-contact printing (liquid delivery). In contact printing an array of either solid pins, split pins or tweezers is dipped into the sample for pickup and a direct contact with the solid surface deposits a drop of sample (Fig. 2.2B). Depending on the application contact printing, droplets of 100–250 μm may be produced. An example of such a spotting pin can be seen in Fig. 2.2C. Non-contact printing involves the ejection of an accurate volume of drops from a dispenser onto the surface. The most common type of non-contact dispensing uses ink-jet printing technologies (similar to printing ink onto paper). cDNA arrays are often fabricated on glass surfaces such as microscope slides. In order to enhance the adhesion of probes, lower the background and prevent spreading of the droplets, the slides are coated with poly-lysine or amino silanes. After spotting, DNA is immobilized (either by UV cross-linking or baking), surface is blocked (by succinic anhydride or sodium borate) and as a final step the DNA on the slide is denatured (by heat or alkali treatment). Processed slides could be stored dry for several months prior to hybridization.

2.2.2.2 Preparation of Targets and Hybridization

In cDNA microarray assay, targets are typically representations of RNA or DNA from two or more biological populations. Most common is the incorporation of modified fluorescent nucleotide during a reverse transcription reaction from an oligo-dT primer. The fluorescent labels Cye3-dUTP and Cye5-dUTP are frequently paired, as they have relatively high incorporation efficiency with reverse transcriptase, good photostability and yield, and absorb and emit light at distinct and separable wavelengths. Either total RNA or mRNA may be used as starting material for labeling, although the use of mRNA provides the best hybridization results. For proper fluorescence, 10–50 μg of total RNA or 1–5 μg of mRNA is used per target per array.

Mixes of equal amounts of both labeled targets (between 5 and 50 μl total volume) are applied to the array most often under coverslips sealed in reaction

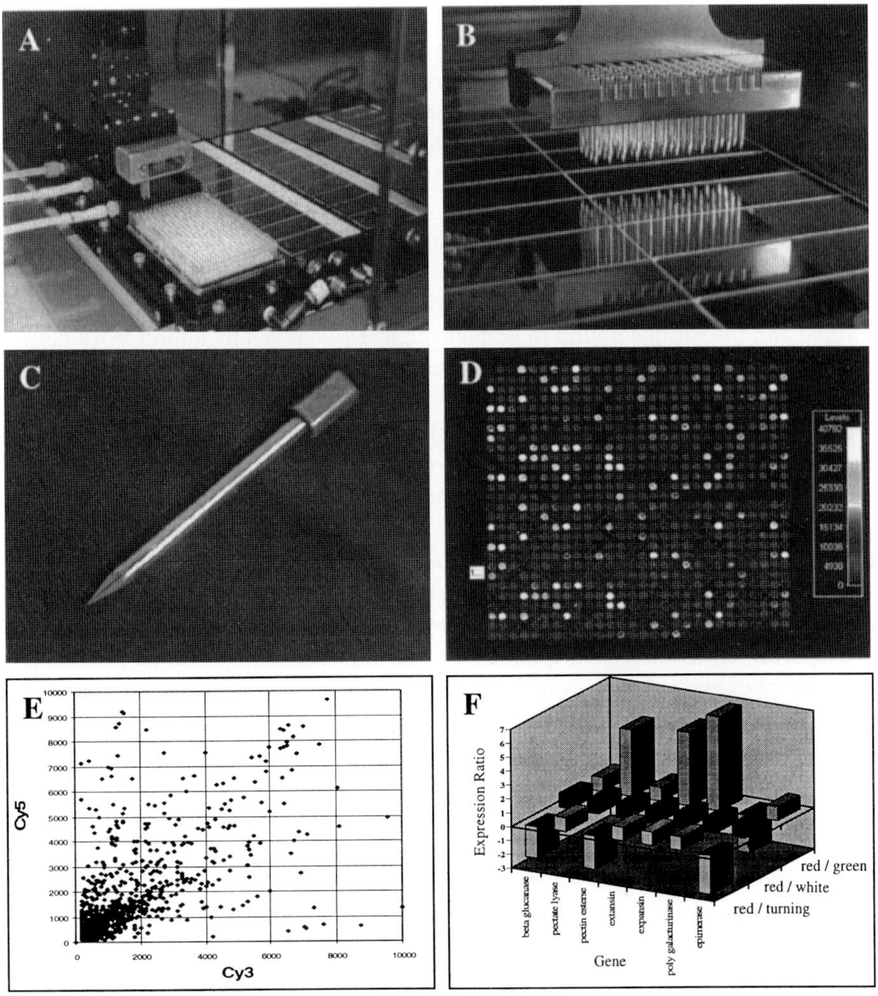

Fig. 2.2. From spotting to analysis of cDNA microarrays. **A** Motion control system spotting (printing) a sample of each probe onto multiple microscope slides in a serial operation [Cartesian Technologies PixSys array printing system with "ChipMaker" microspotting print head (TeleChem International, Inc)]. **B** Print head used for contact printing containing an array of 48 split pins (ChipMaker, TeleChem International, Inc.). **C** Closer look at a spotting split pin. **D** Extracting the signal intensity of each individual probe after microarray hybridization experiments. Measurements are conducted by superimposing a grid on the image, as can be seen in the lower half of the array (*white circles*). **E** Graphic illustration of a typical raw data set obtained from a single microarray experiment expressed as the ratio Cy3/Cy5. Each *spot* in the graph corresponds to an individual array element. **F** Expression ratios of seven strawberry cell-wall-related genes in three microarray experiments comparing stages of strawberry fruit development. *Positive values* represent higher expression in the red stage and *negative values* represent higher expression in green or white or turning stages compared to the red stage. (**B** and **C** are courtesy of T. Martinsky, TeleChem International, Inc.)

chambers. Hybridization conditions such as ionic strength, temperature and target concentration depend on the application. Hybridization temperatures of 42–70 °C are typically used for cDNA array experiments, performed for several hours or overnight. A higher concentration of probes ensures first-order reaction kinetics so that hybridization rate is determined by target concentration.

2.2.2.3 Image Analysis and Data Extraction and Mining

Once the fluorescent sample is reacted with the microarray, unbound material is washed and the sample bound to each element is visualized by fluorescence detection. Both confocal scanning devices and CCD cameras may be used for this purpose. In the next step fluorescence emission from the microarray is converted into a digital output, and signal from each probe on the array is quantified. Measurements of individual array elements are conducted by superimposing a grid on the image and obtaining an intensity value for all probes (Fig. 2.2D). Prior to the actual comparison between the intensities of test and reference targets (performed for each probe) the data are normalized. A few methods may be employed for this purpose such as, for example, using a set of control samples expected to be unchanged in intensity when hybridized with different targets (Hegde et al. 2000). A graphical illustration of a typical data set obtained from a single microarray experiment expressed as the ratio Cy3/Cy5 is shown in Fig. 2.2E. Our results show that ratios of four- to fivefold can be detected with the probability of 95% (van Hal et al. 2000).

The tremendous amount of data obtained by microarray experiments requires construction of databases allowing management of information on probes arrayed and raw data results. Other databases and algorithms are further required for storage and examination of the different data sets. When multiple experiments are performed manual sorting is not sufficient for proper evaluation of the data and therefore several statistical methods for clustering and data representation are frequently used (Eisen et al. 1998). These types of analysis using large data sets can provide novel perspectives on cellular regulatory mechanisms and associate expression of unknown genes with a putative function (Fig. 2.2F).

2.2.3 Key Microarray Applications

2.2.3.1 Monitoring Gene Expression (mRNA Abundance)

The focus of current microarray studies is monitoring RNA expression levels. A very powerful application is the use of array-based technologies to monitor genome-wide gene expression. Both cDNA arrays and oligonucleotide arrays are most suited for large-scale expression monitoring. Apart from small genomes such as yeast and *Escherichia coli* currently being investigated using microarrays (Lashkari et al. 1997; Richmond et al. 1999), exploration of gene expression in larger and more complex genomes (such as humans, containing approximately 35,000 genes) is expected to be a routine in the future. Microarray assays may be directly integrated into functional genomic approaches aimed at both assigning function to identified genes and studying the organization and control of genetic pathways acting together to make up the functional organism. A common approach is to cluster genes into groups based on their expression profiles scored in multiple experiments (Brown and Botstein 1999). The assumption is that genes showing similarity in expression may be functionally related. In most cases, gene clusters comprise both known and unknown genes allowing researchers to associate a putative function to the unknown genes by employing the concept of "guilt by association".

Expression monitoring using arrays may be used to analyze any kind of variability in gene expression between given samples. These variations can be either naturally occurring or induced. Natural variation may arise from, for example, the use of different plant cultivars, different tissues and different stages of development. Induced variability may arise from, for example, exposure to different environmental conditions and mutagenesis. In plants, microarrays are most suited to the analyses of mutant populations generated by a range of methods such as chemical or physical mutagenesis, knockouts (T-DNA tags, transposon tagging) and activation tagging (Pereira 2000). Other variants containing more specific mutations induced by means of site-specific deletions, gene silencing and the introduction of modified genes (induction of dominant mutations) may also be used.

Regulatory circuits may also be unraveled using expression arrays. Monitoring expression in a plant containing a loss of function mutation in a transcription factor may result in the identification of genes acting downstream in the cascade. Data from genome-wide gene expression monitoring can be used to identify new *cis*-regulatory elements in promoter regions and categorize genes in similar regulatory circuits according to the elements identified (Bucher 1999). Another interesting application of gene expression arrays is for the identification of secreted and membrane-associated gene products, as described by Diehn and colleagues (2000). Although applications of gene expression arrays are far more extensive than described above one should realize that factors other then mRNA level alone such as, for example, protein modification and interaction and metabolic flux, determine the actual activity of a gene product. It is therefore very important to couple gene expression data to other information gathered at both the protein and metabolic level.

2.2.3.2 DNA Variation

Microarrays are also well suited to the detection of novel DNA variants. Sequence variations serve as genetic markers in studies of diseases and traits with complex inheritance patterns and strong environmental interactions. Using oligonucleotide arrays one can scan a target sequence for mutation by constructing a "tiling" array.

In a "tiling" array each overlapping 25-mer in the sequence is covered by four complementary oligonucleotide probes that differ only by having A, T, C or G substituted at the central position. The high sensitivity of the assay allows detection of alteration in hybridization pattern when a sequence variation exists. This is actually a most accurate and sensitive way of DNA re-sequencing. The tiling arrays have been used successful in the detection of single nucleotide polymorphisms (SNPs) in humans (Wang et al. 1998), although SNPs can also be detected by spotted arrays (with oligonucleotides) using a method called mini-sequencing (Syvanen 1999). In the mini-sequencing procedure a primer is hybridized to DNA and extended with DNA polymerase with the appropriate dideoxyribonucleoside triphosphate that matches the nucleotide at the variable site. This single base extension reaction can be performed on arrays and the differentially labeled nucleotides incorporated may be scored and used for the detection of point mutations and SNPs.

Large regions may also be scored for DNA discrepancies such as allelic variation. Due to the complexity of genomic DNA compared to total RNA, present methods do not allow the direct hybridization of genomic DNA of complex organisms. An exception, as demonstrated in yeast (Winzeler et al. 1998), is allelic variation which could be detected by hybridizing genomic DNA from two strains to an oligonucleotide array and analyzing the hybridization differences (thus circumventing the requirement for specific target amplification using PCR).

2.2.3.3 Arrays Containing Other Types of Bio-molecules

Microarray technologies are not restricted to the analyses of DNA but might be applied to other biomolecules. These may include peptides, proteins, antibodies, small molecules, mRNAs, tissues, cells and possibly even small multicellular organisms. Recently, protein microarrays fabricated and analyzed in a similar fashion to cDNA arrays were used successfully for screening protein–protein interactions, identifying the substrates of protein kinases and identifying the protein targets of small molecules (MacBeath and Schreiber 2000). Antibody arrays have also been produced that allow isolation of high-affinity antibodies to target proteins and antigens (de Wildt et al. 2000). The array elements were composed of selected clones from a phage display library and these were arrayed as live bacterial cells, and screened on antigen-coated filters. Antibody arrays may also be used for the detection of differentially expressed proteins.

2.3 Strawberry and Flavor Formation

2.3.1 Strawberry Fruit

Strawberry is deemed a false fruit in the respect that what is commonly called the fruit originates from the expansion of the flower base (the receptacle) as a pseudocarp, whereas in reality the real one-seeded fruits (achenes) are actually on the epidermal layer (Perkins-Veazie 1995). The achenes are a combination of seed and ovary tissue and originate at the base of each pistil. They are embedded in the receptacle epidermis and fibrovascular strands connect them to the interior of the receptacle.

Strawberry is classified as a non-climacteric fruit since it does not show the typical increase in ethylene levels and respiration preceding the ripening process (Manning 1994). Ripening is an aspect of fruit development that is initiated after seed maturation has almost been completed. Strawberry displays the same dramatic changes during ripening as other fruits. Fruit ripening is characterized by a rise in soluble solids content in the receptacle, the production of natural aroma and flavor compounds and alterations to fruit shape, size, texture and pigmentation (Woodward 1972).

Despite the long history of genetic selection of fruit such as strawberry, we still lack valuable information on the molecular, cellular and physiological events that control important processes such as flavor formation. In recent years molecular studies in strawberry have resulted in the identification of genes upregulated in expression during ripening (Wilkinson et al. 1995; Medina Escobar et al. 1997a; Nam et al. 1999). Of main interest were genes encoding putative cell-wall-related enzymes such as pectate lyase (Medina Escobar et al. 1997b), expansin (Civello et al. 1999) and various members of the endo-beta-1, 4-glucanase gene family (Harpster et al. 1998; Llop Tous et al. 1999; Trainotti et al. 1999). Genes putatively encoding enzymes involved in other ripening-related processes such as pigmentation, sucrose, lipid metabolism, protein metabolism, respiration and stress have also been identified (Manning 1998; Moyano et al. 1998; Aharoni et al. 2000).

To elucidate the genetic events underpinning fruit flavor formation requires a comprehensive understanding of fruit developmental processes. During fruit development both physical and morphological changes are often a result of changes in protein levels and activities, which may reflect shifts in overall mRNA abundance (Manning 1994). For this reason we used cDNA microarrays and strawberry as our model system to genetically dissect developmental programs, which are central to flavor formation.

2.3.2 Main Flavor and Aroma Components in Strawberry and Their Biosynthesis

In general, fruit flavor compounds develop during ripening when the metabolism of the fruit changes towards catabolism. In similarity with other fruit, a complex mixture of hundreds of compounds (Zabetakis and Holden 1997) determines strawberry flavor and aroma (Fig. 2.3). More than 300 compounds have been identified that can contribute to the unique aroma/flavor of ripe strawberry fruit. The components identified may be grouped into several chemical classes, which include acids, aldehydes, ketones, alcohols, esters and lactones. Other contributing groups are sulphur compounds, acetals, furans, phenols, epoxides and

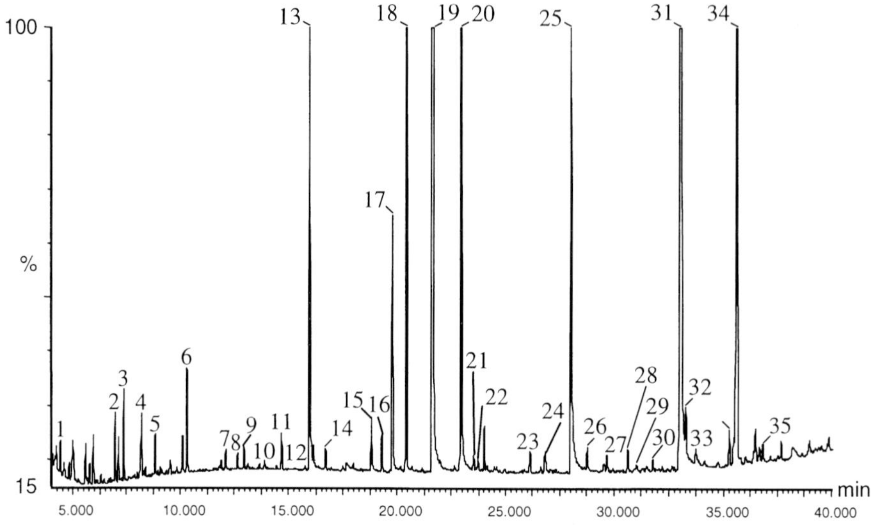

Fig. 2.3. Volatile constituents of ripe strawberry fruit (cv. *Elsanta*) detected by GC-MS after XAD solid-phase extraction. *Peak numbers* correspond to: *1* butyl acetate; *2* 4 methyl-2-butenal; *3* methyl hexanoate; *4* E-2-hexanal; *5* ethyl hexanoate; *6* 3-hydroxy-2-butanone; *7* E-2-hexenyl acetate; *8* E-rose oxide; *9* hexanol; *10* Z-3-hexenol; *11* E-2-hexenol; *12* E-linalooloxid, furanoid; *13* acetic acid; *14* Z-linalooloxid, furanoid; *15* propanoic acid; *16* linalool; *17* 2-methyl propionic acid; *18* methoxyfuraneol; *19* butanoic acid; *20* 2-methyl butanoic acid; *21* gamma-hexalactone; *22* 4-ethyl benzaldehyde; *23* delta-hexalactone; *24* 4-methyl pentanoic acid; *25* hexanoic acid; *26* benzylalcohol; *27* gamma octalactone; *28* benzothiazole; *29* delta-octalactone; *30* furaneol acetate; *31* furaneol; *32* trans nerolidol; *33* octanoic acid; *34* gamma-decalactone; *35* delta-decalactone. (Courtesy of W. Schwab)

hydrocarbons. Members of these groupings, whilst often present at low levels, may have a significant impact on the overall flavor of strawberry. The compound 2,5-dimethyl-4-hydroxy-3(2H)-furanone (furaneol) and its methyl ether 2,5-dimethyl-4-methoxy-3(2H)-furanone were the only compounds identified in diluted solutions as exhibiting typical flavor/aroma associated with strawberry (Roscher et al. 1997). The biochemical pathway leading to the synthesis of furaneol and its methyl ether is unknown although there is strong evidence that sugars (most likely fructose-1, 6-bisphosphate) supply precursors for furaneol biosynthesis (Schwab 1998). Except for carbohydrates, the diverse volatile constituents of fruit like strawberry are derived from various other primary metabolism precursors such as fatty acids and amino acids (Fig. 2.4).

Volatile esters are quantitatively and qualitatively the most important compounds providing fruity odors, and in strawberry alone more than a hundred ester types have been detected (Maarse 1991). A pathway leading to the biosynthesis of volatile esters in fruits has been proposed previously (Olias et al. 1995). Fatty acids are the most important precursors for the formation of volatile aroma compounds. The oxidative degradation of membrane lipids to linolenic and linoleic acid

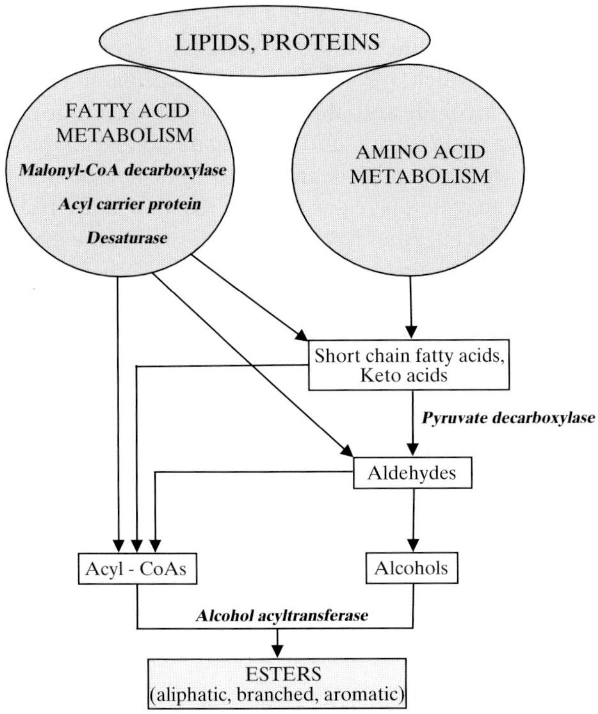

Fig. 2.4. Schematic of possible metabolic routes leading to formation of volatile esters in fruit. Genes identified by our microarray study showing similar expression profile during strawberry ripening are depicted (in *italics*)

catalyzed by lypoxygenase is the main source of precursors for the generation of alcohols, aldehydes, acids and esters found in fruits (Perez et al. 1999). Thiolase is the last enzyme in the β-oxidation of fatty acids. It catalyzes the thiolytic cleavage involving another molecule of CoA. The product of this reaction is acetyl-CoA and acyl-CoA derivatives containing two carbon atoms less than the original acyl-CoA molecule that underwent oxidation. The acyl-CoA formed in the cleavage reaction may be utilized at the final stage of the biosynthetic pathway for ester formation in fruit (Bojorquez and Gomez Lim 1995). Therefore the profile of fatty acid, amino acid and carbohydrate precursors found in each fruit, along with the specificity of the enzymes in the biosynthetic pathway leading to ester formation, could have a key role in determining the type of esters formed.

In addition to fatty acids the transamination and oxidative decarboxylation of amino acids provide the precursors for volatile aroma compounds such as aldehydes, acids, alcohols, esters and thiols (Perez et al. 1992). The transamination of amino acids is catalyzed by aminotransferases. In strawberry alanine is proposed to be the main free amino acid metabolized to flavor compounds (Perez et al. 1992). This was based mainly on the dramatic decrease in its content just before the formation of volatile aroma compounds had commenced in the fruit. In addition, feeding alanine to strawberry cultures resulted in the formation of several esters such as methyl and ethyl hexanoate which are important constituents of strawberry volatiles (Perez et al. 1992). Alcohols and aldehydes are derived from the metabolism of their corresponding amino acids and oxo-sugars. Ketoacids produced by transamination can be enzymatically degraded to the corresponding aldehydes or carboxylic acids. The enzyme that catalyzes this reaction is pyruvate decarboxylase. Alcohol dehydrogenase enzymes have been implicated in the interconversion of the aldehyde and alcohol forms of flavor volatiles (Scharpf and Chandan 1989). Esterification is the result of transacylation from acyl-coenzyme A (CoA) to an alcohol (Perez et al. 1996). The enzyme catalyzing the reaction is termed an alcohol acyltransferase (AAT), and it plays a major role in the biosynthesis of volatile esters.

2.3.3 Alcohol Acyltransferases and Ester Formation

Alcohol acyltransferases (AATs) have been identified from both yeast and fungi (Yamakawa et al. 1978; Fujii et al. 1994, 1996). The yeast AATs showed high affinity to acetyl-CoA and the alcohols ethanol and isoamyl alcohol. The influence of esters (isoamyl and ethyl acetate) on beer flavors renders AAT one of the most important enzymes in the fermentation process performed by microorganisms (Yoshioka and Hashimoto 1984).

In plants ester formation has been studied in both fruit and flowers. In *Clarkia breweri* the ester benzylacetate is an important constituent of the flower scent. The purification of the acetyl-CoA:benzylalcohol acetyl-transferase (BEAT) protein from flowers of *Clarkia* and the isolation of the gene encoding it have been reported (Dudareva et al. 1998). BEAT has a high affinity for aromatic alcohols, such as benzyl alcohol and cinnamyl alcohol.

In fruit, melon, strawberry and banana, AAT proteins have been investigated using crude fruit extracts (Harada et al. 1985; Ueda et al. 1992; Perez et al. 1993, 1996; Olias et al. 1995). Analysis of the substrate specificity of the enzymes from the various sources revealed differences in their affinity to acyl-CoAs and alcohols. Maximum activity for the strawberry (AAT) was obtained using acetyl-CoA and hexyl alcohol as substrates (Olias et al. 1995).

2.4 From Expression to Function: Identification of Strawberry AAT (*SAAT*)

2.4.1 Gene Expression During Development and Ripening

As previously discussed several metabolic pathways are involved in the biosynthesis of flavor and aroma compounds. Some of these pathways have barely been investigated even at the biochemical level and this makes the identification of flavor-related genes more complicated. We anticipated that the dramatic changes related to flavor and aroma are in part a result of changes in abundance of certain mRNAs. Therefore, microarray assay was our method of choice, to discover genes associated with flavor formation in strawberry, as it allows sensitive, quantitative, large-scale, rapid and parallel monitoring of gene expression levels.

cDNA microarrays were used in order to follow gene expression during strawberry fruit development (Aharoni et al. 2000). In total 1,701 cDNAs randomly picked from a strawberry fruit cDNA library were arrayed on glass microscope slides. We hypothesized that genes central to flavor formation would be uncovered by correlating the expression profile of known ripening-related genes over several developmental stages, with candidate genes deduced from biochemical studies on flavor formation. Pigmentation-related genes may serve as good markers since fruit color in the majority of fruits is normally accompanied by the accumulation of flavor and aroma components (Seymour et al. 1993).

Three microarray experiments were performed comparing strawberry fruit developmental stages (1) green with red, (2) white with red and (3) turning with red. In each experiment two mRNA populations fluorescently labeled with cyanine 3 (Cy3) and cyanine 5 (Cy5) respectively were hybridized to the microarray simultaneously. To facilitate proper statistical analysis a second hybridization to an identical microarray was performed in which both populations were inversely labeled. After hybridization the fluorescence of both microarrays was recorded at the specific wavelengths of each fluorescent dye, and calculation of expression ratios was performed for each cDNA (Aharoni et al. 2000).

2.4.2 Identification of the *SAAT* Gene

The expression profile of an unknown gene *(SAAT)* with minimal sequence similarity to genes encoding enzymes with acyltransfer activity matched closely the expression profile of known ripening-regulated genes (e.g. pigmentation- and cell-wall-related genes). Quantitative microarray expression analysis revealed that *SAAT* had a 16-fold higher expression level during the red stage of fruit development when compared to the green stage. As the cDNA library (from which the probes arrayed were selected) was prepared from whole fruit containing a mixture of achene and receptacle tissue, we were capable of performing a fourth experiment to compare gene expression between the ripening stage achene and receptacle tissues. This experiment was directed at studying key processes occurring simultaneously in the two tissue types, and showed that *SAAT* expression was associated with the receptacle tissue. In addition detailed RNA gel blot analysis showed that *SAAT* gene expression was receptacle specific commencing at the white stage of fruit ripening, preceding detectable volatile ester formation in strawberry, and reaching maximal levels between turning and red stages. Gas chromatograph–mass spectrometry (GC-MS) analysis of different stages of strawberry fruit development showed first detectable signs of release of volatile esters during the pink stage of fruit development (between the turning and red stages), with maximal levels attained during the dark-red ripening stage.

2.4.3 *SAAT* Encodes the Ester-Forming Enzyme from Strawberry Fruit

The full-length *SAAT* cDNA clone is 1,618 bp, encoding a polypeptide of 452 amino acid residues with a predicted molecular mass of 50.7 kDa. The presence of several consensus sequences in the SAAT protein sequence assigned it to a super-family of multifunctional acyltransferases responsible for coenzyme A-dependent acyl transfer (St. Pierre et al. 1998). Members of this gene family have a catalytic reaction mechanism related to the ancient chloramphenicol 3-*O*-acetyltransferase (CAT) and dihydrolipoyl *S*-acetyltransferase (DHLAAT) class of enzymes. Three domains were identified in the *SAAT* sequence. One domain, an H-Xaa-Xaa-Xaa-D (aa; amino acid) motif (corresponding to residues 156–160 in SAAT), is the most conserved consensus sequence present even in acyltransferases such as CAT and DHLAAT (Reed and Hackert 1990), carnitine and choline acyltransferases (Brown et al. 1994) from non-plant systems. The H-Xaa-Xaa-Xaa-D motif was initially detected in mammalian systems (Reed and Hackert 1990; Shaw and Leslie 1991). Site-directed mutagenesis experiments with the rat carnitine palmitoyltransferase (CPTII) suggested that both the conserved histidine and the conserved aspartate residues are part of the protein catalytic site mediating acyl transfer and cleavage of free CoA (Brown et al. 1994). A second highly conserved motif was DFGWG (corresponding to residues 388–392 in SAAT), located near the C terminus. A third consensus sequence was located at the N terminus, L-S-Xaa-T-L-Xaa-Xaa-Xaa-Y-Xaa-Xaa-Xaa-G (corresponding to residues 66–78 in SAAT). The first eight residues of the latter domain in SAAT (LSETLTLY) were found to be present in the green alga *Chlorella vulgaris* acetyl-CoA carboxylase, carboxyl transferase â subunit. Acetyl-CoA carboxylase is a biotinylated enzyme that catalyzes the ATP-dependent formation of malonyl-CoA from acetyl-CoA and bicarbonate. This signified the importance of this domain in reactions using acetyl-CoA as co-substrate. In spite of the similarity in catalytic activity between SAAT and the yeast alcohol acyltransferase enzyme, they showed very low sequence identity (4% at the amino acid level). This suggested that there are no direct evolutionary relationships between the two enzymes.

Biochemical evidence that the *SAAT* gene was involved in the formation of fruity esters was provided by the characterization of the recombinant protein expressed in *Escherichia coli* (Fig. 2.5). The SAAT enzyme showed maximum activity with aliphatic medium chain alcohols, whose corresponding esters are major components of strawberry volatiles (Fig. 2.6). It was capable of utilizing short- and medium-chain, branched and aromatic acyl-coenzyme A molecules as co-substrate. The SAAT enzyme was also capable of using the 10-carbon decanoyl-CoA as a co-substrate (Table 2.1). This ability may account for the presence of this type of long chain esters in strawberry (Honkanen and Hirvi 1990). Clear differences in enzyme activity were detected between the four isomers of hexenol tested: *trans-*

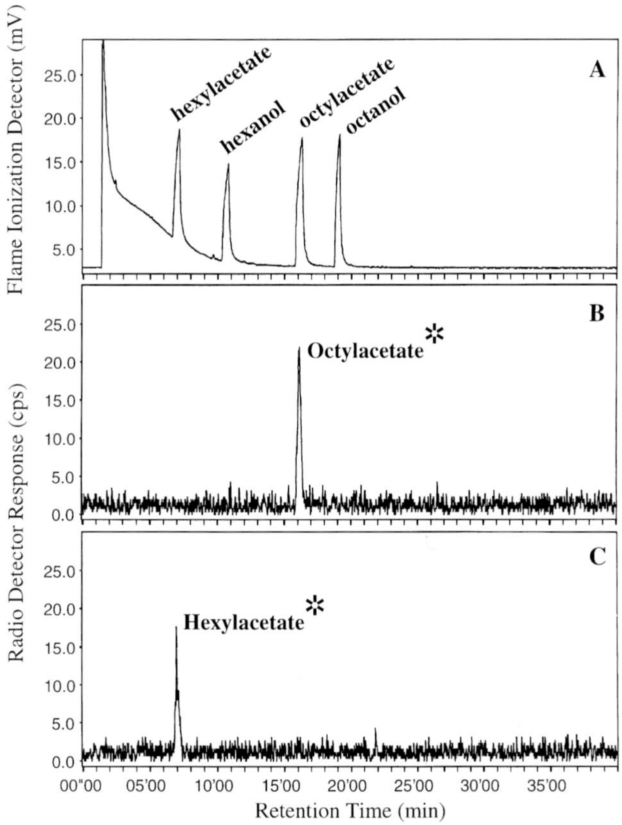

Fig. 2.5. SAAT-catalyzed ester formation from ^{14}C-acetyl-CoA. **A** Flame ionization detector signal of unlabeled standards. **B** and **C** Radiodetector signal of labeled products formed by the SAAT protein from 0.1 mM ^{14}C-acetyl-CoA and alcohols with **B** 2 mM 1-octanol or **C** 2 mM 1-hexanol

2-hexenol was a better substrate than *cis*-2-hexenol, while *cis*-3-hexenol was a better substrate than *trans*-3-hexenol. The effect of the position of the hydroxy moiety of the alcohol on SAAT activity varied: 1-propanol was a better substrate than 2-propanol; however, 2-butanol was a better substrate than 1-butanol. SAAT also accepted the branched primary alcohol isoamyl alcohol. Activity was also detected with aromatic (benzyl- and phenylethyl-) and cyclic (furfuryl-) alcohols though activities were much lower than with 1-octanol (4–10%). In contrast, no activity could be detected with the terpene alcohol linalool. This was in agreement with the absence of linalool esters in strawberry volatile profiles including the *Elsanta* cultivar (Honkanen and Hirvi 1990).

The wide range of alcohols and acyl-CoAs accepted by the SAAT enzyme cannot completely explain the differences in ester composition between the developmental stages and different strawberry cultivars. An important factor is

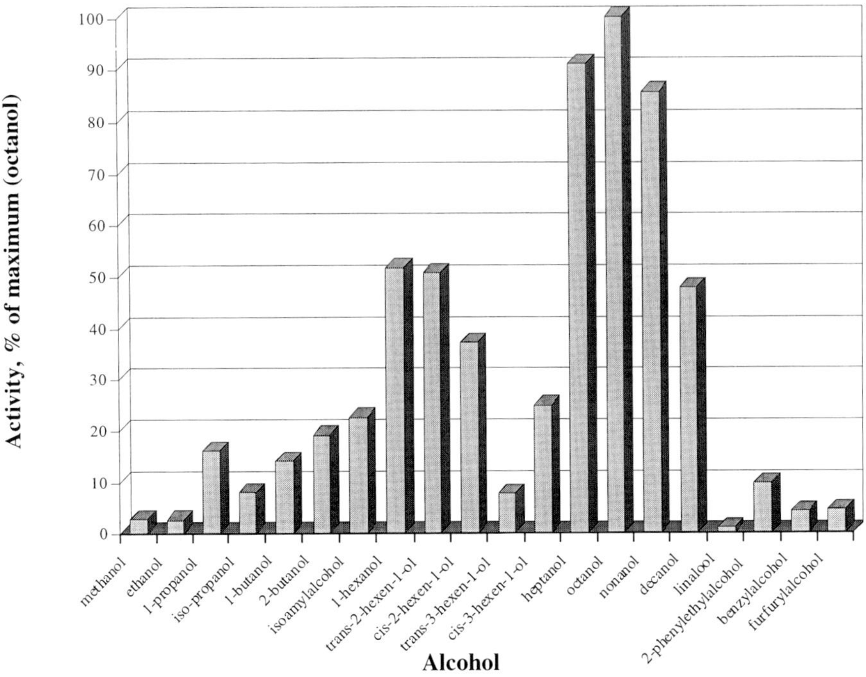

Fig. 2.6. Substrate specificity of SAAT recombinant enzyme towards different types of alcohols. Enzyme activities with different alcohols (20 mM) and ^{14}C-acetyl-CoA (0.1 mM) as acyl donor were assessed by liquid scintillation counting. Activity is expressed as percentage of the maximum activity obtained with octanol

Table 2.1. Different types of acyl-CoAs accepted as co-substrates by the SAAT recombinant enzyme. *ND* Not described

Acyl-CoAs (carbon no.)	Alcohol	Ester formed[a]	Ester properties[c]
n-Propionyl CoA (C3:0)	1-Butanol	1-Butyl propanoate	Ethereal, banana
n-Butyryl CoA (C4:0)	1-Propanol	1-Propyl butyrate[b]	Sharp, pungent, rancid, sweaty, sickening
n-Butyryl CoA (C4:0)	1-Butanol	1-Butyl butyrate[b]	Fruity, pineapple
2-Butyryl CoA (C4:0)	1-Butanol	1-Butyl-2-butyrate	Fruity, apple, banana and pineapple
n-Crotonoyl CoA (C4:1)	1-Butanol	1-Butyl crotoate	ND
n-Hexanoyl CoA (C6:0)	1-Propanol	1-Propyl hexanoate	Wine-like, cheese
n-Decanoyl CoA (C10:0)	1-Butanol	1-Butyl decanoate	Brandy (Whisky-Cognac)-like odor
Benzoyl CoA (C7:3)	1-Butanol	1-Butyl benzoate	Mild floral-balsamic odor

ND, not described

[a] Esters formed after 15-min reaction of the SAAT recombinant enzyme with the substrates indicated were trapped using headspace solid phase microextraction and identified by GC-MS

[b] The ester product was reported to occur in cultivated (*Fragaria* × *ananassa*) strawberries (data from Zabetakis and Holden 1997)

[c] Ester properties are from Arctander (1969)

the availability of substrates, which may be dependent on the activity of catabolic pathways (e.g. lipid breakdown). Another factor, which may influence the ester composition, is the activity of esterases that perform the reverse reaction to AAT and which has been described for fruits, including strawberry (Ueda and Ogata 1976).

The fact that SAAT may use a vast array of acyl CoAs and alcohols as substrates might provide an explanation for the presence of more than a hundred different types of esters present in strawberry and other fruit species during ripening. It is also intriguing that *SAAT* is not a member of a multigene family in strawberry.

2.4.4 Other Candidate Genes Associated with Flavor Formation in Strawberry

The identification of *SAAT* provided us with a valuable expression profile to assist in the identification of other putative flavor-associated genes in strawberry. As described previously, fatty acids serve as the initial precursors for several groups of flavor and aroma compounds present in many fruits including strawberry. Aliphatic C-6 compounds (aldehydes, alcohols, acids and esters) contributing to strawberry fruit flavor are formed from unsaturated aliphatic C-18 fatty acids, linoleic ($C_{18:2}$) and linolenic acid ($C_{18:3}$), through the lipoxygenase/hydroperoxide lyase pathway (Croteau and Karp 1991). Plant fatty acid unsaturation begins with the conversion of 16:0 acyl carrier protein (ACP) and 18:0-ACP into 16:1-ACP and 18:1-ACP, respectively, by Δ9 stearoyl-ACP desaturase (Wang et al. 1996). From our microarray developmental study we observed high expression levels for genes encoding a strawberry Δ9 desaturase, an acyl carrier protein (ACP) and a gene putatively involved in fatty acid biosynthesis (malonyl-CoA decarboxylase) during the white stage of fruit development. The gene expression pattern of a putative strawberry pyruvate decarboxylase (*PDC*) was also investigated, because it may play a role in providing precursors for the formation of ethyl esters. The *PDC*, Δ9 stearoyl-ACP desaturase, malonyl-CoA decarboxylase and ACP expression profile correlated well with the *SAAT* expression profile during fruit development (Fig. 2.7). Interestingly, the activities of two other intermediate enzymes in the metabolism of fatty acids, lipoxygenase and hydroperoxide lyase were previously reported to increase steadily from the white to the red stage of strawberry fruit development (Perez et al. 1999). The similar expression pattern of the newly identified genes putatively related to flavor formation in strawberry with that of *SAAT* suggests that they may be under the same genetic regulation during fruit ripening.

2.5 Conclusion and Future Prospects

Despite the fact that the array technology is still in its infancy, many have demonstrated its power by using it for diverse applications such as global monitoring of

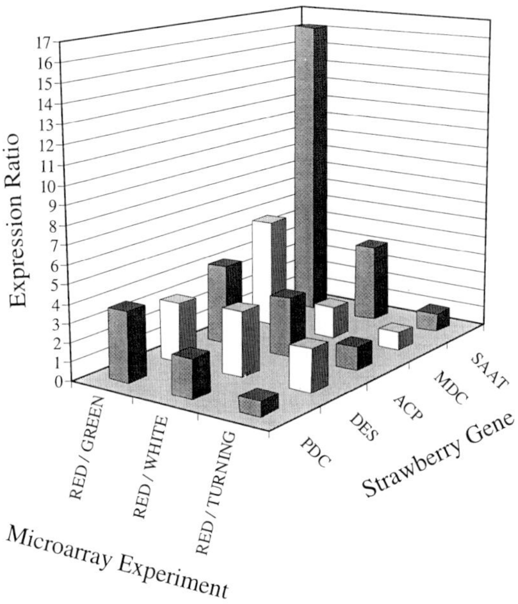

Fig. 2.7. Expression profiles of strawberry *SAAT*, malonyl-CoA decarboxylase (*MDC*), acyl carrier protein (*ACP*), desaturase (*DES*) and pyruvate decarboxylase (*PDC*) during strawberry fruit development as detected in microarray experiments. Expression ratios above 1 correspond to higher expression levels in the red stage compared to green or white or turning stages of development

gene expression, mutation detection and genetic mapping. It is anticipated that array technologies will become more refined in the future, making possible a completely new era of experimental approaches that will affect virtually all fields of biology. The flexible nature of arrays will pave the way for the exploration of more complex genomes (e.g. non-model species such as strawberry) to answer biological questions that simply cannot be answered by analyzing model species such as *Arabidopsis* (e.g. the genetic controls governing biosynthesis of specific secondary metabolites).

We embarked on a research program to genetically dissect development and stress programs in both strawberry achene and receptacle tissue. This study has provided us with a unique and comprehensive overview of signal transduction cascades, regulatory pathways and key metabolic pathways, coinciding in both achene and receptacle tissue, during the development and maturation of strawberry fruit and in response to stress programs (Aharoni et al. 2001, submitted). We hope to couple data arising from this research in the future to data provided by other functional genomic tools (proteomics and metabolomics), for the purpose of generating multicomponent databases linking together sequence, expression and metabolic data.

The identification of the *SAAT* gene involved in fruit flavor biogenesis is a prime example of how microarray technology can be used to aid in the dissection and exploitation of metabolic pathways that lead to the formation of flavor constituents, and as a tool to link gene to function. In recent years the demand for natural flavors/aromas has been steadily increasing. Synthetic organic chemicals constitute more than 80 to 90% (by weight and value) of the raw materials used in flavor and fragrance formulations, and problems often exist concerning production. Extraction from intact plants, conventional fermentation and in-vivo enzyme technology are currently also providing alternative routes for the commercial production of aroma chemicals. However, in many cases demand often outstrips supply, and some sought-after aroma/flavor compounds prelude isolation. An understanding of the precursors and characterization of enzymes involved in the pathways leading to the formation of flavors in fruit is essential for the production of natural flavors.

In this chapter we describe the identification of new candidate flavor-related genes by a similar principal used to identify *SAAT*. Their role in strawberry flavor formation remains yet to be explored. Since the isolation of the *SAAT* gene we have made major strides in the identification of alcohol acyltransferases from a number of other fruits (e.g. banana, apple, melon). The isolation of new fruit alcohol acyltransferases has enabled us to study the significant factors fundamental to the diversity of volatile ester profiles within and between fruit species. Our initial results suggest that substrate availability in each is a major determinant of both the qualitative and quantitative composition of flavor/aroma profiles.

Faced with increasingly sophisticated consumers who are becoming more and more demanding and adventurous, it is vitally important for today's flavor and fragrance manufacturers to create new and innovative flavors. The identification of new fruit alcohol acyltransferases provides a natural alternative route for biotechnology production of bio-flavors, natural aroma chemicals in the form of volatile esters. It will pave the way in the future for natural flavor production and the production of more exotic and novel flavors. Furthermore, it will provide the means for improving flavor characteristics of commercial fresh fruits lost through successive years of breeding and selection.

References

Aharoni A, Keizer LCP, Bouwmeester HJ, Sun ZK, Alvarez Huerta M, Verhoeven HA, Blaas J, van Houwelingen A, De Vos RCH, van der Voet H, Jansen RC, Guis M, Mol J, Davis RW, Schena M, van Tunen AJ, O'Connel AP (2000) Identification of the SAAT gene involved in strawberry flavor biogenesis by use of DNA microarrays. Plant Cell 12:647–661

Alizadeh AA, Eisen MB, Davis RE, Ma C, Lossos IS, Rosenwald A, Boldrick JG, Sabet H, Tran T, Yu X, Powell JI, Yang LM, Marti GE, Moore T, Hudson J, Lu LS, Lewis DB, Tibshirani R, Sherlock G, Chan WC, Greiner TC, Weisenburger DD, Armitage JO, Warnke R, Levy R, Wilson W, Grever MR, Byrd JC, Botstein D, Brown PO, Staudt LM (2000) Distinct types of diffuse large B-cell lymphoma identified by gene expression profiling. Nature 403:503–511

Arctander S (1969) Perfume and flavor chemicals. Steffen Arctander's Publications, Las Vegas

Bojorquez G, Gomez Lim MA (1995) Peroxisomal thiolase mRNA is induced during mango fruit ripening. Plant Mol Biol 28:811–820

Brenner S, Johnson M, Bridgham J, Golda G, Lloyd DH, Johnson D, Luo SJ, McCurdy S, Foy M, Ewan M, Roth R, George D, Eletr S, Albrecht G, Vermaas E, Williams SR, Moon K, Burcham T, Pallas M, DuBridge RB, Kirchner J, Fearon K, Mao J, Corcoran K (2000) Gene expression analysis by massively parallel signature sequencing (MPSS) on microbead arrays. Nature Biotech 18:630–634

Brown NF, Anderson RC, Caplan SL, Foster DW, McGarry JD (1994) Catalytically important domains of rat carnitine palmitoyltransferase II as determined by site-directed mutagenesis and chemical modification. Evidence for a critical histidine residue. J Biol Chem 269: 19157–19162

Brown PO, Botstein D (1999) Exploring the new world of the genome with DNA microarrays. Nat Genet 21(Suppl):S33–S37

Bruce W, Folkerts O, Garnaat C, Crasta O, Roth B, Bowen B (2000) Expression profiling of the maize flavonoid pathway genes controlled by estradiol-inducible transcription factors CRC and P. Plant Cell 12:65–79

Bucher P (1999) Regulatory elements and expression profiles. Curr Opin Struct Biol 9:400–407

Civello PM, Powell ALT, Sabehat A, Bennett AB (1999) An expansion gene expressed in ripening strawberry fruit. Plant Physiol 121:1273–1279

Croteau R, Karp F (1991) Origin of natural odorants. In: Muller PM, Lamparsky D (eds) Perfumes: art, science, and technology. Elsevier, Amsterdam, pp 101–126

de Wildt RMT, Mundy CR, Gorick BD, Tomlinson IM (2000) Antibody arrays for high-throughput screening of antibody–antigen interactions. Nat Biotech 18:989–994

DeRisi JL, Iyer VR, Brown PO (1997) Exploring the metabolic and genetic control of gene expression on a genomic scale. Science 278:680–686

Desprez T, Amselem J, Caboche M, Hofte H (1998) Differential gene expression in *Arabidopsis* monitored using cDNA arrays. Plant J 14:643–652

Diehn M, Eisen MB, Botstein D, Brown PO (2000) Large-scale identification of secreted and membrane-associated gene products using DNA microarrays. Nat Genet 25:58–62

Dudareva N, D'Auria JC, Nam K, Raguso RA, Pichersky E, Nam KH (1998) Acetyl-CoA:benzylalcohol acetyltransferase – an enzyme involved in floral scent production in *Clarkia breweri*. Plant J 14:297–304

Duggan DJ, Bittner M, Chen YD, Meltzer P, Trent JM (1999) Expression profiling using cDNA microarrays. Nat Genet 21(Suppl):S10–S14

Eisen MB, Spellman PT, Brown PO, Botstein D (1998) Cluster analysis and display of genome-wide expression patterns. Proc Natl Acad Sci USA 95:14863–14868

Fodor SPA, Read JL, Pirrung MC, Stryer L, Lu AT, Solas D (1991) Light-directed, spatially addressable parallel chemical synthesis. Science 251:767–773

Fujii T, Nagasawa N, Iwamatsu A, Bogaki T, Tamai Y, Hamachi M (1994) Molecular cloning, sequence analysis, and expression of the yeast alcohol acetyltransferase gene. Appl Environ Microbiol 60:2786–2792

Fujii T, Yoshimoto H, Nagasawa N, Bogaki T, Tamai Y, Hamachi M (1996) Nucleotide sequences of alcohol acetyltransferase genes from lager brewing yeast, *Saccharomyces carlsbergensis*. Yeast 12:593–598

Harada M, Ueda Y, Iwata T (1985) Purification and some properties of alcohol acetyltransferase from banana fruit. Plant Cell Physiol 26:1067–1074

Harpster MH, Brummell DA, Dunsmuir P (1998) Expression analysis of a ripening-specific, auxin-repressed endo-1,4-beta-glucanase gene in strawberry. Plant Physiol 118:1307–1316

Hegde P, Qi R, Abernathy K, Gay C, Dharap S, Gaspard R, Hughes JE, Snesrud E, Lee N, Quackenbush J (2000) A concise guide to cDNA microarray analysis. Biotechniques 29:548–562

Honkanen E, Hirvi T (1990) The flavour of berries. Dev Food Sci. Elsevier, Amsterdam, pp 125–193

Lander ES (1999) Array of hope. Nature Genet 21(Suppl):S3–S4

Lashkari DA, DeRisi JL, McCusker JH, Namath AF, Gentile C, Hwang SY, Brown PO, Davis RW (1997) Yeast microarrays for genome wide parallel genetic and gene expression analysis. Proc Natl Acad Sci USA 94:13057–13062

Lemieux B, Aharoni A, Schena M (1998) Overview of DNA chip technology. Mol Breed 4:277–289

Lipshutz RJ, Fodor SPA, Gingeras TR, Lockhart DJ (1999) High density synthetic oligonucleotide arrays. Nat Genet 21(Suppl):S20–S24

Llop Tous I, Dominguez Puigjaner E, Palomer X, Vendrell M (1999) Characterization of two divergent endo-beta-1,4-glucanase cDNA clones highly expressed in the nonclimacteric strawberry fruit. Plant Physiol 119:1415–1421

Lockhart DJ, Dong H, Byrne MC, Follettie MT, Gallo MV, Chee MS, Mittmann M, Wang C, Kobayashi M, Horton H, Brown EL (1996) Expression monitoring by hybridization to high-density oligonucleotide arrays. Nat Biotech 14:1675–1680

Maarse H (1991) Volatile compounds in foods and beverages. Dekker, New York

MacBeath G, Schreiber SL (2000) Printing proteins as microarrays for high-throughput function determination. Science 289:1760–1763

Manning K (1994) Changes in gene expression during strawberry fruit ripening and their regulation by auxin. Planta 194:62–68

Manning K (1998) Isolation of a set of ripening-related genes from strawberry: their identification and possible relationship to fruit quality traits. Planta 205:622–631

Medina Escobar N, Cardenas J, Valpuesta V, Munoz Blanco J, Caballero JL (1997a) Cloning and characterization of cDNAs from genes differentially expressed during the strawberry fruit ripening process by a MAST-PCR-SBDS method. Anal Biochem 248:288–296

Medina Escobar N, Cardenas J, Moyano E, Caballero JL, Munoz Blanco J (1997b) Cloning, molecular characterization and expression pattern of a strawberry ripening-specific cDNA with sequence homology to pectate lyase from higher plants. Plant Mol Biol 34:867–877

Mir KU, Southern EM (1999) Determining the influence of structure on hybridization using oligonucleotide arrays. Nat Biotech 17:788–792

Moyano E, Portero Robles I, Medina Escobar N, Valpuesta V, Munoz Blanco J, Caballero JL (1998) A fruit-specific putative dihydroflavonol 4-reductase gene is differentially expressed in strawberry during the ripening process. Plant Physiol 117:711–716

Nam Y, Tichit L, Leperlier M, Cuerq B, Marty I, Lelievre JM, Nam YW (1999) Isolation and characterization of mRNAs differentially expressed during ripening of wild strawberry (*Fragaria vesca* L.) fruits. Plant Mol Biol 39:629–636

Olias JM, Sanz C, Rios JJ, Perez AG (1995) Substrate specificity of alcohol acyltransferase from strawberry and banana fruits. ACS Symp Ser 596:134–141

Pereira A (2000) A transgenic perspective on plant functional genomics. Transgenic Res 9:245–260

Perez AG, Rios JJ, Sanz C, Olias JM (1992) Aroma components and free amino acids in strawberry variety Chandler during ripening. J Agric Food Chem 40:2232–2235

Perez AG, Sanz C, Olias JM (1993) Partial purification and some properties of alcohol acyltransferase from strawberry fruits. J Agric Food Chem 41:1462–1466

Perez AG, Sanz C, Olias R, Rios JJ, Olias JM (1996) Evolution of strawberry alcohol acyltransferase activity during fruit development and storage. J Agric Food Chem 44:3286–3290

Perez AG, Sanz C, Olias R, Olias JM (1999) Lipoxygenase and hydroperoxide lyase activities in ripening strawberry fruits. J Agric Food Chem 47:249–253

Perkins-Veazie P (1995) Growth and ripening of strawberry fruit. Hort Rev 17:267–297

Reed LJ, Hackert ML (1990) Structure-function relationships in dihydrolipoamide acyltransferases. J Biol Chem 265:8971–8974

Reymond P, Weber H, Damond M, Farmer Edward E (2000) Differential gene expression in response to mechanical wounding and insect feeding in *Arabidopsis*. Plant Cell 12:707–719

Richmond CS, Glasner JD, Mau R, Jin HF, Blattner FR (1999) Genome-wide expression profiling in *Escherichia coli* K-12. Nucleic Acids R27:3821–3835

Roscher R, Schreier P, Schwab W (1997) Metabolism of 2,5-dimethyl-4-hydroxy-3(2H)-furanone in detached ripening strawberry fruits. J Agric Food Chem 45:3202–3205

Ruan Y, Gilmore J, Conner T (1998) Towards *Arabidopsis* genome analysis: monitoring expression profiles of 1400 genes using cDNA microarrays. Plant J 15:821–833

Scharpf LG Jr, Chandan RC (1989) Biotechnology and natural flavors for yogurt. National Yogurt Association, Mclean, Virginia

Schena M, Shalon D, Davis RW, Brown PO (1995) Quantitative monitoring of gene expression patterns with a complementary DNA microarray. Science 270:467–470

Schwab W (1998) Application of stable isotope ratio analysis explaining the bioformation of 2,5-dimethyl-4-hydroxy-3(2H)-furanone in plants by a biological Maillard reaction. J Agric Food Chem 46:2266–2269

Seymour GB, Taylor JE, Tucker GA (1993) Biochemistry of fruit ripening. Chapman and Hall, London

Shaw WV, Leslie AG (1991) Chloramphenicol acetyltransferase. Annu Rev Biophys Chem 20: 363–386

St Pierre B, Laflamme P, Alarco AM, De Luca V (1998) The terminal O-acetyltransferase involved in vindoline biosynthesis defines a new class of proteins responsible for coenzyme A-dependent acyl transfer. Plant J 14:703–713

Syvanen AC (1999) From gels to chips: "minisequencing" primer extension for analysis of point mutations and single nucleotide polymorphisms. Hum Mutat 13:1–10

Trainotti L, Spolaore S, Pavanello A, Baldan B, Casadoro G (1999) A novel E-type endo-beta-1,4-glucanase with a putative cellulose-binding domain is highly expressed in ripening strawberry fruits. Plant Mol Biol 40:323–332

Ueda Y, Ogata K (1976) Studies of biosynthesis of esters in fruit volatiles. II. Esterification of added alcohol in separated cells from banana, strawberry and melon. J Food Sci Technol Tokyo 23: 288–294

Ueda Y, Tsuda A, Bai JH, Fujishita N, Chachin K (1992) Characteristic pattern of aroma ester formation from banana, melon, and strawberry with reference to the substrate specificity of ester synthetase and alcohol contents in pulp. J Jpn Soc Food Sci Technol 39:183–187

van Hal NLW, Vorst O, van Houwelingen A, Kok EJ, Peijnenburg A, Aharoni A, van Tunen AJ, Keijer J (2000) The application of DNA microarrays in gene expression analysis. J Biotech 78:271–280

Wang CL, Chin CK, Ho CT, Hwang CF, Polashock JJ, Martin CE (1996) Changes of fatty acids and fatty acid-derived flavor compounds by expressing the yeast Delta-9 desaturase gene in tomato. J Agric Food Chem 44:3399–3402

Wang DG, Fan JB, Siao CJ, Berno A, Young P, Sapolsky R, Ghandour G, Perkins N, Winchester E, Spencer J, Kruglyak L, Stein L, Hsie L, Topaloglou T, Hubbell E, Robinson E, Mittmann M, Morris MS, Shen NP, Kilburn D, Rioux J, Nusbaum C, Rozen S, Hudson TJ, Lipshutz R, Chee M, Lander ES (1998) Large-scale identification, mapping, and genotyping of single-nucleotide polymorphisms in the human genome. Science 280:1077–1082

Wilkinson JQ, Lanahan MB, Conner TW, Klee HJ (1995) Identification of mRNAs with enhanced expression in ripening strawberry fruit using polymerase chain reaction differential display. Plant Mol Biol 27:1097–1108

Winzeler EA, Richards DR, Conway AR, Goldstein AL, Kalman S, McCullough MJ, McCusker JH, Stevens DA, Wodicka L, Lockhart DJ, Davis RW (1998) Direct allelic variation scanning of the yeast genome. Science 281:1194–1197

Woodward JR (1972) Physical and chemical changes in developing strawberry fruits. J Sci Food Agric 23:465–473

Yamakawa Y, Goto S, Yokotsuka I (1978) Fractionation and some properties of acetic-ester synthesizing enzyme from *Cladosporium cladosporioides* No. 9. Agric Biol Chem 42:269–274

Yoshioka K, Hashimoto N (1984) Ester formation by brewers' yeast during sugar fermentation. Agric Biol Chem 45:333–340

Zabetakis I, Holden MA (1997) Strawberry flavour: analysis and biosynthesis. J Sci Food Agric 74:421–434

3 Testing for Taste and Flavour of Beer

T. Yonezawa[1] and T. Fushiki[2]

3.1 Introduction

Our knowledge of the taste and flavour of beer has been obtained through two approaches: that of the chemist and that of the taster. In this chapter, we begin by summarizing the methods used in the study of the taste and flavour of beer. The everlasting aim in testing of taste and flavour is the steady production of beer with satisfactory good quality.

In the earliest days when analytical methods did not exist, sensory testing was the only reliable method to obtain good-quality beer. The level of brewing technology has improved little by little at a very slow pace. In contrast to the long history of beer sensory evaluation, the use of chemical methods for measurement of beer constituents has developed only in the last 200 years. Chemical analysis has contributed much to our knowledge, however, because most chemical methods give results with less variability than do sensory panelists. Knowledge of the flavour chemistry of beer, essentially the relationship between beer chemical composition and its sensory characteristics, has evolved only quite recently in the more than 4,000 years during which beer has been brewed.

To make headway in flavour research, one must look at it from two sides: sensory analysis and physico-chemical analysis. Meilgaard (1991) developed a research plan in order to discover the flavour elements of beer. A running inventory of compounds in beer has been discovered (over 900 to date, not counting isomers) by chemists, who have determined the concentration of each compound in various types of beer, and estimated a rough flavour threshold for each so as to discover those likely to be flavour active (perhaps 100). The concentration of the compound itself is less important. Then they obtain or synthesize those compounds likely to be flavour active, and purify each to a state of sensory purity; thus they determine the compound's threshold in a not-too-flavourful beer.

On the other hand, tasters have developed a terminology system for reference, with purification and directions of use for each reference, and methods of analy-

[1] Research Laboratory for Brewing, Technology Development Department, Kirin Brewery Co., Yokohama 230-8628, Japan

[2] Laboratory of Nutrition Chemistry, Division of Applied Life Sciences, Graduate School of Agriculture, Kyoto University, Kyoto 606-8502, Japan

tical taste testing, in particular, threshold tests, descriptive tests and scaling methods for flavour intensity.

The combined approach to relate sensory and chemical properties provides us with knowledge of which beer constituents contribute to its flavour. By taking this approach we can discover how the flavour intensity of added substances varies with the concentration added, and the rules of flavour interaction in mixtures by adding mixtures of compounds. Descriptive testing and detailed chemical analysis of numerous beers are conducted, and the data analyzed by multivariate statistical procedures. Then a hypothesis is developed, and the compounds having the expected type of flavour are tested to confirm or reject the hypothesis.

Here, the characteristics of taste and flavour compounds of beer are reviewed, and sensory testing for beer is introduced from many aspects.

3.2 Characteristics of Taste and Flavour Compounds in Beer

For beer, knowledge concerning flavour is still developing and it is not yet possible to produce a convincing artificial beer flavour. In part, this appears to be because beer flavour is both complicated and subtle. This is related to the diverse kinds of processes necessary for brewing. Barley malt, obtained by germination of barley grain under special conditions followed by a kilning process, is milled and then extracted with hot water according to a temperature program. During this process, called mashing, enzymes from the malt are active in the hydrolysis and release of malt constituents. The resulting extract, which is called sweet wort, is boiled with hops to give the characteristic bitter taste and flavour to the final beer. The cooled wort is fermented with an appropriate yeast strain and stored at a low temperature.

3.2.1 Taste and Flavour Substances in Beer

Meilgaard (1982) compiled a list of approximately 850 compounds reported in beer. There is no key flavour component which hits upon beer itself. Usually, beer contains 30–40 g/l of extract, constituting dextrins, oligosaccharides, maltose, amino acids, glycerin, minerals, hop bitter substances, tannins, organic acids, etc. Dextrins and oligosaccharides, which are the quantitatively dominant components, have a moderate influence on the taste of beer. On the other hand, maltose, amino acids and organic acids, which have a peculiar, strong taste, are limited in finished beer. Except for bitter substances from hops, there are no peculiar taste components in beer.

Many flavours make up the characteristics of beer in complete harmony. Therefore, any off-flavours, which are caused by sunlight and oxidization, severely spoil the harmony of beer flavour.

3.2.2 Threshold

In order to develop meaningful information about flavour chemistry it is necessary to relate information about the concentration of a compound present at a concentration that can be perceived in the product of interest. This is the question: "Is there a perceptible amount of compound, or not?" The concentration of a substance that can just be perceived is called its threshold. There are several different ways to define and measure thresholds.

The "difference threshold" is commonly used and is defined as the extent of change in the stimulus necessary to produce a noticeable difference (Meilgaard et al. 1999). Since the threshold result depends on the medium used for testing, it is important that threshold determinations are made not only in simple beer, but also in a particular type, and perhaps even brand beer, which attract concern.

3.2.3 Flavour Units

Nearly all flavour compounds in beer have different thresholds. It has been found to be convenient to express the amount of compound present in a sample in flavour terms as the flavour unit (FU), where the concentration present is divided by the difference threshold (Meilgaard 1975a). The key question in relating observed concentrations of compounds with their flavour thresholds in beer is "What constitutes proof of a flavour effect?" In other words, given a threshold and a concentration for a compound, what can we conclude about whether the compound may be perceived. Meilgaard (1982) suggested as a rule of thumb that compounds present in amounts greater than 2.0 FU are considered to be highly flavour active. Those with 1.0–2.0 FU are thought to be flavour active. Compounds occurring at slightly lower concentrations may either be flavour active, because they are perceptible to individuals who are particularly sensitive to those compounds, or be perceived through flavour interactions with other similarly flavoured compounds. Concentrations in the range of 0.5–1.0 FU fall into this category. Compounds present between 0.2 and 0.5 FU are possibly flavour active, but in most cases are not important. Compounds present at 0.2 FU and below are unlikely to have an impact on flavour (Meilgaard 1982).

In the case of normal pale lager only three beer components are typically present at concentrations of more than 2.0 FU. These are ethanol, carbon dioxide and the hop bitter substances. A great many other compounds contribute to beer flavour. In a number of cases these other compounds are jointly perceived, although the sum of the similarly flavoured compounds in each group is usually only slightly above 1 FU. Approximately 25 substances in beer are present by themselves at a concentration of more than 0.5 FU (Siebert 1988).

The threshold of detection for the 222 compounds was found to range from 1.3% for ethanol to 0.00007 ppb for tert amyl mercaptan. The latter is the lowest flavour threshold reported for any compound in any medium (Meilgaard 1974).

3.2.4 Bitterness from Hops

The primary bitter taste in beer is from hops. The α- and β-acids are the principal bittering components of hops. The difference between the α- and β-acids is that the β-acids have another prenyl side chain in the bottom position instead of a hydroxyl group. This makes a profound difference in the way the two classes of acids behave during the brewing process. The α-acids, which are only soluble at about 3 ppm in cold water, are isomerized to the much more soluble iso-α-acids during wort boiling. This thermal conversion is necessary not only for solubilization but also for activation of hop bitterness. The iso-α-acids are about nine times more bitter than the α-acids.

The α- and β-acids are oxidized in the hop during storage to other more water-soluble compounds. Kowaka and Kokubo (1977) found at least 90 components contributing to the bitterness of beer. The constitution of beer bitterness was calculated by the bitterness intensity and the contents of divided subfractions. Iso-α-acids are the most bitter compounds, which have a threshold of 4 ppm, and account for 67% of the beer bitterness.

There are three major and several minor α-acids. The only difference between them is the side-chain R (as shown in Fig. 3.1.) The relation between perceived bitterness and the different iso-α-acids has been re-evaluated. The most bitter compound was *cis*-isohumulone, which was 1.82 times more bitter than *trans*-

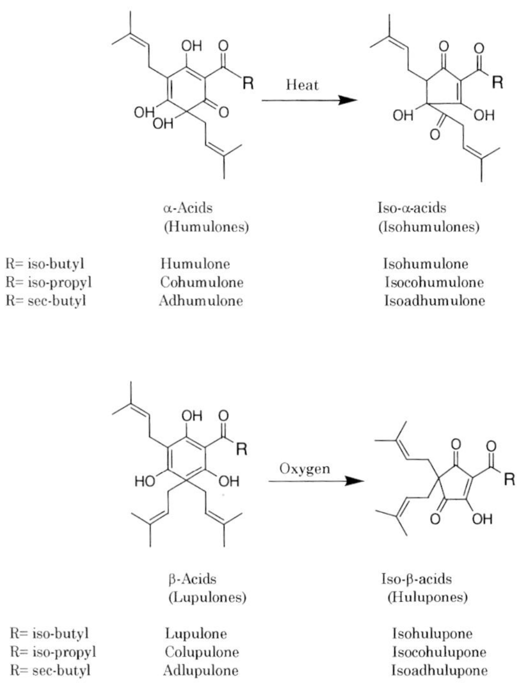

Fig. 3.1. Chemical structures of major hop α- and β-acids, and their transformations

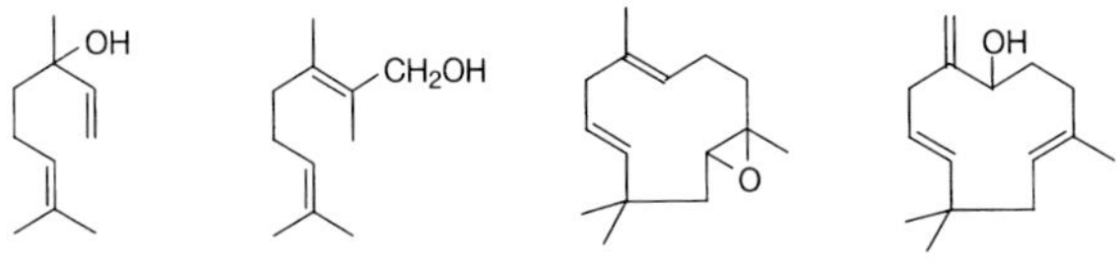

Linalool Geraniol Humulene epoxide I Humulenol II

Fig. 3.2. Structures of some hop-oil-derived compounds

Table 3.1. Typical alcohols in beer. (Meilgaard 1982; Siebert 1988)

Compounds	Threshold (ppm)	Typical concentration (ppm)	FU	Flavour description
Ethanol	13,000	25,000–50,000	1.8–3.6	Alcoholic
Isobutanol	200	4–57	0.0–0.3	Alcoholic
Isoamyl alcohol (3-methylbutanol)	70	25–123	0.4–1.8	Alcoholic, vinous
Amyl alcohol active (2-methylbutanol)	65	7–34	0.1–0.5	Alcoholic, vinous
2-Phenylethanol	125	5–102	0.0–0.8	Rose

isohumulone. The least bitter was *trans*-isocohumulone, which was about 0.74 times as bitter as *trans*-isohumulone (Hughes and Simpson 1996).

3.2.5 Hop Aroma

We believe that a fine hop aroma in beer is as important as a fine bitterness. Especially for a traditional Pilsner beer or a traditional British ale, a fine hop aroma is no doubt "the soul of beer". Several of the hop-oil-derived compounds occur in beer at levels near or above their flavour thresholds. They are linalool, geraniol, humulene epoxide I and humulenol II (Fig. 3.2; Siebert 1994). A fine hop aroma, especially so-called noble hop aroma, is still an elusive quality. We still do not know what compounds cause it.

3.2.6 Alcohols

Among the alcohols, ethanol is present at high concentrations depending on the type of beer. Higher alcohols, often referred to as fusel alcohols, are generally present in high amounts in all beers. The typical alcohols in beer are listed in Table 3.1 (Meilgaard 1982; Siebert 1988). The propyl, butyl and amyl alcohols add primarily an alcoholic flavour to beer with characteristic flavour. 2-Phenylethanol has a pure rose flavour beyond its threshold, but at a normal concentration it is thought to be a positive contribution to beer flavour. The aliphatic alcohols with 6–10

Table 3.2. Typical acids in beer. (Meilgaard 1982; Siebert 1988)

Compounds	Threshold (ppm)	Typical concentration (ppm)	FU	Flavour description
Carbon dioxide	1,000	3,000–5,500	3.0–5.5	Carbonation
Acetic acid	175	30–280	0.2–1.6	Acetic, acidic, vinegar-like
Butyric acid	2.2	0.5–3.3	0.2–1.5	Butyric, butter-like, cheesy
Isovaleric acid	1.5	0.1–3.4	0.1–2.3	Cheesy, old-hop-like, sweaty
Caproic acid	8	1–6	0.1–0.8	Caprylic, vegetable, oil-like, sweaty
Caprylic acid	15	2–15	0.1–1.0	Caprylic
Capric acid	10	0.1–4	0.0–0.4	Tallowy, caprylic

carbon atoms have coconut-like flavours and are probably not flavour-active in beer unless the cumulated FU exceeds 1 U (Meilgaard 1974).

Glycerol occurs in beer at concentrations of around 1.1–2.2 g/l, but it can hardly be regarded as a beer flavour (Meilgaard 1982).

3.2.7 Acids

Carbon dioxide gives the sparkling carbonation effect to taste and mouthfeel of beer. A total of 110 acids have been reported in beer, but the majority of these are not flavour active. The important acids for flavour are caprylic, caproic, isovaleric and possibly phenylacetic, all of which are unpleasant. The two first-mentioned have a goat-like smell. Isovaleric acid appears to be mainly responsible for the flavour of old hops, and phenylacetic acid has a phenolic, astringent flavour (Meilgaard 1974).

The typical acids in beer are listed in Table 3.2 (Meilgaard 1982; Siebert 1988).

3.2.8 Esters

A few esters are very flavour active, especially ethyl caproate (apple-like), isoamyl acetate (banana) and ethyl acetate (fruity, solvent-like). It is interesting that the low-molecular-weight esters have the flavours of temporal fruits, while longer-chain esters show the flavours of subtropical fruits, and the longest-chain members of tropical fruits (Meilgaard 1974).

The typical esters in beer are listed in Table 3.3 (Meilgaard 1982; Siebert 1988).

3.2.9 Ketones and Aldehydes

Ketones were found to be of little importance in beer flavour, with the exception of vicinal diketones. Diacetyl has the lowest threshold and the most unpleasant

Table 3.3. Typical esters in beer. (Meilgaard 1982; Siebert 1988)

Compounds	Threshold (ppm)	Typical concentration (ppm)	FU	Flavour description
Ethyl acetate	30	8–48	0.3–1.6	Solvent-like, fruity, sweet
Isobutyl acetate	1.6	0.01–0.25	0.01–0.15	Banana-like, sweet, fruity
Isoamyl acetate	1.2	0.6–7	0.5–5.8	Banana, estery, solvent-like
Ethyl caproate	0.21	0.1–0.5	0.5–2.4	Apple-like, fruity, aniseed-like, sweet
Ethyl caprylate	0.9	0.1–1.5	0.1–1.7	Apple-like, sweet, fruity
Ethyl caprate	1.5	0.01–1	0.01–0.7	Caprylic, fruity
2-Phenyl acetate	3.8	0.05–2	0.01–0.5	Rose-like, honey-like, apple-like, sweet

Table 3.4. Typical ketones and aldehydes in beer. (Meilgaard 1982; Siebert 1988)

Compounds	Threshold	Typical concentration	FU	Flavour description
Acetaldehyde	10 ppm	1.2–24 ppm	0.1–2.4	Apple-skin-like, green-leaves-like, fruity
Nonanal	18 ppb	1–14 ppb	0.06–0.8	Insect-like, orange-like, aldehydic
trans-2-Nonenal	0.11 ppb	0.0–2 ppb	0.0–18	Cardboard-like, papery, oxidized
Diacetyl (2,3-butanedione)	0.1 ppm	0.01–0.4 ppm	0.1–4	Diacetyl, butterscotch-like, young-red-wine-like

flavour. Aldehydes show a very low threshold and sometimes very unpleasant flavour. It is noteworthy that the short-chain unsaturated aldehydes have a higher threshold and relatively pleasant (grassy or leaf-like) flavour. At longer chain lengths, C-7 to C-10, this is reversed.

The typical ketones and aldehydes in beer are listed in Table 3.4 (Meilgaard 1982; Siebert 1988). In each group (alcohol, esters, aldehydes, etc.), it was found that low-molecular-weight compounds with 1–4 carbon atoms have a high flavour threshold, i.e. weak flavour. With increasing chain length of the molecule, the flavour becomes stronger until a maximum flavour strength is reached, which for most groups of aliphatic compounds occurs at 8–10 carbon atoms. Longer-chain molecules again have a weaker flavour and are fatty in nature. These facts support the hypothesis that a high flavour strength requires a certain solubility in both the lipophilic and the hydrophilic portions of the olfactory membrane.

Nonanal has a threshold of 15–20 ppb and insect-like and orange-like aldehydic flavour, and *trans*-2-nonenal has a threshold of 0.1 ppb and an even more unpleasant cardboard-like flavour. The nonenal flavour becomes strongly noticeable in most beers when they are shelf-stored at a high-temperature. *trans*-2-Nonenal is recognized as a typical oxidized flavour of beer (Meilgaard 1974).

3.2.10 Sulfur Compounds

Some sulfur compounds show extremely low threshold (over three orders of magnitude below our present limit of detection by analysis). For this reason the sporadic or chronic problems of sulfury flavours in breweries are unlikely to be completely resolved by the application of analytical chemistry. We must rely on sensory testing, as we have in the past. The flavour of mercaptans varies from rotten eggs to rotten onions to rotten fish as the chain length of the molecule increases. The dialkyl sulfides, of which dimethyl sulfide (DMS) is common in beer, have a different flavour similar to that of cooked cabbage or sweet corn (Meilgaard 1974).

The correlation coefficient between DMS levels in beers and perceived DMS character, however, was low (Hegarty et al. 1995). The typical sulfur compounds in beer are listed in Table 3.5 (Meilgaard 1982; Siebert 1988).When ultraviolet light penetrates into beer, a skunky, sulfury odour called sun-struck or light-struck flavour can form. Off-flavour of this type is attributed to 3-methyl-2-butene-1-thiol, a product of the degradation of isohumulones from hops in the presence of hydrogen sulfide (Fig. 3.3; Kuroiwa and Hashimoto 1961).

By applying the information presented here the flavour chemist can explain or foresee the alcoholic/aromatic, the estery/fruity and fatty/rancid qualities in beer flavour, just as application of the test for isohumulones allows an estimate of

Table 3.5. Typical sulfur compounds in beer. (Meilgaard 1982; Siebert 1988)

Compounds	Threshold (ppb)	Typical concentration (ppb)	FU	Flavour description
Hydrogen sulfide (free)	8	0.0–20	0.0–2.5	Sulfidic, boiled-egg-like, rotten-egg-like
Dimethyl sulfide	33	10–200	0.6–12	Cooked vegetable, sweet corn
Diethyl sulfide	3	0–5	0.0–1.7	Cooked vegetable, onion
Tert.Butyl mercaptan	0.8	0–1	0.0–1.2	Putrefaction of onion
Ethyl mercaptan	1.7	0–1	0.0–0.6	Putrefaction of leek

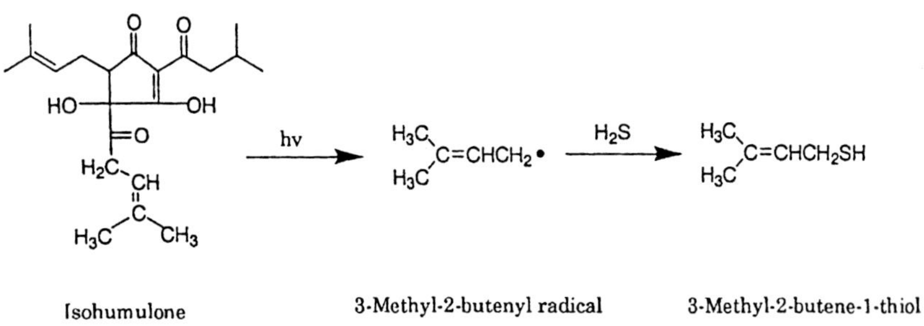

Fig. 3.3. Formation of light-struck flavour and its chemical structure

the bitter qualities. However, such important flavours as the malty, hoppy and oxidized flavours have not yet been fully explained (Meilgaard 1974).

3.2.11 Some Notes on Thresholds

There are great differences between panelists in their individual thresholds for given compounds. Each taster tends to be sensitive to some compounds, average to others and insensitive to some. This means that each of us lives in his or her own world when it comes to flavour. The threshold for compounds known to contribute to beer flavour span a range of over ten orders of magnitude (Meilgaard 1975b). This leads to some potential problems in determining flavour thresholds and attributing the correct sensory character to compounds. Even for compounds that are considered to be reasonably pure chemically, i.e., purer than 99%, it is possible to have a strong flavoured impurity which can contribute more flavour than the major constituent.

Another observation is that the sensory note produced by a given compound may change with its concentration. For example, 1-octen-3-one is perceived as mushroom-like at a low concentration but metallic at a higher level (Swoboda and Peers 1977). At least in theory, two or more compounds may interact in flavour perception. One may mask or hide the flavour of another, or compounds may be jointly or additively perceived. We find it very hard to believe that there is no masking effect in beer because one only has to compare the thresholds of compounds in water to those in beer and one gets very different answers.

Meilgaard and others have carried out experiments to test for the occurrence of synergism of compounds in beer flavour and found that a number of similarly flavoured compounds are approximately additive in flavour. The degree of interaction is measured, and the perceived flavour units of a mixture divided by the sum of the flavour units of the individual components present. It would be 1.0 if the compounds were perfectly additive (Meilgaard et al. 1971). In practice, mixtures of similarly flavoured compounds had degrees of interaction ranging between 0.7 and 1.3. Meilgaard found that substances of similar flavours are perceived as additives, while substances with widely different flavours are perceived as independent.

Two banana-like esters, isobutyl acetate and isoamyl acetate, are approximately additive. The two apple-like esters, ethyl caproate and ethyl caprylate, show a small degree of apparent synergism. The two different flavours, octanoic acid and ethyl acetate, are almost independent. Slight evidence of actual suppression by caproate was seen in the mixture with isoamyl alcohol (Meilgaard 1982).

A survey of commercial lagers showed, however, that a strong negative correlation between perceived DMS and the level of 2-phenyl ethanol existed. The survey showed that the addition of 2-phenyl ethanol to lager suppressed the perception of DMS, explaining the relation as causative (Hegarty et al. 1995).

3.2.12 Effects of Carbonation

We should keep in mind that thresholds of flavours are influenced by carbonation. Sufficient carbonation improves the settlement of each flavour in beer. On the one hand, surplus carbonation stimulates the tongue strongly and hinders the perception of the desirable flavour of beer. A shortage of carbonation lowers the differential threshold value, and therefore the defects could be more tangible.

The results of some tastings of degassed and carbonated lager beer show that the presence of CO_2 raises the threshold level of acetaldehyde, ethylacetate, isoamylacetate and isoamylalcohol by a factor of about 2. However, the diacetyl, methyl nonyl ketone and phenylethyl alcohol threshold levels were not affected by the presence of CO_2 (Harrison and Collins 1968). You could say that draft beer from a cask or keg may be more flavourful than bottle or can beer because the CO_2 has a masking effect.

3.3 Sensory Testing for Taste and Flavour of Beer

We cannot go back to the early time when brewing had just begun, but we can speculate that initially judgements were obviously purely sensory based. The earliest judgement was likely to have been mainly whether the beer was drinkable or undrinkable. Later, other types of judgements developed: "This beer tastes worse than the last one." Eventually this led to the development of descriptive terminology, where attempts at characterizing flavour were made with some flavour description

3.3.1 Descriptive Terminology

When descriptive language first evolved, it was clearly non-standardized and heavily dependent on the experience of the person who produced the description. For example, "This beer tastes like yesterday's milk", or "This beer smells like leaves that I once picked." In fact, this reliance on sensory experience is one of the reasons why working with data from a consumer taste panel is difficult. Different people have had different sensory experiences and make different associations, resulting in non-standard and inconsistent terminology. On the other hand, trained panels have a common sensory experience. They have been given examples of particular flavour notes and trained to call each by a particular descriptive name. In the early days this was done by providing examples of beers that were different in some respect.

3.3.2 Standard Terminology for Beer Flavour

In order to meet the dual needs of enabling brewers to communicate effectively about flavour and naming and defining each separately identifiable flavour note in beer, the joint working group of the European Brewery Convention, the American Society of Brewing Chemists and the Master Brewer's Association of the Americas developed a system of flavour terminology, called the Beer Flavor Wheel.

The system is based on the following principles (Meilgaard et al. 1979). Each separately identifiable flavour characteristic has its own name. There is no duplication of terms for the same flavour characteristic. (In practice, to cover all flavour characteristics leads to conflict with no overlapping on terms, so there are some exceptions.) Similar flavours are placed together. Subjective terms such as good/bad, young/mature, balanced/unbalanced are not included. The meaning of each term is illustrated with readily available reference standards. The flavour terminology system is composed of 14 classes. The 44 objective terms of the first tier and the additional 78 terms of the second tier comprise the Beer Flavour Wheel (Table 3.6; Meilgaard et al. 1979).

Lengthy discussions on the descriptive terms led to a choice of words that have the same meaning in different countries. These discussions revealed some marked differences based on the history of experience of panelists. In North America the "light-struck" flavour was usually called "skunky" because of its resemblance to the defensive aroma chemical produced by a native animal called the skunk. However, this animal is not native to Europe. As a result, the term "light-struck" was agreed upon. Flavour terms cannot be adequately defined other than by use of reference standards. Twenty-six standards had been confirmed already.

Professional training in different beer flavours was available to few people, since it required access to sophisticated laboratory facilities and skilled chemists. With the launch of the Beer Flavour Standards Kit by a new UK company, Flavor ActiV Limited, it has become much more convenient to teach people to identify different beer flavours (Boughton et al. 1999). Each highly purified flavour is stabilized as a free-flowing white powder and measured into individual capsules. This Beer Flavour Standards Kit currently covers 24 typical flavours of beer.

3.3.3 Descriptive Test

The type of sensory test used for characterizing beer flavour is called a descriptive test. The format for a descriptive test can be a blank sheet of paper on which a description of a beer is written or a form on which descriptors of flavours commonly experienced in beer are printed. Often these descriptors are a subset selected from the Flavour Wheel to include the flavours typically encountered in a brewery's products. In some cases panelists are asked to evaluate the intensity of perceived flavour notes.

It is interesting to consider how panelists actually accomplish descriptive evaluations. At any given moment there is a limit to the number of simultaneous

Table 3.6. Flavour terminology in the Flavour Wheel. (Meilgaard et al. 1979)

Class (term)	First tier term	Second tier term
Class 1 (aromatic, fragrant, fruity, floral)	Alcoholic	Spicy, vinous
	Solvent-like	Plastics, can-liner, acetone
	Estery	Isoamyl acetate, ethyl hexanoate, ethyl acetate
	Fruity	Citrus, apple, banana, blackcurrant, melony, raspberry, strawberry
	Acetaldehyde	
	Floral	2-Phenylethanol, geraniol, perfumy
	Hoppy	Kettle-hop, dry-hop, hop oil
Class 2 (resinous, nutty, green, grassy)	Resinous	Woody
	Nutty	Walnut, coconut, beany, almond
	Grassy	Freshly cut grass, straw-like
Class 3 (cereal)	Grainy	Husky, corn grits, mealy
	Malty	
	Worty	
Class 4 (caramelized, roasted)	Caramel	Molasses, licorice
	Burnt	Bread-crust, roast barley, smoky
Class 5 (phenolic)	Phenolic	Tarry, bakelite, carbolic, chlorophenol, iodoform
Class 6 (soapy, fatty, diacetyl, oily, rancid)	Fatty acid	Caprylic, cheesy, isovaleric, butyric
	Diacetyl	
	Rancid	Rancid oil
	Oily	Vegetable oil, mineral oil
Class 7 (sulfury)	Sulfury	
	Sulfidic	H2S, mercaptan, garlic, light-struck, autolysed, burnt rubber, shrimp-like
	Cooked vegetable	Parsnip/celery, DMS, cooked cabbage, cooked sweet corn, cooked tomato, cooked onion
	Yeasty	Meaty
Class 8 (oxidized, stale, musty)	Stale	
	Papery	
	Leathery	
	Moldy	Earthy, musty
Class 9 (sour, acidic)	Acidic	
	Acetic	
	Sour	
Class 10 (sweet)	Sweet	Honey, jam-like, vanilla, primings, syrupy, oversweet
Class 11 (salty)	Salty	
Class 12 (bitter)	Bitter	
Class 13 (mouthfeel)	Alkaline	
	Mouth coating	
	Metallic	
	Astringent	Drying
	Powdery	
	Carbonation	Flat, gassy
Class 14 (fullness)	Body	Watery, characterless, satiating, thick

flavour impressions that a person can hold in his or her mind or attend to. The sensory panelist can focus his/her attention on specific flavour characters in turn to evaluate them. At first he/she focuses on some volatile aroma characteristics, such as estery, acetic and aldehydic, one after another. Then he/she continues focusing on the other characteristics of flavour and taste. This is a strictly analytical manner, which is similar to the perceiver who can evaluate each sound of different instruments in the orchestra at different times by focusing his/her hearing on the sound in turn.

3.3.4 Difference Tests

Difference tests ask "Are these samples different?" Probably the most familiar test is the triangular test. The triangular test at different concentrations is commonly used for threshold determination. Typically, statistics are employed to analyze results from difference tests and a 95% confidence level is often used. Statisticians view each sensory test format as suitable for providing the answer to a single question using a prescribed statistical method. A difference test, like the triangular test, in its purest form thus asks only "Is there a difference between these samples?" without any subsequent questions. The practically used tests are a combination of rating, ranking and descriptive tests which often may ask for an intensity rating for each descriptor.

3.3.5 Bias in Sensory Verdicts

If one is to obtain reproducible verdicts, it is necessary to know about some physiological and psychological factors that cause bias in sensory verdicts (Meilgaard 1991). Typical examples of the physiological factor "adaptation" are as follows: the perceived bitterness declined over time after repeated ingestion of bitter beers whereas it increased for less bitter beers (King and Moreau 1996). It is difficult to detect a slight saltiness of beer because of the presence of sodium already in saliva. After rinsing out one's mouth with water it can be easier to detect a salty taste.

3.3.6 Application of Taste Sensor

Analysis of the quantities of the contained taste substances by conventional chemical sensors is not enough to reproduce the taste sense of humans. It may be difficult to correlate the state mixed from many taste substances with the taste sense. On the other hand, the physical value measured by, for example, a pH meter (to measure sourness) or a refractometer (for sweetness) is affected by extra substances (e.g. starch) which do not elicit taste feeling in humans. In addition, the value cannot express such interactions between taste substances as a synergistic and a suppressing effect, which appear in human taste feeling.

In biological taste reception, the lipids in the biological membrane are suggested to play an important role. Lipid membranes have been investigated in detail as a candidate of transducer of the taste sensor, so far by Toko et al. These workers confirmed that the lipid membranes could respond to taste substances so as to discriminate between basic taste qualities (Toko et al. 1994). Recently, they constructed a multichannel electrode whose transducer is composed of lipid membranes; the electrode had a sensitivity, reproducibility and durability greater than those of humans. The taste was measured in a similar manner to the human gustatory sensation so that the sensor showed a pattern similar to the same group of tastes: similar outputs were obtained for, for example, sucrose, glucose and aspartame (synthetic sweetness) belonging to sweet substances.

The interactions between taste substances were also detected to produce biological taste feeling of, for example, the synergistic effect in umami substances, MSG and IMP, or the suppressing effect between bitter and salty substances (Iiyama et al. 1986; Hayashi et al. 1989). These multichannel taste sensors were applied to 36 brands of beer. Not only different brands of beer but also different production lots of the same brand were easily discriminated by the output patterns. A taste map expressing the human sensation was obtained by using a transformation of the output patterns with a simple relation (Toko et al. 1992). Preference, however, could not be identified by a taste sensor.

3.3.7 Preference Test

Preference tests ask "Which of the samples do you prefer?" For a two-sample test, preference can be expressed simply by a check mark. If more than two samples are involved, a ranking test is usually performed. The results can be calculated to determine whether a difference is sufficiently large to be statistically significant. The ranking tests generally force a preference choice between samples, while with the rating tests, two samples that are perceived not to differ may be given the same score.

It is possible to investigate the consumers' taste preference to the intensity of taste and flavour. Instead of the semi-objective evaluation terms of "strong", "regular" and "weak", the subjective terms of "too strong", "agreeable" and "too weak" are used for this purpose. Through this evaluation method, a uniform study was conducted to investigate regional characteristics of beer preferences in the US, Germany and Japan. The results revealed the differences in acceptability of taste and flavour between these three countries. It was found that more Americans than Germans and Japanese perceived "bitterness" of beer as being "too strong" (Yonezawa et al. 1997).

3.3.8 Drinkability Test

The drinkability test is designed to evaluate which beer is more amenable to consumption in quantity, obviously a parameter of considerable interest to most brewers. For a long time, brewers have paid the closest attention to the quality of beer and toward improving its quality. They have drunk their beer a lot and evaluated whether it tasted as good as before. A good beer possesses the individual characteristics of the beer and gives the desire to drink more even though a lot is drunk every day.

Although a descriptive test can estimate the flavour characteristic of the beer or check the unpleasant defects, this method cannot evaluate its drinkability. In the case of beer, a consumer can make an evaluation after drinking an obviously larger quantity than other alcoholic drinks. This can cause some larger gaps between the actual daily evaluation and the result of the consumer taste test using only one glass of beer. It seems that the importance of the drinkability test is high in beer.

Curin's method for the assessment of drinkability is based on the classification of flavour impressions after the consumption of precisely determined beer volumes. Flavour impression is evaluated after quenching the first thirst, and after consuming 0.5, 1.0 and 1.5 l of beer. After every consumption, the taste panel member renders a judgment chosen from five possible expressions, each with a numerical value. A numerical value is also assigned to each volume of beer consumed. The data on beer drinkability are then presented in numerical form, the value of which may amount to a maximum of 40 (Curin 1978; Ferki and Curin 1979).

Beer drinkability has been examined from the physiological standpoint (Fushiki et al. 1996). Five kinds of beer with a different malt/adjuncts ratio and degree of attenuation were used. Each beer was given to nine healthy volunteers at the rate of 3 ml kg^{-1} body weight every 15 min, and they each recorded the degree of desire to drink, stomach fullness and tastiness every 30 min. With increasing volume drunk, the degree of tastiness and desire to drink were lowered, and the degree of stomach fullness raised. The relaxed cross-sectional area of the pylorus antrum measured by an ultrasonic image analyzer every 30 min was highly correlated with the subjective evaluation of beer drinkability. The beer giving the highest degree of stomach emptiness was appraised to be highly drinkable. These findings suggest that the rate of gastric emptying is one of the factors determining the drinkability of beer (Nagao et al. 1998, 1999).

3.4 Conclusions

Knowledge of taste and flavour of beer has been developed by many devoted chemists and tasters endeavouring to solve the mysteries of beer. We are respectful toward such senior figures and appreciate their tremendous work. Through well-designed experiments coupled with well-trained sensory observers, more precise and more precious knowledge will be compiled without doubt in the future. Moreover, there is no real conclusion to our work; for those of us who are deeply concerned with the taste and flavour of beer, the mysteries of beer will never be exhausted.

References

Boughton RJ, Hadman SI, Simpson WJ (1999) A taste of things to come. Ferment 4:35–41

Curin J (1978) Evaluating drinking quality of beer (in Czech). Kvas Prum 24:169–172

Ferki P, Curin J (1979) Change of organoleptic parameter of beer in relation to the use of adjuncts and shortening of maturing times. Tech Q Master Brew Assoc Am 16:214–217

Fushiki T, Kodama H, Yonezawa T, Morimoto K (1996) Physiological aspects of beer drinkability. In: Edwards JSA (ed) Culinary arts and sciences: global and national perspectives. Computational Mechanics Publications, Southampton, pp 489–497

Harrison GFA, Collins E (1968) Determination of taste thresholds for a wide range of volatile and nonvolatile compounds in beer. Am Soc Brew Chem Proc 26:83–87

Hayashi K, Matsuki Y, Toko K, Murata T, Yamafuji K, Yamafuji K (1989) Sensing of umami taste and synergistic effects with synthetic lipid membrane. Sensors Materials 1:321–334

Hegarty PK, Parsons R, Banforth CW, Molzahn SW (1995) Phenyl ethanol – a factor determining lager character. In: Proc 25th Congr of the European Brewery Convention, pp 515–522

Hughes PS, Simpson WJ (1996) Bitterness of congeners and stereoisomers of hop-derived bitter acids found in beer. J Am Soc Brew Chem 54:234–237

Iiyama S, Toko K, Yamafuji K (1986) Effect of bitter substances on a model membrane systems of taste reception. Agric Biol Chem 50:2709–2714

King BM, Moreau N (1996) A comparison of bitter perception in high-alcohol, low-alcohol and alcohol-free beer. J Inst Brew 102:419–425

Kowaka M, Kokubo E (1977) Composition of bitter substances of hops and characteristics of beer bitterness. J Am Soc Brew Chem 35:16–21

Kuroiwa Y, Hashimoto N (1961) Composition of sunstruck flavor substance and mechanism of its evolution. Am Soc Brew Chem Proc 19:28–36

Meilgaard M (1974) Flavor and threshold of beer volatiles. Tech Q Master Brew Assoc Am 11:87–89

Meilgaard M (1975a) Flavor chemistry of beer, part I: flavor interaction between principal volatiles. Tech Q Master Brew Assoc Am 12:107–117

Meilgaard M (1975b) Flavor chemistry of beer, part II: flavor and threshold of 239 aroma volatiles. Tech Q Master Brew Assoc Am 12:151–168

Meilgaard M (1982) Prediction of flavor differences between beers from their chemical composition. J Agric Food Chem 30:1009–1017

Meilgaard M (1991) The flavor of beer. Tech Q Master Brew Assoc Am 28:132–141

Meilgaard M, Elizondo A, Mackinney A (1971) Beer flavor characterization panel testing and gas chromatography for evaluation of thresholds and interactions of some flavor compounds. Wallerstein Lab Commun 34:95–113

Meilgaard M, Dalgliesh CE, Clapperton JF (1979) Beer flavour terminology. J Inst Brew 85: 38–42

Meilgaard M, Civille GV, Carr BT (1999) Sensory evaluation techniques, 3rd edn. CRC Press, Boca Raton, p 124

Nagao Y, Kodama H, Yonezawa T, Taguchi A, Fujino S, Nakahara K, Fushiki T (1998) Correlation between the drinkability of beer and gastric emptying. Biosci Biotechnol Biochem 62:846–851

Nagao Y, Kodama H, Yamaguchi T, Yonezawa T, Taguchi A, Fujino S, Morimoto K, Fushiki T (1999) Reduced urination rate while drinking beer with an unpleasant taste and off-flavor. Biosci Biotechnol Biochem 63:468–473

Siebert KJ (1988) A data base management system for flavor threshold information and an evaluation of strategies for identifying new flavor-active substances in beer. J Am Soc Brew Chem 46:82–91

Siebert KJ (1994) Sensory analysis of hop oil-derived compounds in beer; flavor effects of individual compounds. Eur Brew Conv Monogr 22:198–215

Swoboda PAT, Peers KE (1977) Metallic odour caused by vinyl ketones formed in the oxidation of butterfat. The identification of octa-1, cis-5-dien-3-one. J Sci Food Agric 28:1019–1024

Toko K, Murata T, Matsuno T, Kikkawa Y, Yamafuji K (1992) Taste map of beer by a multichannel taste. Sensors Mater 4:145–151

Toko K, Matsuno T, Yamafuji K (1994) Multichannel taste sensor using electric potential changes in lipid membranes. Biosensors Bioelectron 9:359–364

Yonezawa T, Ishii T, Go K, Kawaguchi H, Ishibashi T (1997) Comparison of beer preferences in the USA, Germany and Japan. J Jpn Sensory Eval Soc 1:27–139

4 Taste Evaluation for Peptides in Protein Hydrolysates from Soybean and Other Plants

K. MAEHASHI[1] and S. ARAI[2]

4.1 Introduction

A wide variety of peptides derived from proteins have been found in various foodstuffs and fermentation products such as miso (Takeuchi and Yoshii 1969), soy sauce (Oka and Nagata 1974a,b), natto (Yamasaki 1987), katsuobushi (Hosokawa et al. 1990), cheese (Hamilton et al. 1974), and so on. Many kinds of substances are recognized as the principle for taste, some peptides being one of them. Bitter peptides occur in protease-treated proteins as well as in fermented foods. The bitter flavor due to peptides is thus recognized as a flavor produced by food processing. A report presented by Murray and Baker (1952) seems to be the first that deals with this regard. Shortly after, Raadsvelt (1953) reported that a bitter flavor of cheese is due primarily to some peptide. Bitter peptides are formed not only from milk proteins but also from many other proteins by treatment with proteases. Many bitter peptides have been isolated from various sources, and the relationship between bitterness and structure has been well studied (Guigoz and Solms 1976).

Peptides, which have various tastes other than bitterness, also exist widely. However, their tastes are not strong enough for them to be the principal tasty constituents of foodstuffs, although many peptides contribute to foodstuff taste. Aspartame, L-aspartyl-L-phenylalamine methyl ester (Mazur et al. 1969), is intensely sweet. Noguchi et al. (1975) have isolated and identified several acidic oligopeptides from a fish protein hydrolysate which had a flavor-enhancing activity qualitatively resembling that of monosodium glutamate. Yamasaki and Maekawa (1978) found that an octapeptide, which had been purified from beef meat treated with papain, had a delicious taste, and its taste-structure relationship has been studied in detail (Yamasaki and Maekawa 1980; Tamura et al. 1989; Nakata et al. 1995). It is so-called delicious peptide, or BMP, because the peptide (P) is found naturally in beef (B), and it enhances beef's meaty (M) flavor (Wang et al. 1996). Spanier et al. (1997) recently presented a review (1995) in which they named it as STEP (savory taste-enhancing peptide), and proposed a mechanism for perception of umami (savory, delicious) produced STEP.

---

[1] Department of Fermentation Science, Tokyo University of Agriculture, 1-1-1 Sakuragoaka, Setagaya-ku, Tokyo 156-8502, Japan

[2] Department of Nutrition Science, Tokyo University of Agriculture, 1-1-1 Sakuragaoka, Setagaya-ku, Tokyo 156-8502, Japan

In many cases, tasty peptide was found in protein hydrolysate. Since various peptides have different structures formed by degradation of proteins, their tastes also will be various, though most of them are bitter. Therefore, research into peptides in protein hydrolysate may result in the finding of a new tasty or other functional peptide. This paper reviews the present state of knowledge in this field of research, with some emphasis on our finding that several oligopeptides were identified from protein hydrolysate and a mixture of them was recognized as having a function of enhancing a savory flavor.

4.2 Bitterness of Peptides from Soybean Protein

4.2.1 Theory for the Bitterness of Protein Hydrolysate

A bitter taste observed in many enzymatic hydrolysates was first investigated systematically by Murray and Baker (1952). They found a bitter taste in enzymatic hydrolysates from casein and lactalbumin. From a tryptic hydrolysate of casein, Matoba et al. (1969) isolated three bitter peptides. They were all rich in hydrophobic amino acids. Ney (1971) postulated a correlation between the hydrophobicity of the peptides and their bitter taste. He calculated the so-called Q value as an average free energy for the transfer of the amino acid chains from ethanol to water. Ney found that all bitter peptides had Q values above +1,400 cal/mole, whereas all nonbitter peptides had a Q value below +1,300 cal/mole. This empirical correlation between the presence or absence of bitterness and the average hydrophobicity is called the Q rule. Ney's rule is empirically based and reflects the assumption that hydrophobic interaction is essential for the sensation of bitter taste. Ney's Q rule has been critically examined by Guigoz and Solms (1976), who compared the Q value for 206 peptides with reports on their taste. Matoba and Hata (1972) also proposed that hydrophobic amino acid side chains were responsible for the bitter taste. Furthermore, they showed that the bitter taste was intensified when the carboxyl group of the hydrophobic amino acid was blocked, for example by the formation of a peptide bond. These results have all been verified several times and seem generally accepted.

4.2.2 Tastes of Soybean Protein Hydrolysate

There is a worldwide demand for the development of edible protein sources, with emphasis on those of plant origin. Although soybean is regarded as of major importance as a protein source to answer this demand, the soy protein contains unfavorable flavor components that affect its food acceptability. The formation of bitterness during the hydrolysis of soybean protein limits the utilization of this protein. The first report dealing with this topic was presented by Fujimaki et al. (1968a,b). Then, bitterness formation during the hydrolysis of soybean protein

(Fujimaki et al. 1970a) and corn zein (Wieser and Belitz 1975) were studied. Yamashita et al. (1970) described that some of the unfavorable flavor components were removable from the soybean protein concentrate through its proteolysis with pepsin. There was, however, a problem with the appearance of bitterness during the enzymatic proteolysis. Fujimaki et al. (1968b) carried out a series of experiments aimed at characterizing the bitterness of a peptic hydrolysate of soybean protein, and found that at an early stage of the peptic hydrolysis a nondialyzable bitter (NDB) peptide was formed and contributed to the bitterness of the whole hydrolysate.

The cold-insoluble fraction (CIF) of soy protein was hydrolyzed with pepsin under the usual conditions (Arai et al. 1970a). As shown in Fig. 4.1, the bitterness of the nondialyzable fraction of CIF hydrolysate increased slowly through the early period of peptic hydrolysis, reached its maximum intensity after 12 h, and then decreased gradually, while the bitterness of the whole hydrolysate increased over a long period of peptic hydrolysis. This result suggests that the peptic hydrolyze contains a certain NDB-peptic which may be an intermediate product. The nondialyzable part was treated with aqueous ethanol to obtain a 50% ethanol-soluble and 90% ethanol-insoluble fraction. From this fraction, a bitter tetracossapeptide was separated with the aid of column chromatography using Sephadex G-10 and DEAE Sephadex A-25, thin layer chromatography using silica gel G and cellulose powder, and paper electrophoresis. Through these procedures seven bitter peptides were isolated and their structures were proposed. From structural study it was revealed that the several peptides were mostly characterized by leucine as N- or C-terminal amino acid.

Kim et al. (1999) determined the molecular size reduction and the formation of bitterness during a tryptic hydrolysis of soybean 11 S glycinin. Bitterness was formed within 1 h of hydrolysis, and then slowly increased. They observed a degradation of 11 S glycinin at the end of the reaction by SDS-PAGE analysis. They described that the bitterness was mainly contributed by the peptide fractions of molecular weights of 360–2,100 Da based on GP-HPLC analysis.

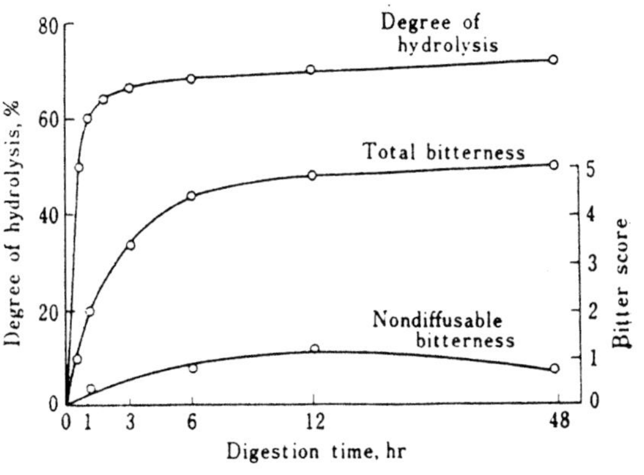

Fig. 4.1. Change in bitterness with increasing peptic hydrolysis of soybean protein. Degree of hydrolysis was shown by 3.3% TCA-insoluble nitrogen per total nitrogen. (Arai et al. 1970a)

4.2.3 Debittering of Peptides

It is widely known that bitterness sometimes is produced in sake (Takahashi et al. 1974), miso, natto (Maekawa and Tamai 1965), and other fermented products, so decreasing their qualities. This bitterness is produced by bitter peptides and their derivatives formed during the aging process of these fermented products. Several debittering methods have been suggested, such as the plastein reaction (Fujimaki et al. 1970b), and the use of exopeptidases (Arai et al. 1970b; Fujimaki et al. 1970a; Umetsu et al. 1983; Hong et al. 1998).

Arai et al. (1970b) found that each of the several bitter peptides isolated from peptic hydrolysate of soybean protein has a leucine residue at the C terminus; therefore, they tried to apply carboxypeptidase A and *Aspergillus* acid car-

Table 4.1. Effects of acidic oligopeptides on masking the bitter flavors of protein hydrolysates. To a protein hydrolysate solution (1%) was added an equal volume of each peptide solution (5 mM) and the resulting mixture was evaluated for bitterness. Scales: +++++ very bitter; ++++ bitter; +++ mildly bitter; ++ weakly bitter; + slightly bitter. The effect of free L-glutamic acid was comparable with that of a peptide in this group. For preparation see Arai (1980)

Acidic oligopeptide	Bitterness after addition of peptide	
	Peptic hydrolysate of soybean protein	Tryptic hydrolysate of casein
Very effective group		
Glu-Asp	±	+
Glu-Glu	±	+
Glu-Gln-Glu	+	++
Glu-Glu-Glu	+	+
Ser-Glu-Glu	+	++
Effective group		
Glu-Ser	++	++
Glu-Thr	++	++
Glu-Gly-Glu	++	++
Glu-Gly-Ser	++	+++
Glu-Glu-Glu-Glu	++	++
Glu-Glu-Glu-Glu-Glu	++	++
Weakly effective group		
Glu-Gly	+++	+++
Glu-Ala-Gly	+++	++++
Glu-Ala-Ala	+++	++++
Ineffective group		
Glu-Arg	++++	+++++
Glu-Ile	++++	+++++
Glu-Leu	++++	+++++
Glu-Lys	++++	+++++
Glu-Phe	++++	+++++
Glu-Trp	++++	+++++
Glu-Tyr	++++	+++++

boxypeptidase to a bitter protein hydrolysate, with the result that its bitter flavor markedly decreases as the enzymatic reaction proceeded to degrade the C-terminal structures of constituent peptides.

Noguchi et al. (1975) carried out an experiment to obtain an acidic oligopeptide fraction in a much higher yield. For this purpose, a glutamic acid-enriched plastein was used as material and hydrolyzed with Pronase. The resulting hydrolysate comprised acidic, basic, and neutral oligopeptides in a weight ratio of 8:3:10, respectively. An aqueous solution of the acidic oligopeptide fraction had no bitter flavor at the concentration of 8 mg/ml, whereas that of the neutral oligopeptide fraction gave a strongly bitter flavor at the concentration of 10 mg/ml. When both solutions were mixed with each other in the same proportion, the resulting mixture no longer showed any bitterness. The acidic oligopeptide fraction used in this experiment was found to consist primarily of di-, tri-, tetra-, and pentapeptides, as listed in Table 4.1. This result indicates that some of these peptides have an ability to mask the bitterness. Arai (1980) confirmed that the addition of each of them to hydrolysates of soybean globulin and casein was more or less effective in masking bitterness. The potent activity for masking the bitter flavor of both proteins well evaluated for the bitterness-masking activity, because each has a bitter flavor of its own.

4.3 Protein Hydrolysates from Soybean and Other Plant Foods

4.3.1 Fermented Foods

Soy sauce (Yokotsuka 1985) is prepared by digesting mould-cultured soybean and wheat in different ratios, usually almost 50:50, in the presence of about 17% salt, and concurrently fermenting the mash with *Lactobacilli* and yeasts. After 6–8 months the well-aged mash is pressed to obtain the liquid part. The total nitrogen content of soy sauce is generally 1.5%, one-half of which is in the form of free amino acids and about 12–15% in peptide form. The nitrogen distribution of dialyzed soy sauce was examined by molecular sieve chromatography using Dowex-50 resins. Most of the nitrogen was in the form of low molecular weight peptides; the size of the major part of the peptides corresponded to oligopeptides having 3.5 amino acid residues on average (Takeuchi et al. 1962; Takeuchi and Yoshii 1967). Oka and Nagata (1974a) fractionated a soy sauce by gel filtration and ion exchange chromatography, and isolated three glycodipeptides and eight dipeptides as the major neutral peptide, four dipeptides and ten glycodipeptides and two tripeptides as the major acidic peptides in soy sauce (Oka and Nagata 1974b). However, it has not been revealed how these peptides contribute to the flavor construction in soy sauce.

Natto, a Japanese traditional fermented food made from soybean with bacterial proteases, often has a bitterness. Yamasaki (1987) separated an *n*-butanol extract from natto and isolated a bitter peptide by Sephadex G-25 gel-filtration, ion-exchange chromatography on CM-Sephadex, and high-voltage paper electrophoresis. He showed a partial sequence of the bitter peptide responsible for bitterness of natto; the N terminus and C terminus of the peptide were both found to be leucine.

Miso (Yokotsuka 1985) is fermented soybean paste, which is popular in Japan, China, Korea, and some other countries. Miso is prepared by enzymatically degrading a mixture of cooked soybeans and molded rice, or barley, or wheat, in the presence of 8–12% salt with a rather small amount of water, along with some lactic and yeast fermentations. There are three main types of miso with respect to the composition of raw materials, i.e., soybean miso (soybean 100%), sendai miso (rice/soybean ca. 5.5), and shiro miso (rice/soybean ca. 22). In all three types of ripened miso, approximately the same distribution pattern of peptides was observed, regardless of the difference in raw materials. The average peptide length is less than 5.0 (Takeuchi and Yoshii 1967).

4.3.2 Other Plant Protein Hydrolysates

Hydrolyzed vegetable protein (HVP) is a savory flavoring product used in a wide variety of food applications. It is produced by hydrolysis of a protein source such as soybean flour, wheat, or maize. Traditionally this hydrolysis is carried out, often by using hydrochloric acid. Aaslyng et al. (1998) produced HVP from soybean by enzymatic hydrolysis and evaluated the chemical and sensory characterization of the eHVP to compare acidic hydrolysate of HVP (aHVP). The small peptides composed mainly of the acidic amino acids or some larger peptides were still present in eHVP compared to aHVP. In the sensory profiling, almost the same score for MSG-like taste (umami) was obtained despite the lower Glu content in eHVP.

Arzu et al. (1972) tested ten commercially available proteolytic enzymes on cottonseed protein in order to increase the protein extraction, and reported that two microbial proteases and bromelain were most active.

Sake is a traditional Japanese alcoholic beverage, which is made from rice by mixed fermentation using fungi, yeast, and bacteria. The unique taste of sake is said to be derived mainly from components of rice, such as hydrolysates of protein, starch, etc. The peptides in sake account for 20–30% of the free amino acid content. Takahashi and Nose (1958) separated peptides in sake by molecular sieve chromatography with Dowex-50 column, and found that 70% of the water-soluble nitrogen in sake consists of amino acids and di- or tripeptides and small amounts of higher molecular weight peptides. Two dipeptides, Val-Gln and Ile-Gln, were isolated (Tajima and Sato 1964) and they were added to artificial sake (Kirimura et al. 1969), and the effect of dipeptides on the taste was evaluated. The dipeptides added to artificial sake improved its body, compared with the control, and caused the taste of artificial sake to come closer to that of natural sake.

4.4 Acidic Oligopeptides

4.4.1 Taste of α-L-Glutamyl Oligopeptides

The soybean protein hydrolyzed with pepsin has a strongly bitter taste; factors responsible for this taste have been elucidated. On the other hand, it has been reported that the plastein reaction is effective in removing the bitter taste from the soybean protein hydrolysate. Arai et al. (1972) explained such an effect based on the fact that the bitter factors, i.e., lower molecular weight peptides involving peculiar sequences of hydrophobic amino acid residues, are polymerized and removed from the system during the plastein reaction, with formation of hydrophobic precipitates of higher molecular weights. Although much has been elucidated on hydrophobic peptides with bitter flavor, as described above, no information appears to be available on the flavor of hydrophilic peptides until about 10 years later.

Arai et al. (1972, 1973) first investigated the flavor of a series of α-L-glutamyl dipeptides in relation to their hydrophobicity. Two independent parameters of hydrophobicity, Δf and ΔW, were obtained. The former was calculated according to the method proposed by Tanford (1962), while the latter was experimentally measured according to the theory of paper partition chromatography (Pardee 1951). Assuming an ideal condition, the work (W) required to move 1 mol of solute from the aqueous phase to the hydrophobic solvent phase under the standard condition is given by

$$W = RT\ 1n\ A_2/A_1[(1/R_f) - 1]$$

where A_1 and A_2 are the cross-sectional areas of the aqueous and the solvent phase, respectively. For two different solvent solutes showing Rf_1 and Rf_2, the difference, ΔW, is expressed as

$$\Delta W = RT\ 1n(1 - Rf_1)Rf_2/Rf_1(1 - Rf_2)$$

since the constants R, T, A_1, and A_2 are cancelled. Applying this method of measurement to glutamyl dipeptides, Arai et al. (1973) obtained a series of ΔW values. Table 4.2 shows the ΔW values of 12 samples as well as their Δf values. It was found that these samples could be classified into three groups according to flavor (Table 4.2). Interestingly, the four dipeptides belonging to the least hydrophobic group do not taste bitter but have a brothy flavor resembling that of monosodium glutamate (MSG), although their threshold values, 0.15–0.30%, are much higher than the threshold value of MSG itself (0.30%).

Arai et al. (1972) reported brothy taste glutamyl oligopeptides including Glu-Asp, Glu-Glu, Glu-Ser, and Glu-Gly-Ser isolated from a proteinase-modified soybean protein. Twelve L-glutamyl dipeptides were prepared and on the basis of taste were classified into three groups: I, brothy taste group; II, flat taste group; and III, bitter taste group. Examination by ion-exchange, thin layer and paper partition chromatographies showed that the dipeptides in I were more acidic, polar, and

Table 4.2. Flavors of α-L-glutamyl dipeptides in relation to hydrophobicity. (Arai et al. 1973)

Dipeptide	Δf[a] (cal/mol)	ΔW[b] (cal/mol)	Flavor
Glu-Asp	<0	−199	MSG-like
Glu-Glu	<0	0	MSG-like
Glu-Ser	40	−47	MSG-like
Glu-Thr	440	43	MSG-like
Glu-Gly	0	0	Flat
Glu-Ala	730	299	Flat
Glu-Pro	2,600	360	Flat
Glu-Val	1,690	480	Flat
Glu-Tyr	2,870	422	Bitter
Glu-Phe	2,650	1,073	Bitter
Glu-Leu	2,420	1,073	Bitter
Glu-Ile	2,970	1,170	Bitter

[a] Relative to the hydrophobicity of Glu-Gly (Tanford 1962).
[b] Relative to the hydrophobicity of Glu-Gly (Pardee 1951).

hydrophilic than those in II, the bitter dipeptides (III) being rather hydrophilic and tasting flat.

4.4.2 Taste Properties of Food Protein Hydrolysates

Fujimaki et al. (1973) found that a similar MSG-like flavor occurred in FPC treated with Pronase under controlled conditions. Noguchi et al. (1975) fractionated a low molecular weight fraction from this FPC hydrolysate and disclosed an acidic oligopeptide fraction, several of which were determined for amino acid sequence by means of mass spectrometry. It was also confirmed that acidic oligopeptides rich in hydrophilic amino acid residues tended to have an MSG-like flavor.

Ishii et al. (1994a–c) reported that acidic oligopeptide fraction obtained from wheat gluten hydrolysate also had potential to enhance an umami. Wheat gluten was hydrolyzed by actinase and deamided. The peptides with a molecular weight of 500–1,000 were divided into acidic, neutral, and basic fractions by ion-exchange chromatographies. The acidic fraction elicited sweet and slightly umami tastes. The addition of an acidic fraction into the niboshi soup stock was shown to enhance its umami and body.

Maehashi et al. (1999) determined the taste properties of several proteolytic protein hydrolysates in order to systematically investigate the effect on taste of the kind of protein digested and of the proteolytic enzymes employed for the digestion. Soybean protein concentrate (Ajipuron KG; Ajinomoto Co. Ltd.), casein (Hammersten; Wako Pure Chemical Ind.), bonito protein (prepared from dorsal meat), and chicken protein (prepared from breast meat) were each hydrolyzed with either one protease or a combination of pepsin, trypsin, α-chymotrypsin, and bromelain (Wako Pure Chemical Ind.). These four proteases are characterized by their specificity for amino acid sequence of substrate protein. The taste of each hydrolysate was examined by sensory evaluation, and the average lengths of the peptides (APL) (Takahashi and Nose 1957) in each hydrolysate were determined. Soybean protein and casein became small peptides only after being treated with the combination of proteases. The taste of each hydrolysate of soybean protein and casein was only bitter, whereas that of bonito and chicken protein was principally bitter, with some associated sour or umami taste. Therefore, the hydrolysates were fractionated in a Sephadex G-25 column to study the tastes of a mixture of the lesser kinds of peptides (Fig. 4.2). With most of the hydrolysates, the bitter peptides were eluted in the first and/or second fractions. Bitterness in soybean hydrolysates became weaker after being fractionated by Sephadex G-25 column. In all fractions of the casein hydrolysates, a bitter taste was found. With the bonito and chicken protein hydrolysates, the fractions eluted at the mid position showed a complex taste of umami, sourness, astringency, and others. Fractions having a sour taste were found in all the chicken hydrolysates. It was revealed that the bitterness masks any other weak tastes; even thought a variety of tasty peptides exists in the hydrolysate, the taste of the hydrolysate was recognized as being bitter.

Maehashi et al. (1999) found that the chicken protein hydrolysate with bromelain possessed the most favorable umami taste in the proteolysates of several foodstuff proteins, soybean protein, casein, bonito protein, and chicken protein. Therefore, they chose the chicken protein hydrolysate with bromelain as the material used for purification of the flavor peptides along with the strategy below:

1. Preparation of chicken protein from fresh meat
2. Enzymatic hydrolysis of chicken protein
3. Separation of hydrophilic fraction by ODS resin
4. Group fractionation on Amberlite IR-120B and IRA-410
5. AG-50 W ion-exchange chromatography
6. Reverse-phase HPLC
7. Sequencing and synthesis of the peptides
8. Taste evaluation of synthesized peptides

The 11 peptides possessing glutamyl residues isolated from chicken protein hydrolysate with bromelain were analyzed for their amino acid sequences by protein sequencer (PPSQ-10, Shimadzu), and synthesized to evaluate their tastes. It was recognized that some of them had a weak umami taste. Furthermore, they found that a combination of these peptides, especially 0.5% each of Glu-Glu, Glu-Val, Asp-Glu-Glu, and Glu-Glu-Asn, with 0.02% IMP in combination, produced a delicious "full" umami taste. This is probably the first report announcing that a mixture of synthetic peptides had a full umami taste in combination with IMP. They proposed that several peptides that have only a weak umami or other taste might contribute to a brothy food taste when they mixed with a small amount of IMP.

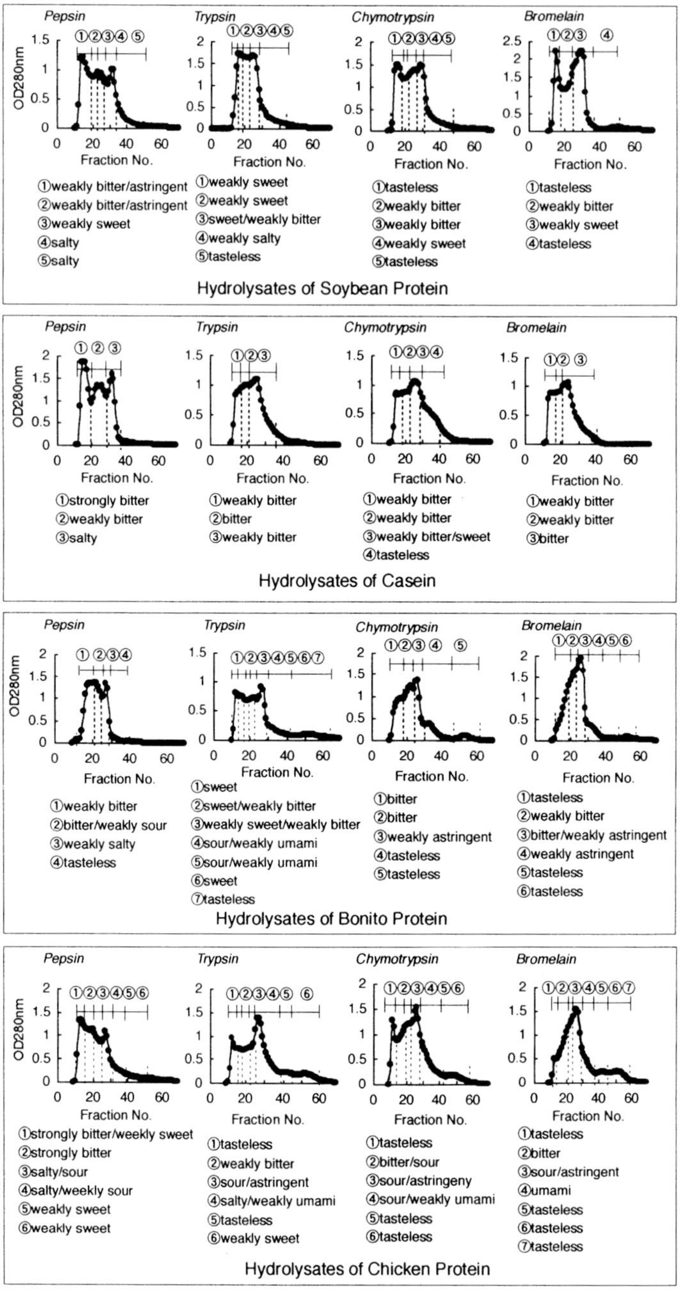

Fig. 4.2. Elution patterns of the various protein hydrolysates from a column of Sephadex G-25. (Maehashi et al. 1999)

4.5 Isolation of Peptides from Protein Hydrolysate

A wide variety of peptides, having almost the same peptide length, ionic strength, or hydrophobicity as each other, exist in small quantities in enzymatic hydrolysate of protein. Therefore, several techniques must be employed in order to separate peptide from enzymatic hydrolysate of protein. Common techniques for protein separation were applied to separate the peptides. In each step, flavor peptides in the fractions were detected by sensory evaluation.

Several procedures for separation of peptides are listed in Table 4.3. It shows typical steps for separation of peptides: primary enzymatic digestion or extraction, secondary gel filtration or ultrafiltration, and then successively ion-exchange, ligand exchange, thin layer chromatography, and/or reverse phase HPLC, etc.

4.5.1 Enzymatic Digestion

Nnatural materials such as foodstuffs contain various flavor substances such as amino acids, nucleotides, and so on. Since these flavor substances often hinder a detection of taste of peptides in separation procedures, the material protein should be purified as far as possible. Since the use of a crude enzyme causes a mixture of the numerous peptides, a commercially available purified enzyme which is known for its specificity for peptide bonding is employed for digestion of protein preparation.

Maehashi et al. (1999) employed four commercially available enzymes (see Table 4.4) in order to systematically investigate the effect on taste of the kind of protein digested and of the proteolytic enzymes employed for the digestion.

4.5.2 Gel Filtration

A number of workers have applied gel chromatography to a food protein hydrolysate. With this technique it was not possible to separate individual peptides; however, this technique is used frequently in the early stage because of its convenience of operation.

Arai et al. (1970a) applied gel filtration to estimate the molecular weight distribution of peptides in soybean protein hydrolysate. Five glass columns (1.2×30 cm) were combined in series, in which Sephadex G-100, 75, 50, 25, and 10 were packed in order of height; each Sephadex had been previously equilibrated with standard buffer (0.0325 M K_2HPO_4–0.0026 M KH_2PO_4–0.40 M NaCl, pH 7.6). The proteolysate sample was applied to the column and developed with the standard buffer of head drop 1 ml. The flow rate was adjusted to 0.2 ml/min, and 3 ml of each was fractionated.

Aristoy and Toldra (1995) fractionated low molecular mass peptides (Mr <3,000 Da) of water-soluble extract from both pork meat and dry-cured ham by

Table 4.3. Procedures for separation of tasty peptides

Peptide	Reference	Material	Procedures
Nondialyzable bitter peptide (bitter) Phe-(Arg, Asp2, Glu2, Gly, Ile, Leu, Lys2, Pro, Ser, Thr)-Trp-(Ala, Arg, Asp, Gly, Val)-Gln-Tyr-Phe-Leu	Arai et al. (1970a)	Soybean protein	Hydrolysis with pepsin Sephadex G-10 DEAE Sephadex A-25 TLC (Sephadex G-25) Paper electrophoresis
Diffusible bitter peptides (bitter) Gly-Leu, Leu-Phe, Ser-Lys-Gly-Leu, Leu-Lys, Phe-(Ile, Leu2)-Gln-Gly-Val, Arg-Leu-Leu, Arg-Leu	Fujimaki et al. (1970a)	Soybean protein	Hydrolysis with pepsin dialysis Sephadex G-10 Dowex 50W-X2 TLC
Octapeptide (BMP) (umami) Lys-Gly-Asp-Glu-Ser-Leu-Ala	Yamasaki and Maekawa (1978)	Beef	Papain treatment Hot water extraction Sephadex G-25 Dowex 50X High-voltage PPC
Acidic oligopeptides (MSG-like) Glu-Glu, Glu-Asp, Glu-Ser, Glu-Asp-Glu, Glu-Gly-Ser, Ser-Glu-Glu, Glu-Glu-Gln, Asp-Glu-Ser, Ile-Glu-Glu	Noguchi et al. (1975)	Fish protein concentrate	Hydrolysate with pronase Removal of nucleic acid Hydrolysis with pronase Activated carbon Amberlite IRC-50 Amberlite IR-45 Amberlite CG-120
Acidic peptide fraction (enhance the umami of *niboshi* soup)	Ishii et al. (1994b)	Wheat gluten	Hydrolysis with actinase Ultrafiltration Amberlite IRC-50 Amberlite IRA-400
Acidic peptides (MSG-like)	Oka and Nagata (1974b)	Soy sauce	Sephadex G-15 QAE-Sephadex A-25
Bitter peptide (bitter)	Yamasaki (1987)	Natto	*n*-Butanol extraction Sephadex G-25 CM-Sephadex
Bitter peptide (bitter) Arg-Gly-Pro-Pro-Phe-Ile-Val	Minamiura et al. (1972)	Casein	Alkaline proteinase (*Bacillus*) *n*-Butanol extraction SE-Sephadex C-25 Sephadex G-15 PPC
Oligopeptides (enhance the taste of 5′-IMP) Glu-Glu, Glu-Val, Ala-Asp-Glu, Ala-Glu-Asp, Asp-Glu-Glu, Ser-Pro-Glu	Maehashi et al. (1999)	Chicken protein	Hydrolysis with bromelain ODS resin Amberlite IR-120B Amberlite IRA-410 AG-50W Reverse-phase HPLC

Table 4.4. Proteases for food protein hydrolysis. (Adler-Nissen 1986)

Protease	Type	Typical pH range	Preferential specificity
Pepsin	Aspartic protease	1–4	Aromatic -COOH and -NH_2, Leu-, Asp-, Glu-COOH
Trypsin	Serine protease	7–9	Lys-, Arg-COOH
α-Chymotrypsin	Serine protease	8–9	Phe-, Tyr-, Trp-COOH
Bromelain	Cysteine protease	5–8	Lys-, Arg-, Phe-, Tyr-COOH

gel filtration chromatography in a Sephadex G-25 column (2.6 × 55 cm). The column was previously equilibrated with 0.01 N HCl in pure water and calibrated with standards of molecular mass ranging between 204 and 6,500 Da. The flow rate was 18 ml/h and 6-ml fractions ware collected. The sample (50 g) was previously deproteinized at 4 °C for 24 h by ethanol addition (1:3), concentrated to dryness, resuspended up to 10 ml, and applied to the column. The fractions were found to have a wide range of flavors: bitter/sour tastes, corresponding to polypeptides with Mr higher than 1,700 Da, while the savory/brothy tastes were found in fractions 2 and 3 corresponding to Mr between 1,500 and 1,700. It should be taken into account that this kind of matrix may present ionic interactions affecting elution volumes. In fact, previous assays revealed the co-elution of salt with several free amino acids such as glutamic acid, methionine, and phenylalanine at 250–270 ml elution volume. On the other hand, tryptophan experiences the greatest retardation and elutes in the peak corresponding to fraction 8.

Maehashi et al. (1999) investigated the elution patterns on Sephadex G-25 column of hydrolysates of soybean and some other protein. After the hydrolysates of proteins were lyophilized, 80 mg of each of them was dissolved in 0.8 ml of 10% ethanol and applied to the column (1.5 × 30 cm) of Sephadex G-25. The peptides were eluted by 10% ethanol at a flow rate of 40 ml/h, and 1.5 ml of each fraction was collected and monitored for its absorbance at 280 nm. The eluates were divided into several peak fractions and evaporated to dryness. Each fraction was then dissolved in 1 ml of distilled water and evaluated for taste(see Fig. 4.2).

4.5.3 Ion-Exchange Chromatography

Ion-exchange chromatography using Dowex, Amberlite, AG, DEAE-Sephadex, CM-Sephadex, etc. is widely applied to separation of peptides from protein hydrolysate. It can separate roughly but effectively acidic, basic, and neutral fractions.

4.5.3.1 Group Fractionation

Maehashi et al. (1999) separated acidic fraction from the hydrolysate of chicken protein. First, the hydrolysate was separated by ODS column to obtain hydrophilic

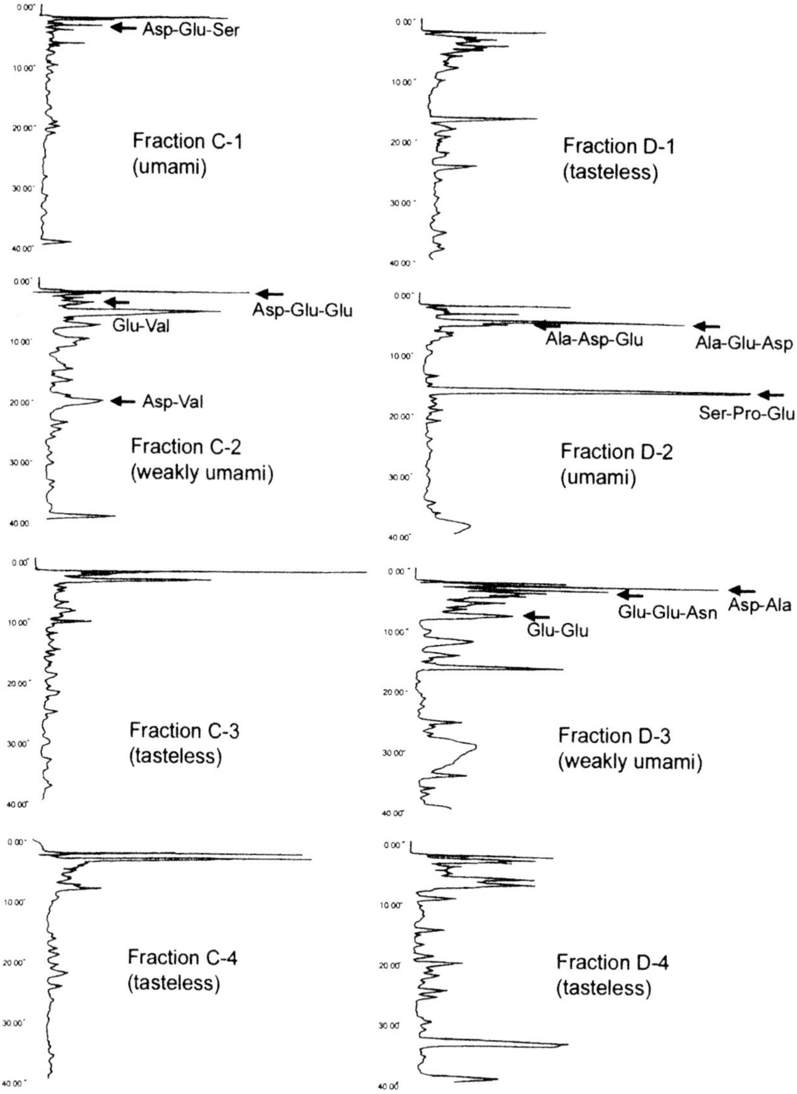

Fig. 4.4. Elution profiles on RP-HPLC of peak fractions from ion-exchange chromatography. *Arrows* indicate the peaks assumed to be peptides, which were collected and analyzed for amino acid sequences

Maehashi et al. (1999) evaluated the taste profiles of each fraction in separations by the following methods:

1. Each fraction is evaporated to dryness to eliminate the volatile buffer, and added to a small amount of distilled water.
2. After rinsing the mouth with distilled water, 0.5 ml of a sample solution is held in the mouth.

3. The taste character is evaluated against a 0.05% NaCl solution for saltiness, 0.05% tartaric acid solution for sourness, 0.2% sucrose solution for sweetness, and 0.1% caffeine solution for bitterness as the standards for average taste.

4.6.2 Determination of Recognition Threshold

Beefy meaty peptide (BMP) was a savory peptide isolated from beef digested with papain. The primary structure of this peptide was proposed as H·Lys-Gly-Asp-Glu-Glu-Ser-Leu-Ala·OH (Yamasaki and Maekawa 1978). Tamura et al. (1989) synthesized the peptide and described the taste as umami and sour with a threshold concentration of 1.41 mM. Spanier (1992) and Spanier et al. (1992) reported that the same octapeptide was found naturally in beef meat and named it beefy meaty peptide (BMP). Wang et al. (1995) considered that this peptide might have application in heat-processed food. To examine the taste properties of BMP under different pH conditions, Wang et al. (1996) determined the recognition threshold of the peptide (BMP) by the following method:

1. Solutions of 0, 0.08, 0.16, 0.32, 0.64, 1.28, and 2.56 mM peptide at pHs of 3.5, 6.5, and 9.5 are prepared by serial dilutions of the highest concentration with double-distilled water.
2. Each panelist was served 0.5 ml of sample solution.
3. Panelists mark the lowest concentration they could recognize and described the perceived sensations.
4. The recognition thresholds of peptide are determined as the concentration at which panelists could recognize the taste sensation of the peptide.

4.6.3 Synergism Among Savory Peptides

When glutamic acid and nucleotides (free forms or salts) coexist, umami is dramatically enhanced. This phenomenon, well known as the synergistic effect of umami substances, has already been investigated and reported (Kuninaka 1960; Yamaguchi et al. 1971).

The synergistic effects of salt and MSG on BMP recognition thresholds were determined by Wang et al. (1996). All synergism experiments were done in duplicate, and the procedures were the same as used for the BMP recognition threshold determination. MSG concentration was kept constant at a threshold value of 1.4 mM. BMP was dissolved in the MSG solution. The highest concentration for BMP was 2 mM.

Table 4.5. Five solutions of basic taste used in sensory evaluation. (Ishii et al. 1994c)

Taste	Substance	Concentration (%)
Sweetness	Sucrose	1.0
Saltiness	NaCl	0.6
Sourness	Citric acid	0.15
Bitterness	1-Phenyl-2-thiourea	0.0002
Umami	5′-IMP	0.03

4.6.4 Effect on Five Basic Tastes

Ishii et al. (1994c) prepared a peptide fraction possessing a molecular weight of 500–1,000 from deaminated wheat gluten hydrolysate, and examined the effect of this peptide fraction on the strength of primary taste. They prepared five basic taste solutions, as shown in Table 4.5, at the minimum concentrations enough to distinguish clearly from each other. The peptide fraction was added at 0.03% of each basic taste to the sample solution. The taste intensity of each basic solution added to the peptide fraction was compared with that with no addition by paired preference test.

4.6.5 Taste Duration-Intensity Curve

The tastes of peptides are sometimes characterized by their lasting aftertastes, suggesting that they became tightly adsorbed to the tongue surface. Ishii et al. (1994c) measured the change of taste with time to evaluate aftertaste of peptides. After tasting 8 ml of 1% sucrose solution added to 0.03% of peptide fraction for 5 s, panelists drank it and scored its taste intensity at the time as 5 points, and change of taste intensity was scored until sweetness disappeared.

4.6.6 Buffer Capacity of Peptide

The addition of some dipeptides to sake resulted in an increased buffering capacity of the product (Kirimura et al. 1969). Gly-Leu and Pro-Gly increased the buffer action at pH > 9, and Val-Gln showed strong buffer action at pH 7–8. In general, peptides have complex dissociation constants, compared to those of amino acids.

It is also reported that the dipeptides naturally exist in fish meat, carnosine, and anserine and possessed strong buffering capacities above pH 6 (Suyama and Shimizu 1982). Ishii et al. (1994c) measured the buffer capacity of peptide fraction from wheat gluten hydrolysate by titration with 0.1 N NaOH or 0.1 N HCl recording pH. The buffering capacity was represented as the value that was the amount of NaOH or HCl divided by pH value.

4.7 Conclusions and Vista

There are many reports on the tasty peptides from animal proteins; however, little is known of peptides derived from plant protein, except for soybean protein. The field of plant protein hydrolysis is dominated by the use of soybean protein as substrate. Most of the more fundamental research into food protein hydrolysis is based on soybean protein. Since many researchers have studied soybean protein for its nutritional or functional use since the 1970s, a considerable knowledge on protein hydrolysis or peptide analysis was accumulated and necessary techniques have developed. This knowledge and techniques have been applied to many researches into hydrolysis of other protein. Recently, the importance of the physiological function of foods involved in the modulation of our physiological systems such as the immune, endocrine, nerve, circulatory, and digestive systems has come to be recognized. A number of functional peptides were found from enzymatically hydrolyzed food proteins (Arai 1996), e.g., hydrophobic peptides from soybean glycinin lowering the plasma cholesterol level (Gatchalian-Yee et al. 1994). Since it is clinically important to supply organoleptically acceptable foods to patients, the knowledge and techniques that developed in protein hydrolysis and taste evaluation should be applied to design and construct acceptable functional foods in the near future.

Acknowledgements. We are most grateful to Professor Shigezo Udaka, Department of Fermentation Science, Tokyo University of Agriculture, who made a number of pertinent suggestions to us at the time of writing.

References

Aaslyng MD, Martens M, Poll L, Nielsen PM, Flyge H, Larsen LM (1998) Chemical and sensory characterization of hydrolyzed vegetable protein, a savory flavoring. J Agric Food Chem 46: 481–489

Adler-Nissen J (1986) Enzymic hydrolysis of food proteins. Elsevier, London

Arai S (1980) The bitterness flavor due to peptides or protein hydrolysates and its control by bitterness-masking with acidic oligopeptides. In: Charalambous G (ed) The analysis and control of less desirable flavors in foods and beverages. Academic Press, New York, pp 133–147

Arai S (1996) Studies on functional foods in Japan. Biosci Biotech Biochem 60:9–15

Arai S, Yamashita M, Kato H, Fujimaki M (1970a) Applying proteolytic enzymes on soybean; a nondialyzable bitter peptide in peptic hydrolyzate of soybean protein and its bitterness in relation to the chemical structure. Agric Biol Chem 34:729–738

Arai S, Noguchi M, Kurosawa S, Kato H, Fujimaki M (1970b) Applying proteolytic enzymes on soybean. 6. Deodorization effect of Aspergillopeptidase A and debittering effect of *Aspergillus* acid carboxypeptidase. J Food Sci 35:392–395

Arai S, Yamashita M, Fujimaki M (1972) Glutamyl oligopeptides as factors responsible for tastes of a proteinase-modified soybean protein. Agric Biol Chem 36:1253–1256

Arai S, Yamashita M, Noguchi M, Fujimaki M (1973) Tastes of L-glutamyl oligopeptides in relation to their chromatographic properties. Agric Biol Chem 37:151–156

Aristoy MC, Toldra F (1995) Isolation of flavor peptides from raw pork meat and dry-cured ham. In: Charalambous G (ed) Food flavors: generation, analysis and process influence. Elsevier, Amsterdam, pp 1323–1344

Arzu A, Mayorga H, Gonzalez J, Rolz C (1972) Enzymatic hydrolysis of cottonseed protein. J Agric Food Chem 20:805–809

Fernandez M, Singh T, Fox PF (1998) Isolation and identification of peptides from the diafiltration permiate of the water-soluble fraction of cheddar cheese. J Agric Food Chem 46:4512–4517

Fujimaki M, Yamashita M, Okazawa Y, Arai S (1968a) Diffusible bitter peptides in peptic hydrolyzate of soybean protein. Agric Biol Chem 32:794–795

Fujimaki M, Kato H, Arai S, Tamaki E (1968b) Applying proteolytic enzymes on soybean. 1 Proteolytic enzyme treatment of soybean protein and its effect on the flavor. Food Technol 22:889–893

Fujimaki M, Yamashita M, Okazawa Y, Arai S (1970a) Applying proteolytic enzymes on soybean. 3. Diffusible bitter peptides and free amino acids in peptic hydrolysate of soybean protein. J Food Sci 35:215–218

Fujimaki M, Yamashita M, Arai S, Kato H (1970b) Plastein reaction. Its application to debittering of proteolyzates. Agric Biol Chem 34:483–484

Fujimaki M, Arai S, Yamashita M, Kato H, Noguchi M (1973) Taste peptide fractionation from a fish protein hydrolysate. Agric Biol Chem 37:2891–2898

Gatchalian-Yee M, Imamura M, Nonaka M, Gu JY, Sugano M (1994) Effect of dietary fats on cholesterol metabolism and eicosanoid production in hamsters fed undigested fraction of soybean protein. J Nutr Sci Vitaminol 40:499–504

Guigoz Y, Solms J (1976) Bitter peptides, occurrence and structure. Chem Senses Flavor 2: 71–84

Hamilton JS, Hill RD, Van Leeuwen H (1974) A bitter peptide from Cheddar cheese. Agric Biol Chem 38:375–379

Hong YS, Lee CH, Lee KY (1998) Effect of weak acid pretreatment on the enzymic hydrolysis against wheat gluten of high concentration. J Kor Soc Food Sci Nutr 27:1110–1116

Hosokawa M, Sakakibara H, Yajima I, Hayashi K (1990) Non-volatile flavor components in dorsal, abdominal and red meat parts of dried skipjack (Katsuobushi). Nippon Shokuhin Kogyo Gakkaishi 37:856–861

Ishii K, Nishimura T, Ono T, Hatae K, Shimada A (1994a) Taste of peptides in wheat gluten hydrolyzate by protease. J Home Econ Jpn 45:615–620

Ishii K, Nishimura T, Hirata Y, Hatae K, Shimada A (1994b) Fractionation and taste of peptides in wheat gluten hydrolyzate by actinase. J Home Econ Jpn 45:791–796

Ishii K, Nishimura T, Hatae K, Shimada A (1994c) Effect of peptides in wheat gluten hydrolyzate on basic tastes. J Home Econ Jpn 45:797–801

Kim MR, Choi SY, Lee CH (1999) Molecular characterization and bitter taste formation of tryptic hydrolysis of 11 S glycinin. J Microbiol Biotechnol 9:509–513

Kirimura J, Shimizu A, Ninomiya T, Katsuya N (1969) The contribution of peptides and amino acids to taste of foodstuffs. J Agric Food Chem 17:689–695

Kuninaka A (1960) Studies on taste of ribonucleic acid derivatives. J Agric Chem Soc Jpn 34: 489–497

Maekawa K, Tamai S (1965) Abstracts of papers, 36th National Meeting of the Agricultural Chemical Society of Japan, Tokyo, April

Maehashi K, Matsuzaki M, Yamamoto Y, Udaka S (1999) Isolation of peptides from an enzymatic hydrolysate of food protein and characterization of their taste properties. Biosci Biotechnol Biochem 63:555–559

Matoba T, Hata T (1972) Relationship between bitterness of peptides and their chemical structures. Agric Biol Chem 36:1423–1431

Matoba T, Nagayasu C, Hayashi R, Hata T (1969) Bitter peptides in tryptic hydrolysate of casein. Agric Biol Chem 33:1662–1663

Matoba T, Hayashi R, Hata T (1970) Isolation of bitter peptides from tryptic hydrolysate of casein and their chemical structure. Agric Biol Chem 34:1235–1243

Mazur RH, Schlatter JM, Goldkamp AM (1969) Structure–taste relationships of some dipeptides. J Am Chem Soc 91:2684–2691

Minamiura N, Matsumura Y, Fukumoto F, Yamamoto T (1972) Bitter peptides in cow milk casein digest with bacterial proteinase – isolation and determination of amino acid sequence of a bitter peptide. Agric Biol Chem 36:588

Murray T, Baker BE (1952) Studies on protein hydrolysis. I Preliminary observations on the taste of enzymic protein-hydrolysates. J Sci Food Agric 3:470

Nakata T, Takahashi M, Nakatani M, Kuramitsu R, Tamura M (1995) Role of basic and acidic fragments in delicious peptides (Lys-Gly-Asp-Glu-Glu-Ser-Leu-Ala) and the taste behavior of sodium and potassium salts in acidic oligopeptides. Biosci Biotech Biochem 59:689–693

Ney KH (1971) Voraussage der Bitterkeit von Peptiden aus deren Aminosäurezusammensetzung. Z Lebensm Unters Forsch 147:64–71

Noguchi M, Arai S, Yamashita M, Kata H, Fujimaki M (1975) Isolation and identification of acidic oligopeptides occurring in a flavor potentiating fraction from a fish protein hydrolysate. J Agric Food Chem 23:49–53

Oka S, Nagata K (1974a) Isolation and characterization of neutral peptides in soy sauce. Agric Biol Chem 38:1185–1194

Oka S, Nagata K (1974b) Isolation and characterization of acidic peptides in soy sauce. Agric Biol Chem 38:1195–1202

Pardee AB (1951) Calculations on paper chromatography of peptides. J Biol Chem 190:757

Raadsvelt CW (1953) Bitter compounds from cheese. In: Proc 13th Int Dairy Congr, Publ 2, p 676

Spanier AM (1992) Current approaches to the study of meat flavor quality. In: Charalambous G (ed) Food science and human nutrition. Elsevier, New York, pp 695–709

Spanier AM, Miller JA, Bland JM (1992) Lipid oxidation: effect on meat proteins. In: St Angelo AJ (ed) Lipid oxidation in foods. ACS Books, Columbus, pp 104–119

Spanier AM, Bland JM, Miller JA, Glinka J, Wasz W, Duggins T (1995) BMP: a flavor enhancing peptide found naturally in beef. Its chemical synthesis, descriptive sensory analysis, and some factors affecting its usefulness. In: Charalambous G (ed) Food flavors: generation, analysis and process influence. Elsevier, New York, pp 1365–1375

Spanier AM, Bland JM, Flores M, Bystricky P (1997) What is the NExT STEP in flavor enhancers? Chemistry of novel foods. In: Proc Symp Int Chemical Congr of Pacific Basin Societies, Honolulu, 17–22 Dec 1995, pp 45–66

Suyama M, Shimizu T (1982) Buffering capacity and taste of carnosine and its methylated compounds. Bull Jpn Soc Sci Fish 48:89–95

Tajima O, Sato J (1964) Studies on peptide in sake brewed from refined defatted soybean meal. On the method of isolating peptide from acidic peptide fraction and the nature of peptide isolation. Rep Inst Phys Chem Res 40:259–264

Takahashi A, Nose K (1957) Study on peptides in sake, 1. Group fractionation of peptides in sake using Dowex-50. J Ferment Technol 35:318–321

Takahashi A, Nose K (1958) Study on peptides in sake, 4. Changes in amount of peptides during sake brewing. J Ferment Technol 36:248–251

Takahashi K, Tadenuma M, Kitamoto K, Sato S (1974) L-Prolyl-L-leucine anhydride. A bitter compound formed in aged sake. Agric Biol Chem 38:927–932

Takeuchi T, Kato H, Yoshii H (1962) Studies on the peptides in miso and soy sauce. (I) Changes of peptides in amounts during soybean treatment in the process of miso brewing. J Ferment Technol 40:375–379

Takeuchi T, Yoshii H (1967) Studies on the peptides in miso and soy sauce (IV). Relationship between the C/N ratio of the raw materials and the distributive patterns of peptides in miso. J Ferment Technol 45:29–33

Takeuchi T, Yoshii H (1969) Studies on peptides in miso and soy-sauce (VI). Group separation of lower peptides in mame-miso and their properties. J Ferment Technol 47:496–501

Tamura M, Nakatsuaka T, Tada M, Kawasaki Y, Kikuchi E, Oka H (1989) The relationship between taste and primary structure of "delicious peptide" (Lys-Gly-Asp- Glu-Glu-Ser-Leu-Ala) from beef soup. Agric Biol Chem 53:319–325

Tanford C (1962) Contribution of hydrophobic interactions to the stability of the globular conformation of proteins. J Am Chem Soc 84:4240–4247

Umetsu H, Matsuoka H, Ichishima E (1983) Debittering mechanism of bitter peptides from milk casein carboxypeptidase. J Agric Food Chem 31:50

Yamaguchi S, Yoshikawa T, Ikeda S, Ninomiya T (1971) Measurement of the relative taste intensity of some L-α-amino acids and 5′-nucleotides. J Food Sci 36:846–849

Wang K, Maga JA, Bechtel PJ (1995) Stability of beefy meaty peptide to pasteurization and sterilization temperature. Food Sci Technol/Lebensmitt Wiss Technol 28:539–542

Wang K, Maga JA, Bechtel PJ (1996) Taste properties and synergisms of beefy meaty peptide. J Food Sci 61:837–839

Wieser H, Belitz HD (1975) Bitter Peptide aus dem Maisprotein Zein durch Hydrolyse mit Pepsin. Z Lebensm Unters Forsch 159:329–336

Yamasaki Y (1987) The bitter peptide in natto. J Home Econ Jpn 38:93–97

Yamasaki Y, Maekawa K (1978) A peptide with delicious taste. Agric Biol Chem 42:1761–1765

Yamasaki Y, Maekawa K (1980) Synthesis of a peptide with delicious taste. Agric Biol Chem 44:93

Yamashita M, Arai S, Gonda M, Kato H, Fujimaki M (1970) Enzymatic modification of proteins in foodstuffs. 2. Nutritive properties of soy plastein and its bio-utility evaluation in rats. Agric Biol Chem 34:1333–1337

Yokotsuka T (1985) Fermented protein foods in the Orient, with emphasis on shoyu and miso in Japan. In: Wood BJB (ed) Microbiology of fermented foods. Elsevier, London, pp 197–247

5 Hop Aroma Extraction and Analysis

G. Lermusieau and S. Collin

5.1 Introduction

Hops (*Humulus lupulus* L.) are dioecious (male and female flowers on separate plants), perennial, climbing plants of the hemp (Cannabis) family and Urticales order. As shown in Fig. 5.1, the female cone, of interest to brewers, is a close grouping of flowers arranged on a zigzagging central axis (strig). At each bend, there are florets with large covering bracts (leaves). Epidermal hair grows on the bract basis, forming a beaker-like gland, called the lupulin gland, secreting resins (bitter substances) and oils (hoppy flavours) (Davies 1960; Moerman 1982; O'Rourke 1994). Picked hops contain 75 to 80% water. Hops are carefully dried at a maximum temperature of 60°C to a water content of 10%. They are then stored in a cool, dark room until used within the year (Stevens 1966; Hughes and Simpson 1993). The composition of the whole hop cone is depicted in Table 5.1.

In the brewing industry, hops are processed in several different ways, yielding cones, pellets, stabilised or isomerised pellets, organic solvent extracts, CO_2 extracts (supercritical or liquid CO_2), or isomerised extracts. Each type of processing has its advantages and disadvantages (Hupe 1981; Moir 1988; Hughes and Simpson 1993; Lewis 1994; Wilson and Fincher 1995). Liquid CO_2 is the most selective solvent used commercially for hops, producing the purest whole resin and oil extract. It solubilises none of the hard resins or tannins, much lower levels of plant waxes, no plant pigments, and less water and water-soluble materials (Moir 1988; Hughes and Simpson 1993).

Mainly used in the boiling kettle, hop imparts bitterness to beer because of the isomerisation of humulones (α-acids) to isohumulones (Fig. 5.2) at this stage of the process (Tatchell 1955; Kuroiwa et al. 1972; Sandra 1976; Verzele and De Keukeleire 1991; Deinzer and Yang 1994). Since hop varieties contain bitter compounds in variable amounts and proportions, hop cones are usually classified into two categories (Table 5.2) on the basis of their α-acid content: (1) aromatic hops, containing low levels of α-acids (from 3.8 to 5.1%) (Moll 1991); and (2) bittering hops, essentially used to impart bitterness to beers (from 8.6 to 12% α-acids) (Kenny 1990).

Université Catholique de Louvain, Unité de Brasserie et des Industries Alimentaires, Croix du Sud, 2 Bte 7, 1348 Louvain-la-Neuve, Belgium

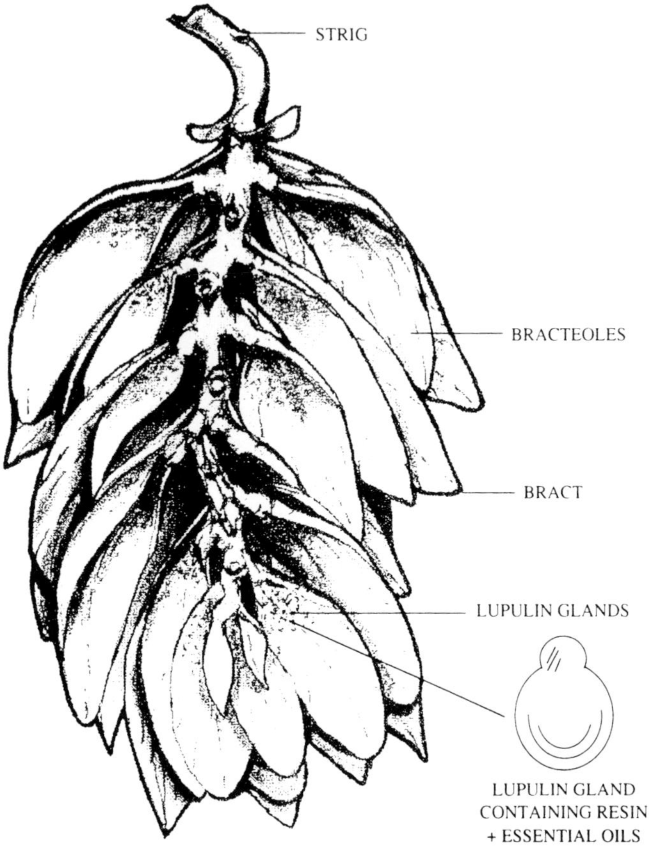

Fig. 5.1. Hop cone (O'Rourke 1994)

Table 5.1. Hop cone composition (in %; resin and oil content vary markedly with variety). (De Keukeleire 1993)

Substances			(w/w)%
Proteins			15
Total resins	Soft resins	α-acids	2–16
		β-acids	2–10
	Hard resins		3
Water			8–12
Essential oils			0.1–2
Tannins			4
Polyphenols			2–5
Monosaccharides			2
Amino acids			0.1
Lipids and waxes			<1–25
Pectin			2
Ash			10
Cellulose and lignin, etc.			40–50

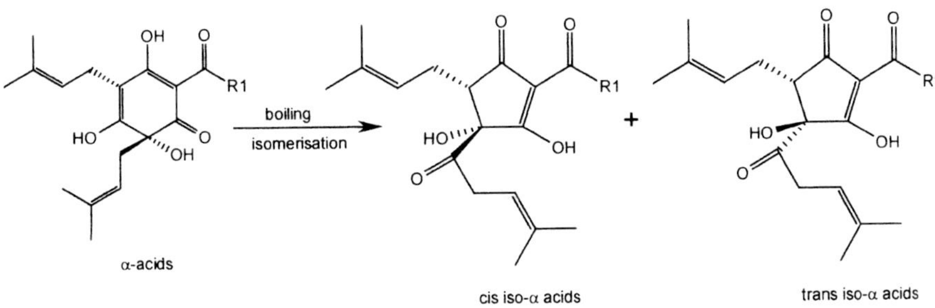

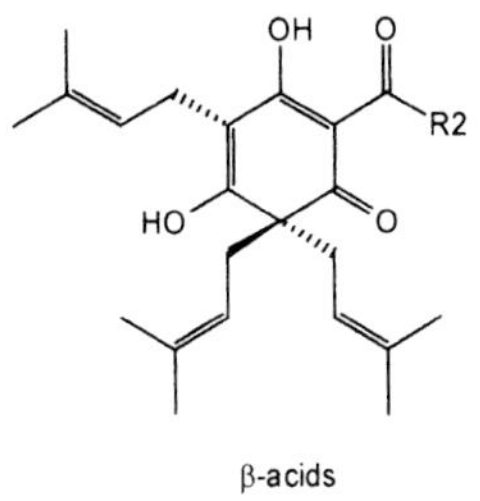

	R_1	R_2
$CH_2CH(CH_3)_2$	humulone	lupulone
$CH(CH_3)_2$	cohumulone	colupulone
$CH(CH_3)CH_2CH_3$	adhumulone	adlupulone
$(CH_2)_2CH(CH_3)_2$	prehumulone	prelupulone
CH_2CH_3	posthumulone	postlupulone

Fig. 5.2. Hop humulones and isohumulones

Table 5.2. Several "aromatic" (low-bitter) and bitter varieties. (Moll 1991; O'Rourke 1994)

Aromatic hops		Bitter hops	
Backe (E)	Progress (E)	Aquila (US)	Nugget (US)
Bramling (US)	Saaz (E)	Banner (US)	Olympic (US)
Cascade (US)	Spalt (E)	Bullion (US)	Orion (E)
Crystal (E)	Spalt Select (E)	Brewers Gold (E)	Pride (AUS)
East Kent Goldings (E)	Trisselspalt (E)	Centennial (US)	Pride of Ringwood (AUS)
Fuggle (E+US)	Styrian Goldings (E)	Challenger (E)	Record (E)
Golding (E)	Styrie (E)	Chinook (US)	Ringwood (AUS)
Hallertau Gold (E)	Tettnang (E+US)	Cluster (US)	Southern Brewer (AFR)
Hallertau Hersbrucker (E)	Whit bread Goldings (E)	Comet (US)	Star (E)
Hallertau Mittelfrüh (E+US)	Willamette (US)	Eroica (US)	Talisman (US)
Hallertau Tradition (E)		Galena (US)	Target (E)
Hersbruck Spät (E)		Hallertau Magnum (E)	Wye Target (E)
Hüller bitterer (E)		Hallertau Perle (E+US)	Wye Challenger (E)
Lublin (E)		Hueller (E)	Yeaman (E)
Mont Hood (US)		Northdown (E)	Yakima Cluster (US)
Perle (E+US)		Northern Brewer (E)	

E, European varieties; US, American varieties; AUS, Australian varieties; AFR, African varieties

Hop is also responsible for the appearance in beer of the lightstruck, sunstruck, or skunky flavour (3-methyl-2-buten-1-thiol) caused by photochemical degradation of iso-humulones in the presence of cysteine and riboflavin (Kuroiwa and Hashimoto 1961; Gunst and Verzele 1978; Goldstein et al. 1993; Lewis 1994). Other compounds derived from the isohumulone lateral chain can impart a cheesy (volatile fatty acids) or wine-like (the corresponding ethylic esters) flavour to beer. The soporific agents 2-methyl but-3-ene-2-ol and 3-methyl but-2-ene-1-ol also derive from humulone degradation. Lastly, bitter compounds exhibit high bacteriostatic activity, very useful in the brewing process, and contribute to beer foam stability (Derdelinckx and Maudoux 1992).

5.2 Hop Aroma

For brewers, qualitative and quantitative analyses of hop aroma are of prime importance in assessing hop quality (variety and freshness) (Moerman 1982; De Keukeleire 1993). To date, over 300 components have been positively or tentatively identified in hop oils: mainly hydrocarbons, ketones, aldehydes, esters, carboxylic acids, alcohols, oxygen heterocycles, and sulfur compounds (Buttery et al. 1965; Naya and Kotake 1972; Tressl et al. 1978b; Kralj et al. 1991; Moir 1992).

Probably the most reliable method for recovering hop oils is steam distillation liquid/liquid extraction (Fig. 5.3; Likens and Nickerson 1964; Nickerson and Likens 1966; Bouseta and Collin 1995; Perpète et al. 1998). The extraction time, the condenser temperature, and the flow rate of the nitrogen ensuring an inert atmosphere in the microextractor are crucial to good recovery (Bouseta and Collin 1995). With a 45-min extraction time, a temperature of −10°C in the cold finger, and a 2-mL/min nitrogen flow, recovery factors as high as 97 to 99% are obtained for terpenic compounds such as terpineol, verbenone, bornyl acetate, and *trans*-caryophyllene. After extraction, the Kuderna-Danish flask allows a high concentration factor without flavour loss (Bouseta and Collin 1995).

The above method has enabled us to determine easily more than a hundred molecules in hop pellets (Table 5.3 and Fig. 5.4).

5.2.1 Terpenic Compounds

Terpenic compounds are hydrocarbons representing 40 to 80% (by weight) of the total hop essential oils. Large amounts of three terpenes are present in all hop varieties (Fig. 5.5). The most abundant monoterpene ($C_{10}H_{16}$) is myrcene (about 30% of the total oil content) while α-humulene (18 to 33%) and β-caryophyllene (4 to 22%) are sesquiterpenes ($C_{15}H_{24}$) (Jenard 1966; Tressl et al. 1978b; Sharpe and Laws 1981; Moir 1992). According to Kralj et al. (1991), only eleven compounds repre-

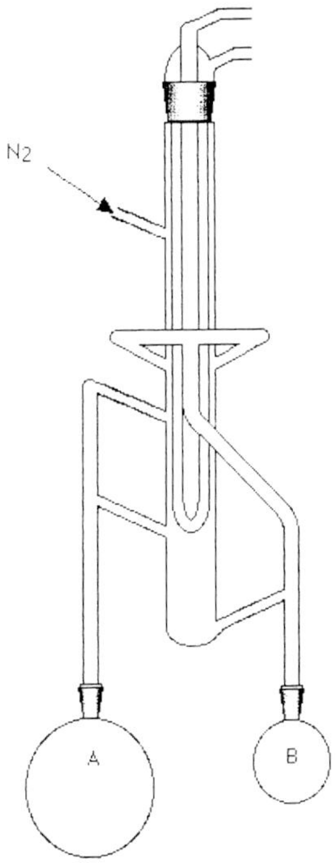

Fig. 5.3. Microextractor used for simultaneous steam-distillation solvent extraction (Perpète et al. 1998). *A* 50 ml ultrapure deoxygenated water + 0.5 g hop; *B* 1.5 ml dichloromethane

sent the fundamental hop oil composition: myrcene, α-humulene, β-caryophyllene, α-muurolene, β-farnesene, carvone, methyl-deca-4-enoate, 2-undecanone, and three unidentified compounds. Other hydrocarbon terpenes such as β-selinene, γ-cadinene, or tricyclic sesquiterpenes could be specific to some varieties (Davies and Menary 1982; Tressl et al. 1983).

Terpenic compounds are derived from isopentenyl diphosphate units also called activated isoprene. Located in the chloroplast, these units assemble to create unsaturated hydrocarbon chains of two or more units. The chains are further modified by oxidation, reduction, or carbon removal. The enzyme *isopentenyl diphosphatase-Δ-isomerase* (Fig. 5.6) catalyses the isomerisation of a first isopentenyl diphosphate molecule to 3,3-dimethyl allyl diphosphate.

The corresponding C_5 carbocation is further transferred to a second isopentenyl diphosphate molecule, yielding either geranyl diphosphate (*trans* compound) or its isomer, neryl-diphosphate (*cis* compound). The latter is a precursor of monoterpenes such as myrcene (Fig. 5.7). Cyclic monoterpenes are also usually created from neryl diphosphate or from linalyl diphosphate (Fig. 5.8).

Diphosphate removal from geranyl diphosphate again yields an allylic carbocation, allowing further isoprene additions. After removal of the diphosphate anion, farnesyl diphosphate is a precursor of various other sesquiterpenes such as α-humulene or β-caryophyllene (Fig. 5.9).

Figure 5.10 shows a series of cyclic sesquiterpenes also derived from farnesyl diphosphate.

Table 5.3. List of compounds detected in 11 hop varieties (Saaz, Lublin, Styrian Goldings, Hallertau, Mount Hood, Northdown, Northern Brewer, Galena, Nugget, Challenger, Pride of Ringwood). *n* Pic number, see Fig. 5.4; *RT* retention time (min); I_k retention index on a CP-Sil5 column; *Identification* mass spectrometric identification (*MS*) and/or pure compound elution (*GC*)

Compounds	*n*	RT	I_k	Identification
Furan,2,3-dihydro,4-methyl	1	7.70	659	MS
Unknown 95	95	12.76	805	MS
Unknown 96	96	13.30	813	MS
Isobutyl propanoate	2	15.15	844	GC
Isobutyl isobutyrate	3	18.56	898	GC/MS
Methyl hexanoate	4	19.03	904	GC/MS
3,3,6-Trimethyl,1,5-heptadiene	5	19.23	906	MS
Unknown 97	97	19.70	911	MS
Terpene 6[a]	6	21.96	936	MS
α-Pinene	7	22.57	942	MS
2-Methylbutyl propanoate	8	23.20	949	GC/MS
β-Pinene	9	25.80	978	GC/MS
Myrcene	10	26.42	985	GC/MS
2,7-Dimethyl,2,6-octadiene	11	26.70	988	MS
3-Methylbutyl isobutyrate	12	27.42	996	GC/MS
2-Methylbutyl isobutyrate	13	27.88	1,001	GC/MS
Methyl heptanoate	14	28.37	1,005	GC/MS
Limonene	15	30.74	1,025	GC/MS
Unknown 16	16	31.33	1,030	MS
γ-Terpinene	17	31.95	1,036	MS
Heptanoic acid	18	32.84	1,044	GC/MS
1-Octanol	19	33.39	1,049	MS
Unknown 20	20	35.50	1,067	MS
Unknown 21	21	36.36	1,074	MS
Linalool oxide	22	36.78	1,078	MS
Nonanal	23	37.36	1,082	GC/MS
Linalool	24	37.46	1,084	GC/MS
3-Methyl,2-methylbutyl butanoate	25	38.04	1,089	MS
2-Methylbutyl pentanoate	26	38.35	1,092	MS
Methyl octanoate	27	39.90	1,104	GC/MS
Unknown 28	28	43.24	1,130	MS
3-Hexen-1-ol formate	29	43.93	1,136	MS
Octanoic acid	30	44.96	1,143	GC/MS
1,14-Tetradecanediol	31	47.43	1,162	MS
2-Decanone	32	48.48	1,171	GC/MS
Unknown 33	33	49.00	1,175	MS
α-Terpineol	34	49.55	1,179	MS
Octyl propanoate	35	50.23	1,184	MS
Methyl,4-nonenoate	36	51.12	1,191	MS
Methyl nonanoate	37	52.91	1,205	GC/MS
IST (carvone)		54.91	1,220	GC/MS
Unknown 38	38	56.24	1,230	MS
Geraniol	39	56.84	1,235	GC
Unknown 40	40	57.27	1,238	MS
Nonanoic acid	41	58.03	1,244	GC/MS
Unknown 42	42	59.64	1,256	MS
Unknown 43	43	60.79	1,265	MS
4,6-Dimethyl,methyl octanoate	44	61.65	1,272	MS
2-Undecanone	45	62.13	1,276	GC/MS
Unknown 46	46	63.72	1,288	MS
Methyl, deca,4-enoate	47	64.05	1,291	MS
Methyl, deca,4,8-dienoate	48	64.24	1,292	MS
Methyl geranate	49	65.45	1,302	GC/MS

Table 5.3. *Continued*

Compounds	n	RT	I_k	Identification
Methyl decanoate	50	65.84	1,306	GC/MS
Unknown 51	51	67.88	1,328	MS
Unknown 52	52	68.43	1,334	MS
Unknown 53	53	69.38	1,345	MS
Geranyl acetate	54	71.22	1,365	GC
Decanoic acid	55	71.76	1,371	MS
2-Dodecanone	56	72.38	1,378	GC/MS
Ylangene	57	72.79	1,382	MS
Copaene	58	73.18	1,387	MS
Unknown 59	59	73.61	1,391	MS
Unknown 60	60	73.94	1,395	MS
7-Methyl,methyl nonanoate	61	74.63	1,403	MS
β-Caryophyllene	62	76.76	1,435	GC/MS
Sesquiterpene 63[b]	63	77.22	1,441	MS
Bergamotene	64	77.22	1,441	MS
β-Farnesene	65	78.18	1,455	GC/MS
Unknown 66	66	78.22	1,456	MS
α-Humulene	67	79.22	1,470	GC/MS
2-Tridecanone	68	79.82	1,479	GC/MS
α-Amorphene	69	80.09	1,483	MS
Methyl,3,6-dodecadienoate	70	80.40	1,488	MS
β-Selinene	71	81.08	1,498	MS
Sesquiterpene 72[c]	72	81.36	1,503	MS
Sesquiterpene 73[d]	73	81.46	1,504	MS
α-Selinene	74	81.57	1,505	MS
Sesquiterpene 75[e]	75	81.98	1,513	MS
γ-Cadinene	76	82.46	1,522	MS
Unknown 77	77	82.79	1,527	MS
δ-Cadinene	98	82.92	1,529	MS
Sesquiterpene 78[f]	78	83.05	1,532	MS
Sesquiterpene 79[g]	79	83.50	1,540	MS
α-Muurolene	80	83.81	1,545	MS
1,13-Tetradecadiene	81	84.50	1,558	MS
Humuladienone	82	85.77	1,581	MS
Caryophyllene oxide	83	86.50	1,593	GC/MS
Humulene epoxide I	84	87.17	1,606	GC/MS
Humulene epoxide II	85	87.82	1,619	GCMS
Humulene epoxide III	99	88.66	1,639	MS
Unknown 16	86	89.13	1,646	MS
Humulenol II	87	89.66	1,657	MS
Unknown 88	88	90.10	1,666	MS
Unknown 100	100	90.68	1,678	MS
Unknown 89	89	90.90	1,684	MS
Unknown 90	90	92.06	1,707	MS
Humulene diepoxide A	91	92.74	1,722	GC/MS
Humulene diepoxide B	92	93.34	1,736	GC/MS
Humulene diepoxide C	93	94.94	1,773	GC/MS
Humulene diepoxide D	94	95.46	1,785	GC/MS

[a] Terpene 6: bicyclo [3.1.1] hep-2-ene,3,6,6-trimethyl

[b] Sesquiterpene 63: 1H-cyclopenta [1,3] cyclopropa [1,2] benzene, octahydro-7-methyl-3-methylene-4-(1-methylethyl)-

[c] Sesquiterpene 72: naphtalene,1,2,3,5,6,7,8,8a-hexahydro-4,7-dimethyl-1-(1-methylethyl)-

[d] Sesquiterpene 73: cyclohexene,1-methyl-4-(5-methyl-1-methylene-4-hexenyl)-

[e] Sesquiterpene 75: naphtalene,1,2,3,5,6,8a-hexahydro-4,7-dimethyl-1(1-methylethyl)-

[f] Sesquiterpene 78: cyclohexene,6-ethenyl-6-methyl-1-(1-methylethyl)-3-(1-methylethylidene)-

[g] Sesquiterpene 79: naphtalene,1,2,3,4,4a,7-hexahydro-1,6-dimethyl-4-(1-methylethyl)-

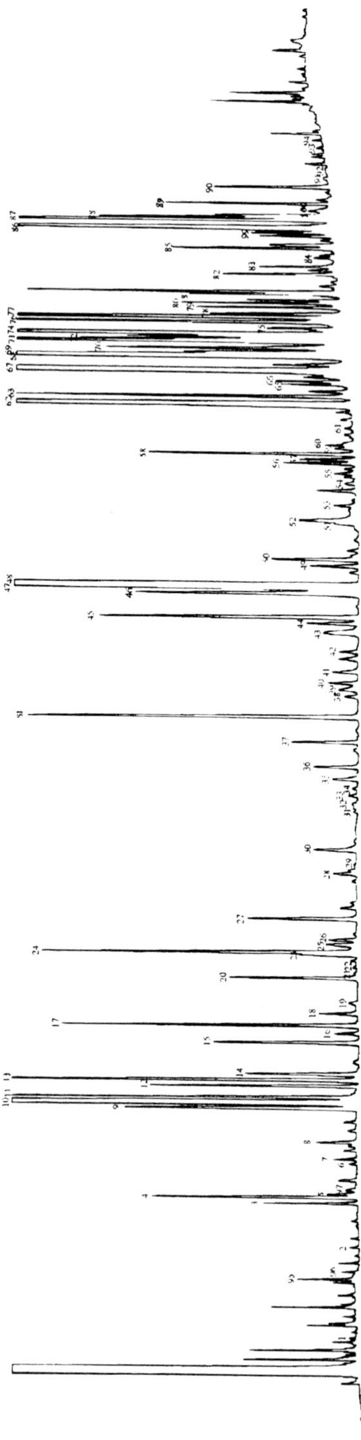

Fig. 5.4. Chromatographic profile corresponding to analysis of hop oil by GC

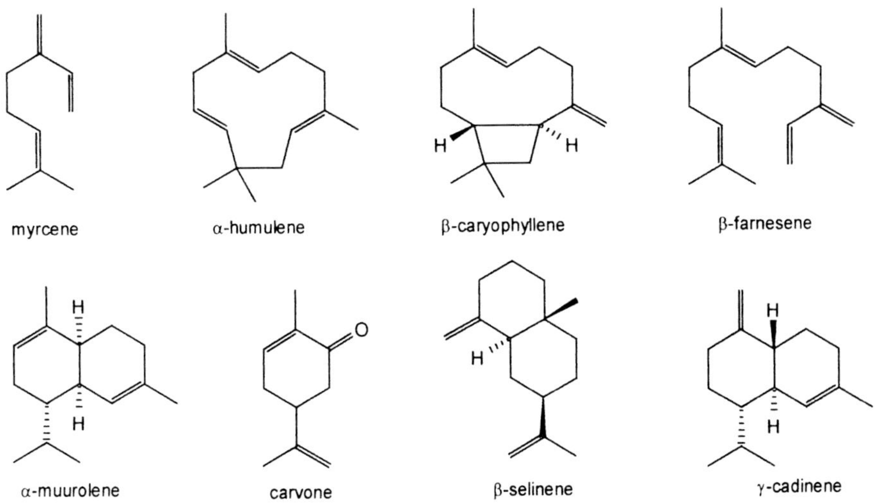

Fig. 5.5. Myrcene, α-humulene, β-caryophyllene, β-farnesene, α-muurolene, carvone, β-selinene, and γ-cadinene

5.2.2 Oxidation and Hydrolysis Products from Sesquiterpenes

Several authors (Tressl et al. 1978a, 1983; Peacock and Deinzer 1981) have shown that sesquiterpenes can be easily oxidised during hop storage to yield the corresponding epoxides (Fig. 5.11). Caryophyllene epoxide, three humulene epoxides, and five humulene di-epoxides have been evidenced in aged samples (Peacock et al. 1980; Sharpe and Laws 1981; Peacock and Deinzer 1989; Moir 1992). Subsequent hydrolysis during storage, boiling, or fermentation yields various alcohols and ketones such as humulol II, humulenol II, and humuladienone (Fig. 5.12; Tressl et al. 1978a; Fukuoka and Kowaka 1983; Lam et al. 1986).

The terpene oxidation rate depends on the storage temperature, light exposure, and hop reducing power (carotenoids, polyphenols, etc.) (Stevens 1966; Liégeois et al. 2000).

5.2.3 Alcohols, Carbonyles, Acids and Esters

As depicted in Fig. 5.13, geraniol, a monoterpenic alcohol derived from geranyl diphosphate, can yield either geranial or esters such as geranyl acetate and geranyl isobutyrate. In the former case, an additional oxidation can occur, yielding geranic acid or methyl geranate. Hop esters can be hydrolysed or transesterified by yeast during beer fermentation, provided no α,β-unsaturated esters are concerned (e.g. methyl geranate; Palamand et al. 1971). Thus, geraniol resulting from yeast hydrolysis of geranyl acetate and geranyl isobutyrate may contribute to the floral

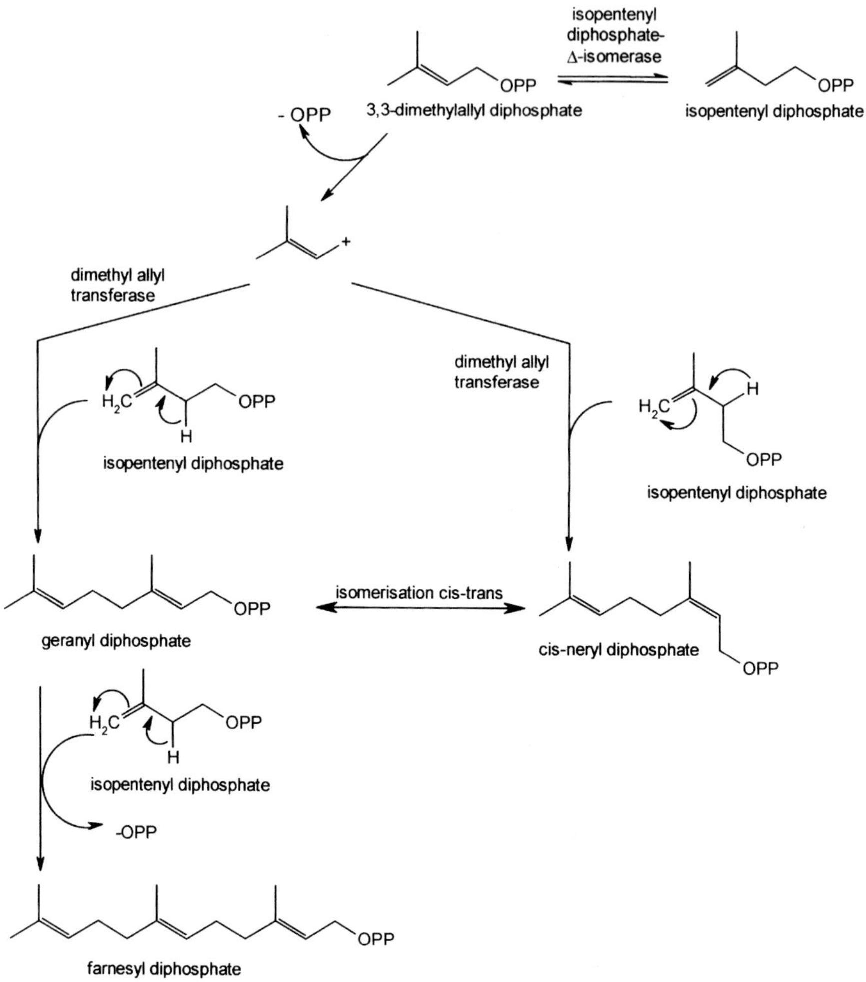

Fig. 5.6. Biosynthesis of monoterpenes and sesquiterpenes precursors

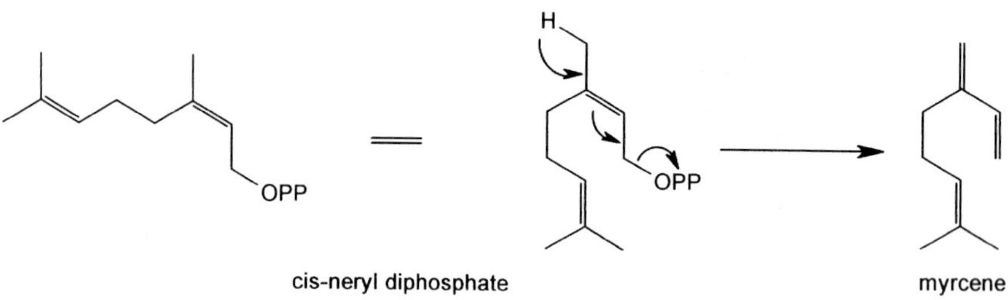

Fig. 5.7. Myrcene biosynthesis

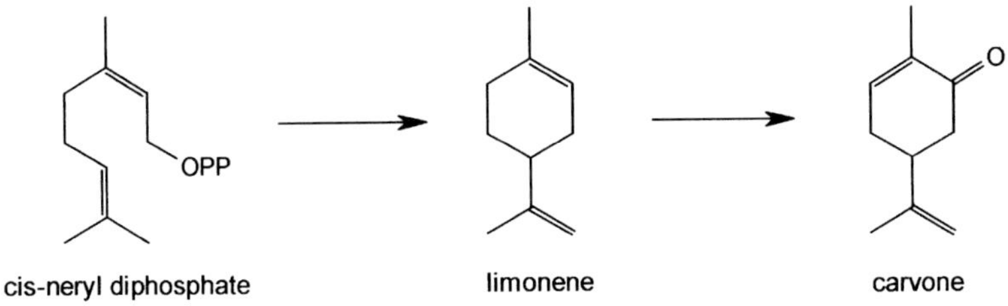

Fig. 5.8. Limonene and carvone synthesis from *cis*-neryl diphosphate

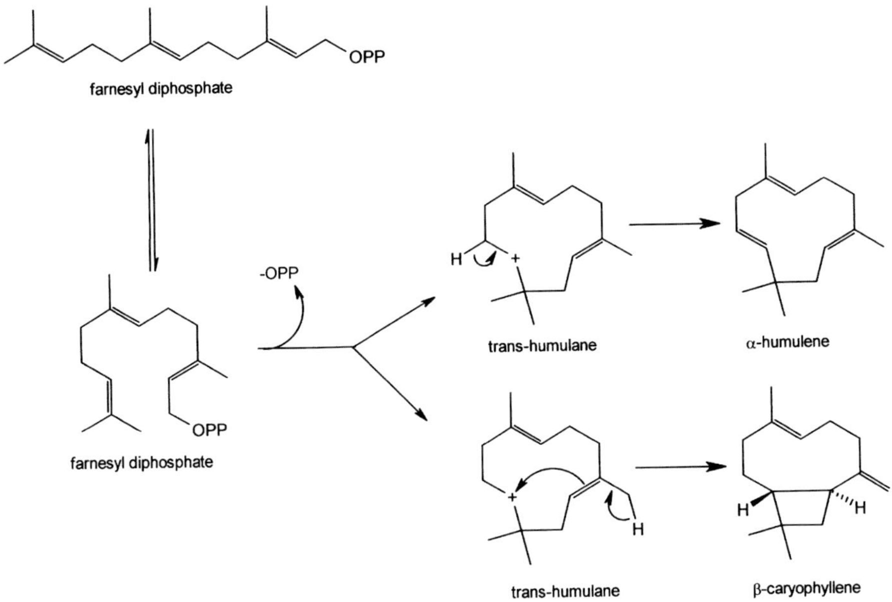

Fig. 5.9. α-Humulene and β-caryophyllene synthesis

hop aroma of beer (Peacock et al. 1981). Geraniol, nerol, and citronellol may also arise through yeast reduction of citral and citronellal.

Other terpenic alcohols such as linalool (the most abundant), α-terpineol, and farnesol are present in hop oils (Fig. 5.14). These terpenic alcohols can be viewed as end products of plant biosynthesis. Levels of most terpenic alcohols tend to drop during hop storage whereas allylic alcohol levels tend to increase (Tressl et al. 1978b; Moir 1992).

Linear ketones (10 to 13 carbon atoms) and 6 methyl-5-hepten-2-one (hoppy aroma) are present in hop essential oils (Maier 1966; Markl and Palamand 1973; Davies and Menary 1982; Kralj et al. 1991). Unsaturated linear esters have also been identified in hops (Fig. 5.15; Maier 1966; Shimazu et al. 1978; Haley and Peppard

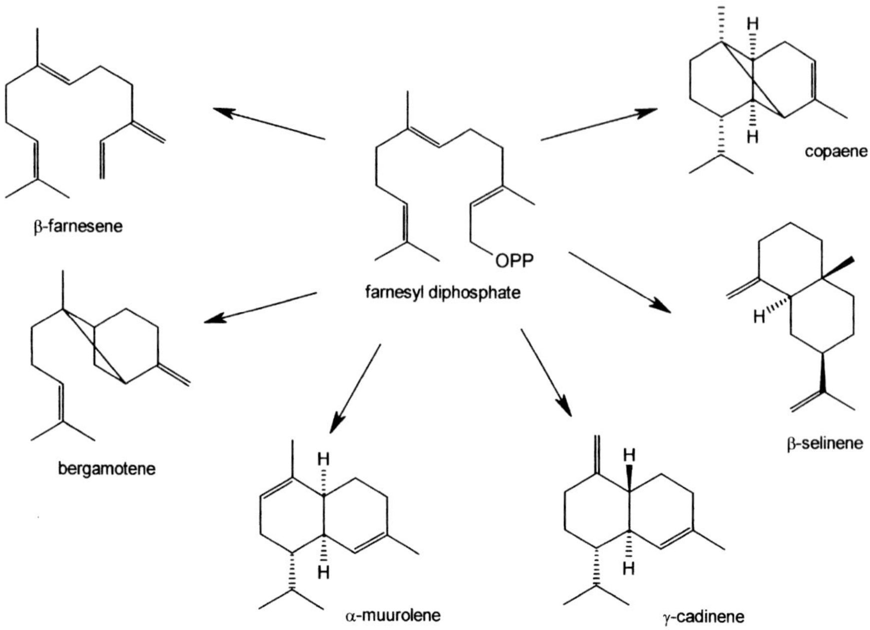

Fig. 5.10. Six sesquiterpenic compounds frequently present in hop essential oils

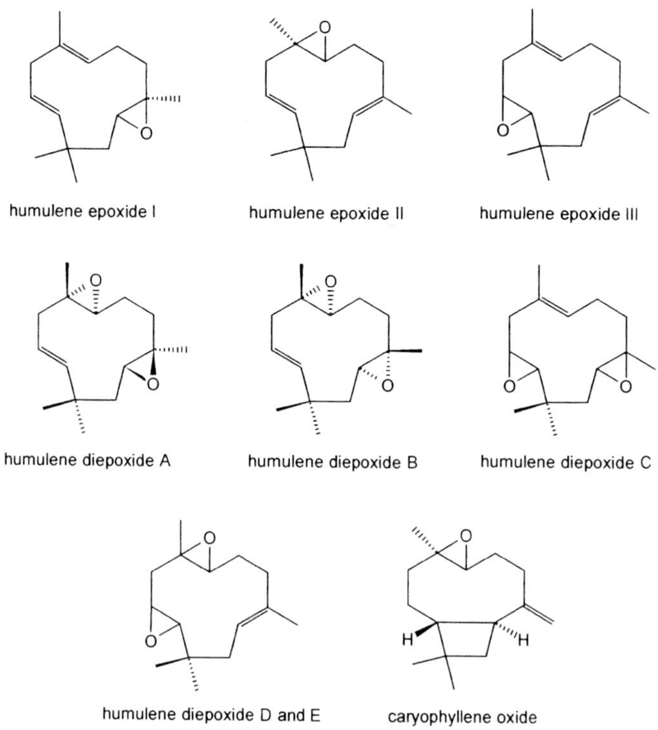

Fig. 5.11. Main epoxides and diepoxides in hop

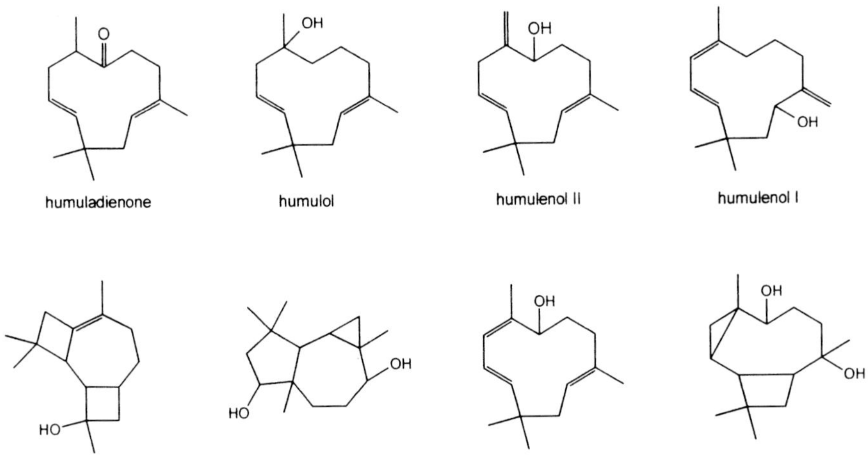

Fig. 5.12. Some humulene hydrolysis products. (Yang et al. 1991, 1993)

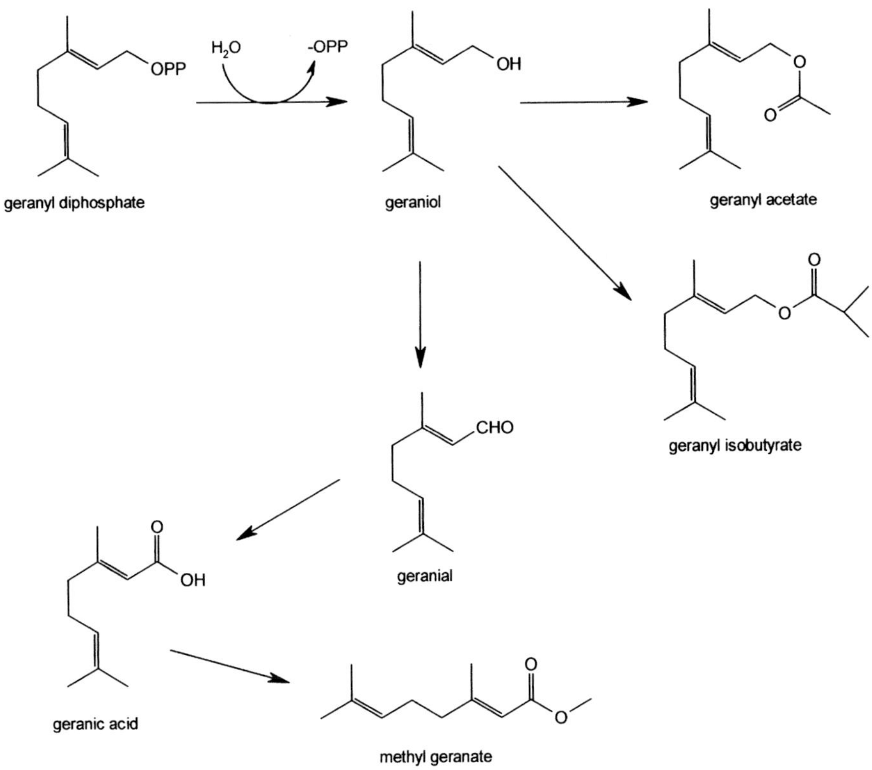

Fig. 5.13. Methyl geranate synthesis from geranyl diphosphate

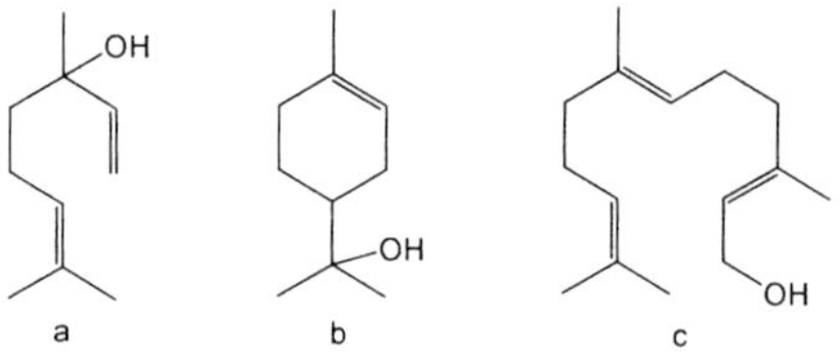

Fig. 5.14. **a** Linalool, **b** α-terpineol, **c** farnesol

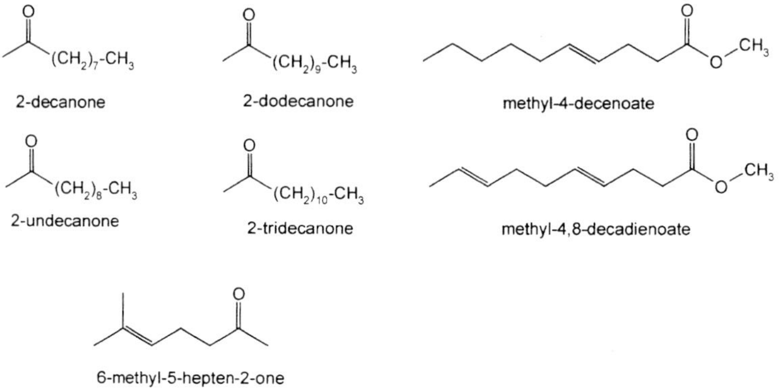

Fig. 5.15. Other important ketones and esters present in hop oils

1983; Kralj et al. 1991). Ethyl lactate is another ester found in cones, mainly after ageing.

Tressl et al. (1978b) identified β-ionone and β-damascenone in hops. According to Strating and Van Eerde (1973), damascenone could impart a blackcurrant odour to aged beers. Both of these compounds are known as oxidative degradation products of β-carotene and analogues.

5.2.4 Hop Aroma Glycosides

After liquid carbon dioxide extraction, which removes essential oils and soft resins, hop solids still contain a water-soluble fraction with several aroma glycosides (Goldstein et al. 1999). Hydrolysis of hop solid with *β-glucosidase* releases a large number of aromatic compounds (Table 5.4). Direct LC-MS analyses of glycosides have confirmed the structures of several aglycones: grasshopper ketone, 3-methyl-2-pentanol, 4-methyl-2-pentanol, citronellol, aceto vanillonol, 2,6-dimethyl-2,7-octadiene-1,6-diol, and 3-hydroxy-7,8-dihydro-β-ionone.

Table 5.4. Identified molecules derived from treatment of hop solids with β-glucosidase (Goldstein et al. 1999)

2-Methyl-3-buten-2-ol	1-Phenylethanol	2-Hydroxy-5-methyl acetophenone
3-Methyl-2-butanol	2-Phenylethanol	Vanillyl acetone
2-Methyl-1-propanol	p-Hydroxyphenylethanol	1-Ethyl-2-pyrrolidone
3-Pentanol	4-Methyl-2-pentanone	2,4,5-Trimethyl acetophenone
2-Pentanol	3-Methyl-2-pentanone	2-Hexenal
4-Methyl-2-pentanol	4-Methyl-3-penten-2-one	Geraniol
3-Methyl-2-pentanol	2-Octen-4-one	α-Terpineol
3-Methyl-1-butanol	6-Methyl-5-hepten-2-one	Linalool
2-Methyl-1-butanol	3-Hydroxy-2-butanone	c- and t-Linalool oxide
1-Pentanol	Phenylacetaldehyde	Benzaldehyde
3-Methyl-2-buten-1-ol	Vanillin	Loliolide
1-Hexanol	4-Hydroxybenzaldehyde	2,6-Dimethyl-2,7-octadien-1,6-diol
3-Hexen-1-ol	Coumaran	4-Hydroxy-3-methoxy-acetophenone
2-Hexen-1-ol	Benzylic alcohol	2-Methyl-6-methylene-2,7-octadien-4-ol
1-Octen-3-ol	2-Methoxy-3-isopropyl-pyrazine	2-Methyl-6-methylene-3,7-octadien-2-ol
6-Methyl-5-hepten-2-ol	Raspberry ketone	2,2-Methylene bis[6-(1,1-dimethylethyl)-4-ethylphénol]
2-Ethyl-1-hexanol	4-Terpineol	5-Methyl-3-hexen-2-one
1-Octanol	5,5-Dimethyl-2-(5H)-furanone	
Ipsdienol	Acetophenone	

5.3 Varietal Discrimination of Hop Cultivars According to Their Oil Content

Buttery and Ling (1967) and Likens and Nickerson (1967) have shown that the chemical composition of hop essential oil, genetically determined, is only slightly influenced by external conditions. As this composition depends on the hop variety, some compounds can be taken as markers for varietal discrimination (Buttery and Ling 1967; Likens and Nickerson 1967; Naya and Kotake 1972; Kralj et al. 1991; Forster and Schmidt 1994; Perpète et al. 1998).

A first noteworthy difference between hop cultivars is their relative content in three terpenic hydrocarbons: α-humulene (H), β-caryophyllene (C) and myrcene (M) (Peacock et al. 1980).

- H > C > M in Saaz, Styrian, and Target
- M > H > C in Lublin, Galena, and Nugget
- H > M > C in Northern Brewer
- M > C > H in Pride of Ringwood

For Deinzer and Yang (1994), low humulene/caryophyllene (H/C) ratios are also typically associated with non-European hop varieties (e.g., H/C = 2.1 for Cluster) whereas high ratios (H/C = 3 for Hallertau Mittelfrüh and 2.9 for Tettnang) may be associated with the European hoppy aroma.

According to Davies and Menary (1982), the myrcene level is higher in bitter hops. Yet because myrcene is too volatile and too readily oxidised, it is hard to use it for varietal discrimination. Myrcene is usually not desired although partially responsible, once oxidised, for the hoppy aroma in beer.

Aromatic hops such as Saaz, Tettnang, Styrian Goldings, and Lublin can also be distinguished from bitter varieties by a higher β-farnesene content (150 to 1,140 ppm) (Moerman 1982; Perpète et al. 1998). As with myrcene, unfortunately, this conjugated diene readily undergoes Diels-Alder degradation, leading to a significant loss in aged hops or after pelleting (Peacock and McCarty 1992).

Bergamotene (Fig. 5.10) is closely associated with farnesene (Peacock and McCarty 1992; Perpète et al. 1998). In fresh cones, its concentration is usually about 5% of the level of farnesene. Perpète et al. (1998) found concentrations between 25 and 74 ppm in all tested pellets produced from aromatic cultivars and less than 5 ppm in bitter varieties. Bergamotene is much more stable than farnesene as it cannot undergo a Diels-Alder reaction. Hops nearly devoid of farnesene but containing 20 ppm bergamotene should thus be classified as high-farnesene hops.

Additional sesquiterpenes and their degradation products (α-amorphene, β-selinene, α-selinene, humulene epoxide I, humulene epoxide II, etc.) also prove very useful for varietal discrimination (Kralj et al. 1991; Perpète et al. 1998). In the "European bitter hops" group of Perpète et al. (1998) (Fig. 5.16), Northdown and Challenger are characterised by large amounts of α- and β-selinenes (>200 ppm, caryophyllene equivalent), as compared to Target and Northern Brewer. Northdown is distinguishable from Challenger by the concentration of a sesquiterpene that may be α-amorphene (>125 ppm, caryophyllene equivalent). The same terpene also occurs as a significant constituent distinguishing Mount Hood (>125 ppm, caryophyllene equivalent) from Hallertau in the "low bitterness" group. The most interesting marker of Northern Brewer hops emerging from their study is the humulene-to-farnesene ratio (above 220). All other cultivars, and especially Target with its low humulene content, show a ratio below 220. The compound 2-undecanone is an additional marker for distinguishing Target (>100 ppm) from Northern Brewer hops. High 2-undecanone concentrations can also occur in Pride of Ringwood samples (Perpète et al. 1998). This methylketone could give rise to an organoleptically active reduced compound in beer, where the estimated flavour threshold of 2-undecanol is 70 ppb (Meilgaard 1975).

Perpète et al. (1998) also used esters to establish their varietal classification flowchart (Fig. 5.16). Non-European and European bitter hops were discriminated from aromatic cultivars by quantification of 4-decenoic acid methyl ester (>150 ppm in the former) and 3-methylbutyl isobutyrate (<11 ppm in the latter). A high level of 3,6-dodecadienoic acid methyl ester, furthermore, can help to recognise the Galena cultivar.

Using 30 hop oil constituents as markers, Kralj et al. (1991) were able to classify 95 different hops in 14 groups. The European hoppy aroma correlates positively with the presence of 8 of these 30 compounds and negatively with the presence of the others, responsible for unwanted flavours (Kralj et al. 1991; Deinzer and Yang 1994; see Table 5.5). Among them, methyl decanoate, neryl acetate, α-copaene, α-humulene, and β-caryophyllene emerged as determining. According to Kralj et al. (1991), methyl decanoate, neryl acetate, and α-copaene are responsible for the non-European aroma.

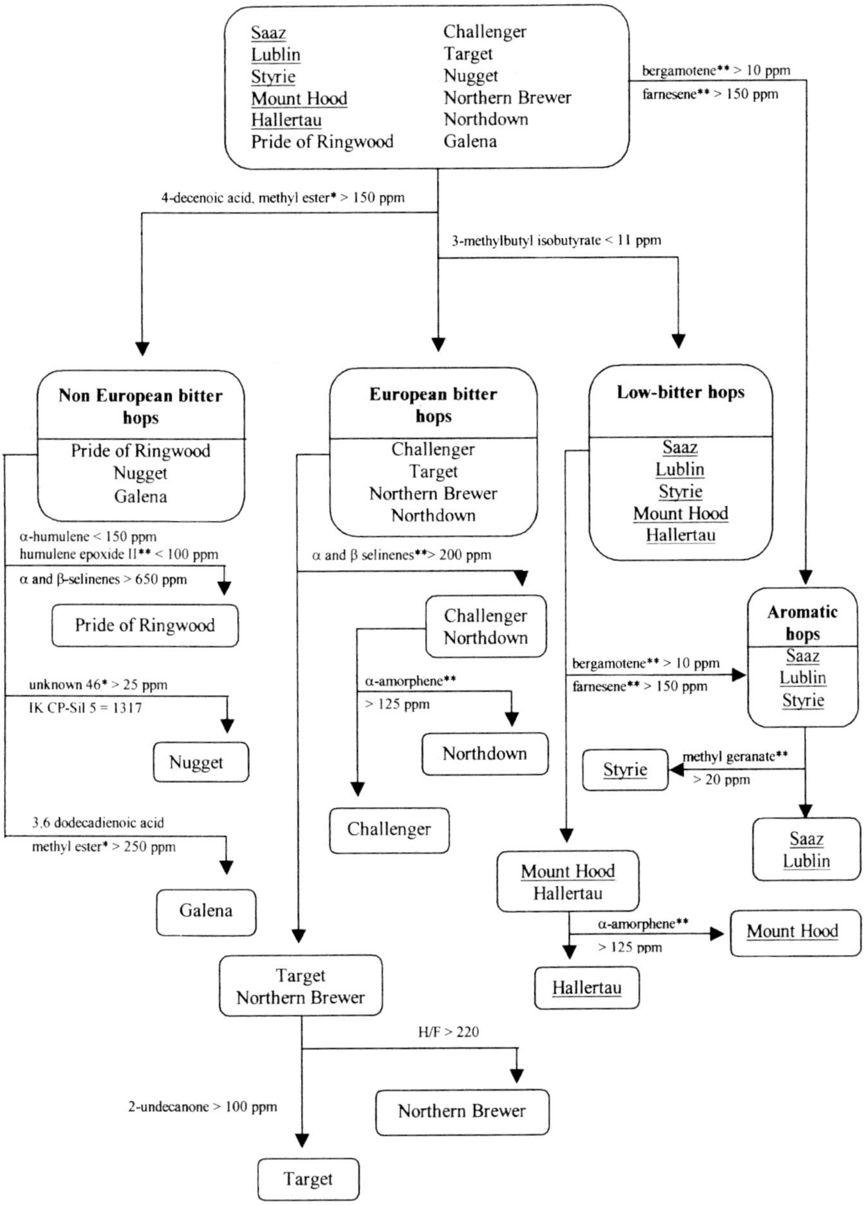

Fig. 5.16. Discrimination flowchart proposed by Perpète et al. (1998) for distinguishing 12 hop cultivars. * Carvone equivalent for quantification; ** caryophyllene equivalent for quantification

Table 5.5. Correlations between some hop compounds [selection results from analysis of 95 local varieties and cultivars from different regions and grown in the ecological environment of Slovenia, 1986–1988 (Kralj et al. 1991)] and the European hoppy aroma. (Deinzer and Yang 1994)

Positive correlation	Negative correlation	
α-Humulene[a,b]	Myrcene[c]	Methyl decanoate
β-Caryophyllene[b]	α-Muurolene	Neryl acetate
Carvone	Germacrene-D	γ-Copaene
Methyl, deca,4-enoate	Curcumene[c]	γ-Cadinene
2-Undecanone	β-Selinene	δ-Cadinene[c]
Farnesene[a]	α-Selinene	Selina-3,7(11)-diene[c]
Humulene epoxide I[a]	o-Cymene	Selina-3(14),7(11)-diene
Unidentified compound[a]	Methyl caprylate	Santalene[c]
	Geranyl acetate[c]	Calamenene
	Geraniol	Two unidentified compounds
	α-Phellandrene	

[a] Improved European aroma with increasing concentration
[b] Humulene/caryophyllene ratio must be greater than 3
[c] Increasing negative influence on aroma as concentration increases

Acknowledgements. Guillaume Lermusieau is grateful to the Interbrew – Baillet Latour Foundation for financial support. We also thank the *Journal of the American Society of Brewing Chemists*, *Cerevisia*, the *European Brewery Convention*, *The New Brewer*, and the authors Philippe Perpète, Denis De Keukeleire, Max Deinzer, Henry Goldstein, and Tim O'Rourke for having allowed the reproduction of tables and figures.

References

Bouseta A, Collin S (1995) Optimized Likens-Nickerson methodology for quantifying honey flavors. J Agric Food Chem 43:1890–1897

Buttery RG, Ling L (1967) Identification of hop varieties by gas chromatography patterns and analysis of hop oils from American grown varieties. J Agric Food Chem 15:531–535

Buttery RG, Black DR, Guadagni DG, Kealy MP (1965) A study of the volatile oxygenated constituents in different hop varieties. J Am Soc Brew Chem 23:103–111

Davies E (1960) Visual analysis of hop samples. Wall Commun 23:19–24

Davies NW, Menary RC (1982) Essential oils of Tasmanian grown hops. J Inst Brew 88:80–83

Deinzer M, Yang X (1994) Hop aroma: character impact compounds found in beer, methods of formation of individual compounds. In: Proc European Brewery Convention, Zoeterwoude, Monogr 22, pp 1–17

De Keukeleire D (1993) The effects of hops on flavour stability and beer properties. Cerevisia 18: 33–46

Derdelinckx G, Maudoux M (1992) Importance de la variété de houblon et de son type de conditionnement sur la qualité de la bière. In: Chaire J. De Clerck V (eds) Wort boiling and clarification: a challenging operation. Laboratoire de Brasserie, Université catholique de Louvain, Louvain-la-Neuve

Forster A, Schmidt R (1994) Characterisation and classification of hop varieties. Brauwelt Int 108–124
Fukuoka Y, Kowaka M (1983) Identification of compounds imparting hoppy flavour to beer. Rep Res Lab Kirin Brew Co 26:31–36
Goldstein H, Rader S, Murakami AA (1993) Determination of 3-methyl-2-butene-1-thiol in beer. J Am Soc Brew Chem 51:70–74
Goldstein H, Ting P, Navarro A, Ryder D (1999) Water-soluble hop flavor precursors and their role in beer flavor. In: Proc of the European Brewery Convention. IRL Press/Oxford University Press, Oxford, pp 53–62
Gunst F, Verzele M (1978) On the sunstruck flavour of beer. J Inst Brew 84:291–292
Haley J, Peppard TL (1983) Differences in utilisation of the essential oil of hops during the production of dry-hopped and late-hopped beers. J Inst Brew 89:87–91
Hughes PS, Simpson WJ (1993) Production and composition of hop products. MBAA Tech Q 30:146–154
Hupe J (1981) Le houblonnage de la bière: aspects qualitatifs et économiques. Cerevisia 3:129–136
Jenard H (1966) Constituants volatils des houblons. Echo Brass 69:672–674
Kenny ST (1990) Identification of US-grown hop cultivars by hop acid and essential oil analyses. J Am Soc Brew Chem 48:3–8
Kralj D, Zupanec J, Vasilj D, Kralj S, Psenicnik J (1991) Variability of essential oils of hops, *Humulus lupulus* L. J Inst Brew 97:197–206
Kuroiwa Y, Hashimoto H (1961) Composition of sunstruck flavour substance and mechanism of its evolution. J Am Soc Brew Chem 19:28–36
Kuroiwa Y, Kokubo E, Hashimoto N (1972) Advanced hop chemistry in connection with beer flavor. In: Proc 4th IFS, Ferment Technol, pp 633–637
Lam KC, Foster RT, Deinzer ML (1986) Aging of hops and their contribution to beer flavor. J Agric Food Chem 34:763–770
Lewis GK (1994) Kiss of the hops. New Brewer 11:11–19
Liégeois C, Lermusieau G, Collin S (2000) Measuring antioxidant efficiency of wort, malt, and hops against the 2,2'-azobis(2-amidinopropane) dihydrochloride-induced oxidation of an aqueous dispersion of linoleic acid. J Agric Food Chem 48:1129–1134
Likens ST, Nickerson GB (1964) Detection of certain hop oil constituents in brewing products. J Am Soc Brew Chem 22:5–13
Likens ST, Nickerson GB (1967) Identification of hop varieties by gas chromatographic analysis of their essential oils. Constancy of oil composition under various environmental influences. J Agric Food Chem 15:525–530
Maier J (1966) Zusammensetung des Hopfenöls in Beziehung zur Sorte und Qualität des Hopfens und seinen Umwandlungsprodukten. Brauwissenschaft 19:425–434
Markl KS, Palamand SR (1973) New flavor compounds and their relationship to beer aging. MBAA Tech Q 10:184–189
Meilgaard MC (1975) Flavor chemistry of beer. II. Flavor and threshold of 239 aroma volatiles. MBAA Tech Q 12:5–13
Moerman E (1982) Etude bibliographique concernant l'identification des varieties de houblon. Cerevisia 4:197–204
Moir M (1988) Developments in hop usage. Ferment 3:49–56
Moir M (1992) Hop aromatic compounds. In: Chaire J. De Clerck V (eds) Wort boiling and clarification: a challenging operation. Laboratoire de Brasserie, Université Catholique de Louvain, Louvain-la-Neuve
Moll M (1991) Bières et coolers. Tec Doc-Lavoisier, Paris
Naya Y, Kotake M (1972) The constituents of hops (*Humulus lupulus* L.). VII. The rapid analysis of volatile components. Bull Chem Soc Jpn 45:2887–2891
Nickerson GB, Likens ST (1966) Gas chromatographic evidence for the occurrence of hop oil components in beer. J Chromatogr 21:1–5
O'Rourke T (1994) Making the most of your hops. New Brewer 11:20–33
Palamand SR, Markl KS, Hardwick WA (1971) Trace flavor compounds of beer. J Am Soc Brew Chem 29:211–218
Peacock VE, Deinzer ML (1981) Chemistry of hop aroma in beer. J Am Soc Brew Chem 39: 136–141

odorant in the effluent of a gas chromatograph (Fuller et al. 1964). Of course, the first forms of olfactometry were very simple, and consisted of the sensory description of the effluent from the chromatographic column, as an aid in identification. The first reference listed in Table 6.1 dates from 1975, although it is almost sure that previous GCO work on wine had been done. The pioneers very soon realized what Nelson and Acree stated in 1978, that there is no correlation between volatile and aroma analysis, and there is no doubt that these results encouraged the subsequent development of Charm analysis (Acree et al. 1984; Acree and Cotrell 1985; Cunningham et al. 1986). A large part of the works published in the 1970s and 1980s suffered from the lack of knowledge about the nature of the wine impact aroma compounds. In addition, as they did not use quantitative techniques, they failed to assess whether an odor was really a key one.

Some of the first successes of olfactometry came in the 1980s, when olfactometry was used satisfactorily as an aid to identify some off-odors. The references by Rapp et al. (1980) and Heiman et al. (1983) are worth mentioning. In the first work, Rapp et al. identified furaneol as responsible for the strawberry odor of some hybrid wines; in the second one, Heiman et al. found some sesquiterpenes that could be related to the "corky" off-flavor. Yet, these authors did not used quantitative olfactometry, and they were able to succeed because most off-odors are easy to detect in the effluent of a GC. Quantitative olfactometry was first reported by Acree and Lavin in 1990 and by McDaniel et al. (1990; Miranda-López et al. 1992). The first authors demonstrated the importance of o-aminoacetophenone in Labruscana grapes through the measurement of Charm peaks. The second group of researchers introduced the technique known as Osme to measure the influence of grape maturity and of the vintage on the wines from Pinot Noir. These works were brilliant and introduced new concepts. However, the conclusions reached by the authors are not very meaningful because they failed in the identification of most of the odor peaks registered and were not able to establish a potential hierarchy of odorants. Also from this time is the work by Darriet et al. (1991), which used a rather original approach (measuring the width of the odor peak) to get quantitative data from an odorant which, at that moment, they had not been able to identify. Their work demonstrates the potential of olfactometry as a way to get quantitative data from odor-active compounds, regardless of whether they have been identified or not.

Another very active research group in wine olfactometry was the group of Chisholm who used Charm analysis and was able to develop a quite rigorous approach that should be known to all the practitioners of GCO. They demonstrated that GCO is a powerful way to detect differences between wines. Particularly, they were able to determine the differences between some French–American hybrids and white Riesling and to study the changes observed during wine aging. Very interesting as well is the work carried out by several researchers working at Dijon (France) who introduced strategies of data treatment to look for correlation between the aromatic profile and the aromagram (Schlich and Moio 1994; Escudero and Etievant 2000a). The work from these authors poses several questions about the relationship between Charm values and the number of odor

Table 6.1. Studies concerning olfactometry in wine. *NIF* Nasal impact frequency

Year	Authors	Technique	Subject
1975	Wildenradt et al.	Just odor evaluation	Is a pioneer paper that uses the evaluation of the gas chromatographic column as a way of detecting grassy and leafy compounds from vine leaves
1975	Stern et al.	Just odor evaluation	Is a pioneer paper that uses the evaluation of the gas chromatographic column as a way to gather information about the changes in aroma during aging of wine
1978	Nelson and Acree	Just odor evaluation	Compares Concord wines from different maturities or winemaking strategy. States that there is no correlation between volatile and aroma analysis
1980	Rapp et al.	Just odor evaluation	Identifies furaneol as the key compound responsible for the strawberry-like character of some interspecific grape-wine breedings
1980	Slingsby et al.	Just odor evaluation	Uses olfactometry to try to detect methoxypyrazines in Cabernet Sauvignon wine extracts
1982	Augustyn et al.	Just odor evaluation	A complete study about the aroma compounds of must from Sauvignon Blanc
1983	Criddle et al.	Just odor evaluation	A complete study about the aroma compounds of Sherry wine
1983	Etievant et al.	Just odor evaluation	A complete study about the aroma compounds of muscat wine
1983	Heiman et al.	Just odor evaluation	Identifies several sesquiterpenes as possible cause of a corky off-flavor of microbiological origin
1984	Simpson and Miller	Just odor evaluation	Gives a complete qualitative study of the aroma of Chardonnay wine, and for each compound or chromatographic peak gives the odor detected. Uses this as a tool for the discussion of results
1985	Acree and Cotrell	Charm	Presents Charm analysis as bioassay for measuring wine quality
1985	Reglero	A precursor of NIF	Uses a group of tasters to identify the odors from the column, concluding that some ethyl esters may be important since they are most frequently detected in young wines
1986	Baumes et al.	Just odor evaluation	Makes a complete qualitative and quantitative determination of the odorants of monovarietal wines, and gives for each compound the odor perceived during its elution

Table 6.1. *Continued*

Year	Authors	Technique	Subject
1990	Rapp and Pretorius	Just odor evaluation	Presents some new data about the role of some compounds detected by GCO in wine off-flavors
1990	Acree and Lavin	Charm	Uses Charm analysis to demonstrate the odor activity of o-aminoacetophenone in Labruscana grapes
1990	McDaniel et al.	Osme	Semiquantitative evaluation of the odorants formed in Pinot Noir wines as a function of grape maturity
1991	Darriet et al.	A kind of Osme	Semiquantitative measurement of two box-tree odorants from Sauvignon Blanc wines
1992	Miranda-López et al.	Osme	Semiquantitative evaluation of the odorants formed in Pinot Noir wines as a function of grape maturity
1992	Chisholm and Samuels	Charm	Determines the sensory impact of some products derived from sorbic acid in a spoiled red wine
1994	Moio et al.	Charm	Olfactometric study of Chardonnay wine
1994	Schlich and Moio	Charm	A study of the correlation between sensory and olfactometric data
1994	Chisholm et al.	Charm	Compares the aroma profiles of Riesling wines with some others made with French–American hybrid grapes
1995	Moio and Etievant	Charm	Uses Charm analysis to propose ethyl anthranilate, ethyl cinnamate, ethyl 2,3-dihydrocinnamate and methyl anthranilate as important aroma contributors of Pinot Noir from Burgundy
1995	Chisholm et al.	Charm	Studies the changes in the aroma profiles of Vidal Blanc wines during aging, and makes some comparison with their sensory characteristics
1995	Moio et al.	Charm	Uses sensory and olfactometric analysis to study the representativity of different types of extract from wine
1995	Darriet et al.	Just odor evaluation	Uses it to isolate and identify the powerful odorant 4-methyl-4-mercaptopentanone as a key odorant of Sauvignon Blanc wines
1997	Priser et al.	Sensory analysis	Uses quantitative descriptive sensory analysis to study the representativity of different types of extract from champagne

Table 6.1. *Continued*

Year	Authors	Technique	Subject
1997a	Guth	AEDA	Uses AEDA to classify in order of aromatic power the most important odorants of Schreube and Gewürtztraminer wines
1997	Aubry et al.	OUV	Uses quantitative analysis to verify the role played in Burgundy Pinot Noir by some important odorants previously identified through Charm analysis as key odors
1997	Guedes de Pinho et al.	Just odor evaluation	Uses olfactometry just as a guide to identify important aromatic sulfur-containing odorant linked to the aroma of hybrid grape wines
1997	Cutzach et al.	Just odor evaluation	Uses olfactometry just as a guide to identify important aromatic compounds with toasty aroma from heated wood
1997a,b	Guth	OUV	Determines Aroma values of Schreube and Gewürtztraminer wines and makes reconstitution studies
1997	Baek et al.	AEDA	Uses AEDA to classify in order of aromatic power the most important odorants of muscadine grape juices
1998	Le Fur and Etievant	Charm	Identifies the most important odorants of Chardonnay wines and studies the potential importance of guaiacol, maltol, ethyl cinnamate and cyclotene in some key descriptors
1998	Schneider et al.	Just odor evaluation	Uses olfactometry just as a guide to identify some important odorants linked to the aroma of Natural Sweet Wines
1998	Bouchilloux et al.	Just odor evaluation	Uses olfactometry just as a guide to identify some important aromatic thiols odorants linked to the aroma of Cabernet Sauvignon and Merlot wines
1998a	Ferreira et al.	AEDA	Uses AEDA to classify in order of aromatic power, the most important odorants of Grenache red wines
1998	Lavigne et al.	Just odor evaluation	Uses olfactometry just as a guide to identify some important aromatic thiols odorants linked to the toasted aromas of some wines
1998b	Ferreira et al.	AEDA/OUV	Studies the relationship between flavor detection values and OUV in hydroalcoholic solutions
1998	Guth	AEDA/OUV	Uses AEDA, and later determines aroma values of Schreube and Gewürtztraminer wines and makes reconstitution studies

Table 6.1. *Continued*

Year	Authors	Technique	Subject
1999	Escudero and Etievant	NIF	Determines the effect of some antioxidants on the olfactometric profiles of champagne extracts. Gets quantitative results
1999	Ong and Acree	Charm	Compares the aroma profiles and the aroma spectra of canned lychees and Gewürtztraminer wines
1999	López et al.	AEDA	Uses AEDA to classify in order of aromatic power the most important odorants of Grenache, Merlot and Cabernet Sauvignon red wines
2000a	Escudero et al.	NIF	Uses this quantitative technique of olfactometry to determine the differences in aroma composition caused by the ageing of champagne
2000a	Ferreira et al.	OUV	Determines Aroma values of a large number of monovarietal red wines
2000	Kotseridis and Baumes	AEDA	Uses AEDA to classify in order of aromatic power the most important odorants of Merlot and Cabernet wines from Bordeaux,
2000c	Escudero et al.	AEDA	Uses AEDA to compare the aromatic profiles of wines before and after oxidation in order to characterize the most important odors
2000	Charles et al.	NIF	Determines the most important odorants of two red wine vinegars
2000b	Ferreira et al.	AEDA/OUV	Uses AEDA to determine the most important odorants of Rioja red wines. Olfactometric data are compared with the sensory evaluation of flavor fractions and with OUV

units (concentration/threshold) of an odor. By using Charm analysis, they proposed four esters as key odorants of Pinot Noir wines from Burgundy (Moio and Etievant 1995), but later on (Aubry et al. 1997) they found that the concentrations at which those compounds were found in wine did not support the conclusions extracted by Charm analysis. This group also pioneered the critical analysis of olfactometry and the study of the different steps involved in the olfactometric operation. They compared Charm and aroma extract dilution analysis (AEDA) (Abbot et al. 1993a; Guichard et al. 1995), and studied in depth the representativeness of the extracts (Abbot et al. 1993b; Etievant et al. 1994; Moio et al. 1995; Priser et al. 1997), the effect of antioxidants (Escudero and Etievant 1999), and the artifact formation during analysis (Etievant 1996). In recent years they have also been very active in the use of the frequency of citation of an odor to get quantitative data (Escudero and Etievant 1999; Charles et al. 2000; Escudero et al. 2000a).

AEDA was firstly applied to wine by Guth (1997a, 1998) who studied in depth the aroma composition of some German white wines. He was the first one to give a complete picture of the hierarchy of odorants of a wine and to complement olfactometric data with a complete quantitative study (Guth 1997b, 1998) to determine the odor unit value (OUV) of each odorant, according to the general strategy proposed by Grosch (1994). After this, he used a technique called recombination and suppression tests that allowed him to find out which is the sensory effect of a given compound by studying how much the suppression of that compound from a mixture of odorants imitating wine makes the mixture differ from the wine. AEDA was also used by Ferreira et al. (1998a) and by López et al. (1999) to determine the most important odorants from young wines made with Grenache, Tempranillo, Merlot and Cabernet Sauvignon. These authors also present an extensive quantitative study (Ferreira et al. 2000a) about the number of odor units of Spanish single-variety young red wines and about the role played by the variety of grape the wine is made with. Concerned about the differences between the number of odor units of a compound in wine and its flavor dilution (FD) factor estimated in an AEDA experiment, they studied the relationship between these two concepts in a wine-like medium (Ferreira et al. 1998b). In some recent reports, they combined AEDA and the determination of the odor units of the odorants with the sensory evaluation of the effluent of a reversed-phase HPLC column (Ferreira et al. 1999, 2000b; Aznar et al. 2001). Kotseridis and Baumes (2000) presented another very complete AEDA research. They studied the impact odorants of Bordeaux Cabernet Sauvignon and Merlot must and wines, and of the yeast used for their fermentation.

Finally, it is not possible to end a revision about wine olfactometry without mentioning the works developed by several researchers over the last years concerning the isolation and identification of powerful wine odorants. Most of these works are not, properly speaking, about olfactometry, but they have largely contributed to the understanding of wine aroma. The work developed in Bordeaux by Professor Dubourdieu and collaborators is particularly outstanding. They have been the first researchers reporting the presence of very powerful thiol compounds in wine aroma responsible for several wine aroma nuances. The first compound they identified was 4-mercapto-4-methyl-2-pentanone (Darriet et al. 1995), and later 3-mercaptohexylacetate (Tominaga et al. 1996) and several other compounds (Bouchilloux et al. 1998; Tominaga et al. 1998a,b). Other important areas of research have been the aroma composition of the Vins Doux Naturels (Schneider et al. 1998; Cutzach et al. 1999) and the nature of some important odorants from wood which could impart different flavor nuances to wine (Cutzach et al. 1997; Lavigne et al. 1998).

6.3 Wine Olfactometry: An Overview

Table 6.2 summarizes all the odor regions detected by the different researchers working in wine olfactometry. They are arranged following the elution order in a

Carbowax 20 M column, and, when available, the elution index in a 5% phenylmethylpolixyloxane phase is also given. The table lists nearly 140 odor regions eluting from the column, 110 of them being well correlated with a known chemical compound. Some of the odor regions reported in the table will be found only in some particular kinds of wine, but most of them are general and will be found in almost every wine sample. Of course, in a normal experiment only between 40 and 80 odor regions will be registered, which means that some of those odor regions are very weakly perceived, or that several of them coelute and are registered as a single odor peak. It must be taken into account that almost all the volatile molecules have some smell, and that if they are concentrated enough, they will be perceived in an olfactometric experiment. In fact, Peña (1997) was able to detect more than 200 different odor regions in very concentrated fractions from a Chardonnay wine extract. This does not mean that those 200 compounds have an effective role in the Chardonnay wine aroma, but that one should be careful when injecting very concentrated samples.

Figures 6.1 and 6.2 show the typical flame ionization detector (FID) chromatograms obtained from the injection of a crude extract of wine aroma in both columns. In order to make it easier to correlate data in Table 6.2 with the figures, we have included the retention index and the identification of some of the prominent peaks in the chromatograms. It should be noted that slight drifts in the retention indexes must be expected, and that retention index data in Table 6.2 must be taken as approximate. Of course, nobody should expect a clear correlation between the FID peaks and the odor regions presented in Table 6.2, except for those major compounds such as ethyl esters, fatty acids and fusel alcohols.

Finally, Table 6.3 summarizes the aroma compounds that have been found to be at concentrations above their threshold, or at least very close to it, in some kind of wine. Data in Table 6.3 show that at least 58 different odorants can be odor-active in wine. Of them, about 20 will be found at concentrations higher than their threshold in all wines, and thereby these compounds constitute the sensory background of wine. Some of them are responsible for specific odor nuances if they are concentrated enough, as is the case of ethyl esters in white and rosé wines (Ferreira et al. 1995b), or in red wines made with Pinotage or Tempranillo grape varieties (Van Wyk et al. 1978; Ferreira et al. 2000a). Some of these compounds can also be found at abnormally high concentrations, which, probably, will constitute an off-odor. It is thus with acetic acid and ethyl acetate in the case of wine spoiled by acetobacter, and with diacetyl in the case of wine spoiled by lactic bacteria. Some well-known compounds are strongly linked to variety, and only will be found to be odor-active in wines made with some particular grapes. This is the case with terpenols, which are only really odor-active in muscat and other related grape varieties, and methoxypyrazines. We would like to remark that 15 of the compounds listed in Table 6.3 are present in wines at concentrations well below the 1 ppb. This fact explains the lack of success of some past researches, because it is only recently that these compounds have been identified and quantified in wine.

All these data are gathered here for two purposes: (1) because it will be very useful for anybody working in wine olfactometry to have at hand that qualitative

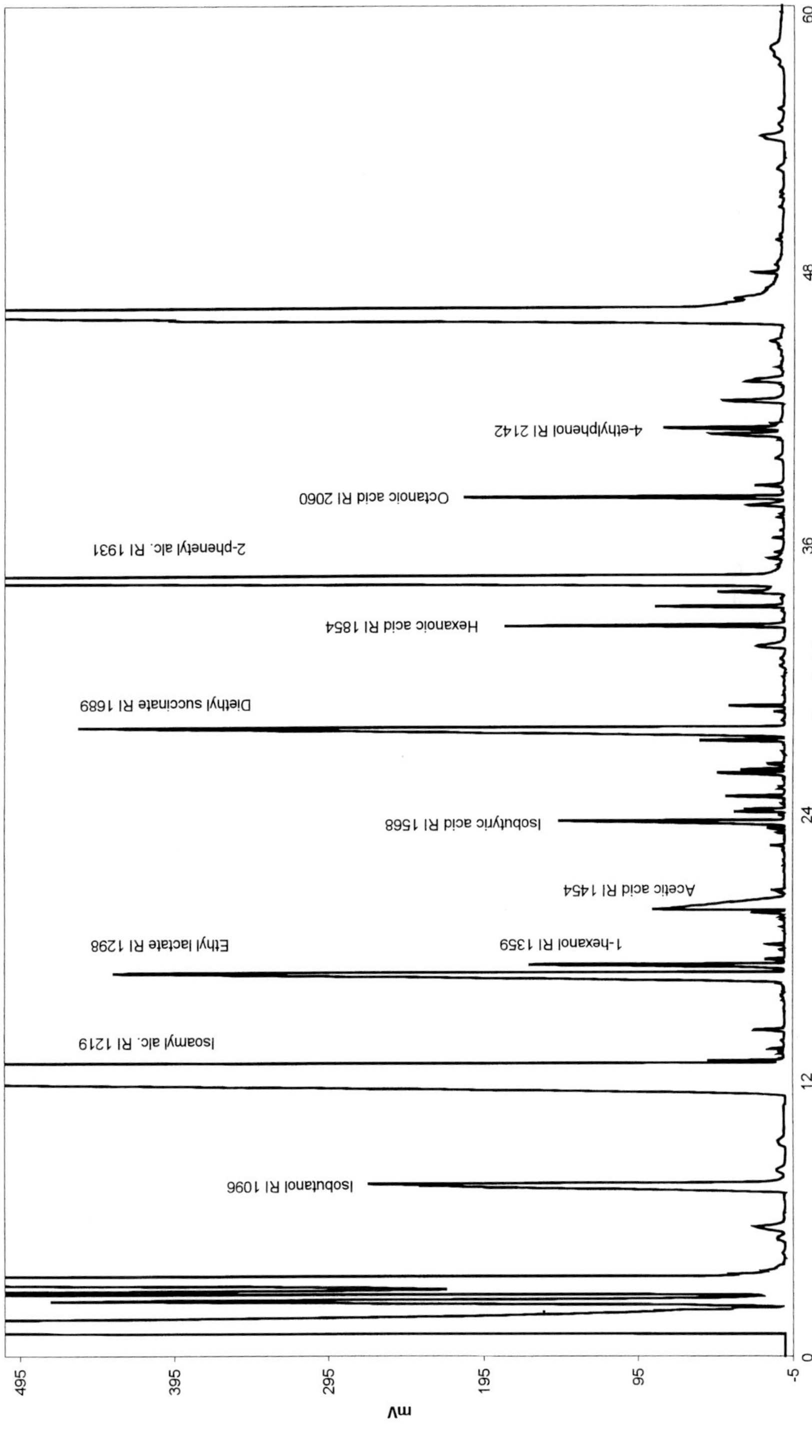

Fig. 6.1. Gas chromatogram (FID detection) of a dichloromethane wine extract in a polar column (Carbowax 20 M)

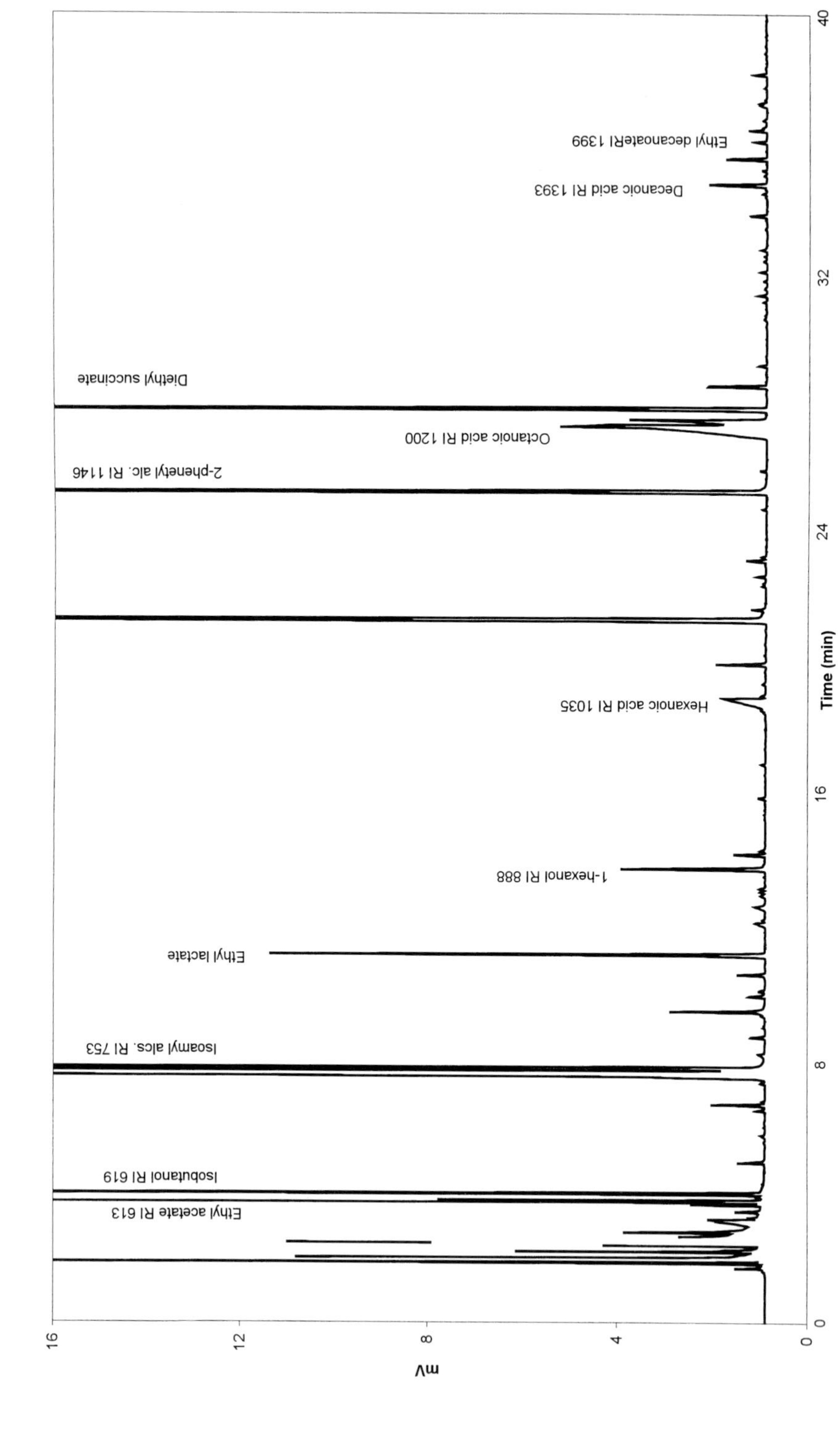

Fig. 6.2. Gas chromatogram (FID detection) of a Freon 113 wine extract in a non-polar column (5% phenyl-methylpolysiloxane)

Table 6.2. Odor regions detected in wine extracts by different researchers in effluent of the gas chromatograph. *R.I. pol* Retention index in a polar column (Carbowax 20M); *R.I. npol* retention index in a non-polar column (5% phenyl polymethylsiloxane) intensity

R.I. pol	R.I. npol	Odor description	Identity	Observations	Reference
750		Apple, fruit	Acetaldehyde		
885	<700	Fruity, solvent	Ethyl acetate		
900	730		1,1-Diethoxyethane		Guth (1997a)
900		Green, aldehyde	2,4,5-Trimethyldioxolane	Highest in oxidized wines	Escudero et al. (2000c)
929	757	Fruity, pineapple	Ethyl 2-methylpropanoate		
955	770	Fruity	Isobutyl acetate	Highest in young wines	
969	<700	Cream	2,3-Butanodione		
1,008		Orange peel, citric			
1,043	812	Acid fruit	Ethyl butyrate	Highest in young wines	
1,052	910	Cabbage	Dimethyl sulfide	Highest in white and red wine	Guth (1997a); Kotseridis and Baumes (2000)
1,061	860	Sweet fruit	Ethyl 2-methylbutyrate	Highest in aged wines	
1,075	864	Berry, blackberry	Ethyl isovalerate	Highest in aged wines	
1,079	800	Green	Hexanal	Found in muscadine grape juice	Chisholm et al. (1995);
1,095		Green, grass	2,5-Dimethyl-1,4-dioxane		Baek et al. (1997)
1,096	<700	Fusel, alcohol	Isobutanol		
1,128	886	Banana	Isoamyl acetate	Highest in young wines	
1,141	908	Fruity, ester	Ethyl pentanoate		
1,202	857	grass	(E)-2-Hexenal	Found in aged Vidal Blanc wine	Chisholm et al. (1995)
1,219	753	Alcohol, harsh	Isoamyl alcohol	Highest intensity	
1,238	1,004	Green apple	Ethyl hexanoate		
1,296	1,010	Green, unpleasant, bitter	Octanal		

Table 6.2. *Continued*

R.I. pol	R.I. npol	Odor description	Identity	Observations	Reference
1,302	976	Earthy, mushroom	1-Octen-3-one	Found in muscadine grape juice	Baek et al. (1997)
1,310	977	Onion, sunflower seeds	2-Methyl-3-furanthiol	Highest in Cabernet and Merlot wines	Kotseridis and Baumes (2000)
1,312	728	Cream	Acetoine		
1,337	1,112	Roses	(Z)-Roseoxide	Key odor of Gewürtztraminer	Guth (1997a)
1,359	888	Green, grass	1-Hexanol		
1,360	970	cabbage	Dimethyl trisulfide	Only detected in headspace	Guth (1997a)
1,374	940	Patchuli, green, bitter	4-Mercapto-4-Methylpentan-2-one	Key odor of Sauvignon Blanc and Schreube	Guth (1997a)
1,386		Pleasant			
1,389	870	Freshly cut grass	(Z)-3-Hexenol		
1,393		Incense			
1,408		Wet, green, grass	2-Buthoxyethanol		Escudero et al. (2000c)
1,439	1,201	Sweet, soap	Ethyl octanoate		
1,439		Coffee, toasted			Aznar et al. (2001)
1,454	<700	Vinegar	Acetic acid		
1,459		Citrus zest	4-Mercapto-4-Methylpentan-2-ol	Found in Sauvignon Blanc	Bouchilloux et al. (1998)
1,459	1,099	Pleasant, grass, alcoholic	3-Isopropyl-2-methoxy-Pirazine		
1,474		Food, seed			
1,477	985	Cereals, potato, soup-like	3-Methylthiopropanal	Highest in some oxidized wines	Escudero et al. (2000c)
1,478	853	Incense	Furfural		
1,495		Caramel, bakery		Found in aged champagne	Escudero et al. (2000c)

1,495		Plastic, brand new car			Aznar et al. (2001)
1,517		Green wood	Decanal		Kotseridis and Baumes (2000)
1,520		Plastic, gas oil	2-Methyltetrahydrotiophenone	Found in Rioja wine	Aznar et al. (2001)
1,522		Bitter, cherry	Benzaldehyde	Found in oxidized wines	Escudero et al. (2000c)
1,524	935	Burnt marshmallow, muscadine	Ethyl 3-hydroxybutyrate	Found in muscadine grape juice	Baek et al. (1997)
1,528		Fruity, toasted			
1,532	1,158	Wet	(E)2-Nonenal	Found in oxidized wines	Escudero et al. (2000c)
1,542		Toasty, caramel	Unknown m/z 144	Found in toasted oak wood	Cutzach et al. (1997)
1,543	1,315	Pleasant, complex	Vitispirane		
1,544	1,186	Green pepper (capsicum)	3-Isobutyl-2-Methoxypyrazine		
1,551	1,106	Flowery	Linalool		
1,566		Lavender			
1,568	796	Acid	Isobutyric acid		
1,572		Meat			
1,582	1,156	Cucumber	(E,Z)-Nona-2,6-dienal	Found in muscadine grape juice and in red wines	Baek et al. (1997); Kotseridis and Baumes (2000) Aznar et al. (2001)
1,613		Toasted			
1,620		Cooked leeks	3-Mercapto-3-methylbutanol	Found in Sauvignon Blanc	Bouchilloux et al. (1998)
1,620		Orange	Methyl benzoate	Found in Rioja wine	Ferreira et al. (2000b)
1,623	796	Cheese	Butyric acid		
1,624	1,062	Woody, oily	Ethyl furoate		
1,634	1,399	Pleasant, soap	Ethyl decanoate		
1,635	1,047	Honey	Phenylacetaldehyde	Found in muscadine grape juice and in red wines	Baek et al. (1997); Kotseridis and Baumes (2000)
1,654	1,449	Flower, fruity	Isoamyl octanoate		

Table 6.2. *Continued*

R.I. pol	R.I. npol	Odor description	Identity	Observations	Reference
1,655	878	Blue cheese	Isovalerianic acid		
1,662	1,179	Ripe fruit	Ethyl benzoate		
1,678		Fruity, violets, fruit candy			
1,684	1,199	Pleasant, sweet	α-Terpineol		
1,686		Fruit candy			
1,699		Anise			
1,708		Acid, root, grass			
1,714	994	Garlic	3-Methylthiopropanol		
1,725		Grapefruit, banana	3-Mercaptohexyl acetate		Kotseridis and Baumes (2000)
1,748		Tree fruit			
1,758		Coconut			
1,763	1,227	Citric	Citronellol		Guth (1997a)
1,771	1,167	Cucumber, melon	(E,Z)-2,6-Nonadien-1-ol	Found in muscadine grape juice	Baek et al. (1997)
1,772		Flower, liquorice			
1,775		Flowery			
1,793		Vegetables			
1,837		Caramel	Ciclotene	Found in toasted oak wood, proposed as important in Chardonnay from Burgundy	Cutzach et al. (1997); Le Fur and Etievant (1998)
1,850	1,393	Bark	β-Damascenone		
1,852	1,265	Pleasant, flowery	Phenylethyl acetate		
1,854	1,035	Cheese	Hexanoic acid		
1,854	1,034	Vegetable, dry, grapefruit	3-Mercaptohexanol	Maximum in Cabernet and Merlot, found first in Sauvignon Blanc	Bouchilloux et al. (1998); Kotseridis and Baumes (2000)

1,868		Toasty, caramel	Dihydromaltol	Found in toasted oak wood	Cutzach et al. (1997)
1,869	1,267	Citric	Geraniol		
1,875		Strawberry, sulfur	4-Mercapto-2,5-dimethylthiophen-3-one	Key odor of some non-*Vitis vinifera* wines	Guedes de Piho et al. (1997)
1,879	1,434	Sweet fruit	α-Ionone		
1,882	1,103	Phenolic, chemical	Guaiacol		
1,889		Nice, flowery			
1,900	1,357	Strawberry, plum, flowery	Ethyl dihydrocinnamate		
1,916		Citric, pleasant, jam	2-Phenylethyl isobutyrate		
1,931	1,146	Roses	β-Phenethylalcohol		
1,952	1,489	Pleasant, berry	β-Ionone		
1,961		caramel	Maltol	Found in toasted oak wood, proposed as important in Chardonnay from Burgundy	Cutzach et al. (1997); Le Fur and Etievant (1998)
1,965		Pleasant, lactone-like			
1,973	1,332	Peach, coconut	(E)-Whiskylactone	Proposed as important in Chardonnay from Burgundy, maximum in Rioja red wines	Le Fur and Etievant (1998); Aznar et al. (2001)
1,985		Lactone-like, green, apple jam	δ-Octalactone		
2,025	1,288	Phenolic	4-Ethylguaiacol	Maximum in Rioja red wines	Aznar et al. (2001)
2031	1,081	Cotton candy	Furaneol		
2,038	1,372	Coconut	γ-Nonalactone		
2,047		Fruit candy, vanilla			
2,060	1,200	Rancid, harsh	Octanoic acid		
2,074	1,175	Cotton candy, peach	Homofuraneol	Maximum in Cabernet and Merlot	Kotseridis and Baumes (2000)

Table 6.2. *Continued*

R.I. pol	R.I. npol	Odor description	Identity	Observations	Reference
2,088		Woody, smoky	m-Cresol		
2,122	1,474	Strawberry cream	Ethyl cinnamate		
2,130	1,475	Peach	γ-Decalactone		
2,140	1,369	Clove	Eugenol	Highest in oxidized wines and in Rioja red wines	Escudero et al. (2000c)
2,142	1,188	Phenolic	4-Ethylphenol	Maximum in Rioja red wines	Aznar et al. (2001)
2,142	1,188	Phenolic	3-Ethylphenol		Guth (1997a)
2,169	1,124	Curry, black pepper (spice)	Sotolon		
2,173	1,328	Phenolic, pleasant	4-Vinylguaiacol	Highest in white wines	
2,180	1,505	Peach	δ-Decalactone		
2,192	1,455	Coconut	Wine lactone	Highest in Schreube and Gewürtraminer	Guth (1997a)
2,211		Toasty, caramel	2,3-Dihydro-3,5-dihydroxy-6-methyl-4(H)-piran-4-one	Found in toasted oak wood	Cutzach et al. (1997)
2,218		Farm feed, phenolic	Unknown	Highest in oxidized wines	Escudero et al. (2000c)
2,221		Leather, coffee		Detected in aged champagne	Escudero and Etievant (1999)
2,223	1,309	Sweet	o-Aminoacetophenone	Found in muscadine grape juice	Baek et al. (1997)
2,223		Honey		Proposed as important in Chardonnay from Burgundy	Le Fur and Etievant (1998)
2,229	1,393	Fatty, unpleasant	Decanoic acid		

2,233		Toasty, caramel	Hydroxymaltol	Found in toasted oak wood	Cutzach et al. (1997)
2,238		Coconut, flowery	γ-Undecalactone	Tentatively found in aged champagne	Escudero and Etievant (1999)
2,239	1,343	Coconut, flowery	Methyl anthranilate	highest in Rioja red wines	Chisholm et al. (1995); Moio and Etievant (1995); Aznar et al. (2001)
2,240	1,360	Floral	4-Phenyl-3-hydroxybutan-2-one		Kotseridis and Baumes (2000)
2,249	1,345	Incense, phenolic, chemical	2,6-Dimethoxyphenol		Kotseridis and Baumes (2000)
2,251		Carnation			
2,255		Incense			
2,261		Flowery, rose			
2,280	1,425	Sweet, fruity	Ethyl anthranilate	Found in Pinot Noir wines	Moio and Etievant (1995)
2,308		Medicinal, phenolic, gouache	4-Vinyl-phenol	Highest in white wines	Chatonnet et al. (1993)
2,352	1,438	Woody, sweet	Isoeugenol		Kotseridis and Baumes (2000)
2,396	1,657		(Z)-6-Dodeceno-γ-lactone		Guth (1997a)
2,415		Incense, almond, cypress			Aznar et al. (2001)
2,566		Oak			Aznar et al. (2001)
2,580	1,249	Pollen, flowery, roses	Phenylacetic acid		Kotseridis and Baumes (2000)
2,585	1,389	Vanilla, candy	Vanillin		
2,600	1,579	Pollen, flowery,	Ethyl vanillate		
2,600		caramel	Acetovanillone	Found in oxidized wines	Escudero et al. (2000c)
2,600		Cinnamon, fruity, flowery			This work
	913	Roasted coffee	2-Furanmethanethiol	Found in sweet wines	Tominaga et al. (2000)

Table 6.3. Wine odorants that have been found at concentrations higher or very close to their olfactory threshold. *VDN* Vins Doux Naturels(=Sweet Natural Wine); *Hyb* wine from hybrid grapes

Odorant	Concentration range (ppb)	Olfactory threshold[a] (ppb)	OUV range	References
Ethyl isobutyrate	30–480	15 (G)	2–32	Baumes et al. (1986); Guth (1997b); Ferreira et al. (2000a)
Ethyl 2-methylbutyrate	1–30	1 (G); 18 (F)	0–30	Guth (1997b); Ferreira et al. (2000a)
Ethyl isovalerate	2–36	3 (G, F)	0.7–12	Guth (1997b); Ferreira et al. (2000a)
Ethyl acetate	2–150 ppm	7.5 ppm (G)	0.27–20	Maarse and Visscher (1989)
Ethyl butyrate	69–2,194	20 (G)	3.5–110	Ferreira et al. (1995b, 2000a); Guth (1997b)
Ethyl hexanoate	153–2,731	5 (G), 14 (F)	11–546	Baumes et al. (1986); Ferreira et al. (1995b, 2000a); Guth (1997b)
Ethyl octanoate	138–2,636	2 (G), 5 (F)	28–1,318	Ferreira et al. (1995b, 2000a); Guth (1997b)
Ethyl decanoate	14–821	200 (F)	0.1–4	Ferreira et al. (1995b, 2000a)
Isoamyl acetate	118–7,354	30 (G)	4–245	Ferreira et al. (1995b, 2000a)
Phenylethyl acetate	0.5–744	250 (G)	0–3	Ferreira et al. (1995b, 2000a)
Ethyl dihydrocinnamate	0.21–3	1.6 (F)	0.1–2	Aubry et al. (1997); Ferreira et al. (2000a)
Ethyl cinnamate	0.11–8.9	5.1 (F)	0–2	Aubry et al. (1997); Ferreira et al. (2000a)
Ethyl lactate	0.2–382 ppm	150 ppm	0–2.5	Simpson and Miller (1984)
Isobutyric acid	430–4,160	2,300 (F)	0.2–2	Guth (1997b); Ferreira et al. (2000a)
Isovaleric acid	300–1,150	33.4 (F)	9–34	Ferreira et al. (2000a)
Isobutanol	25,700–108,900	40,000 (G)	0.6–2.7	Guth (1997b); Ferreira et al. (2000a)
Isoamyl alcohol	72,000–318,000	30,000 (G)	2.4–10.6	Baumes et al. (1986); Ferreira et al. (2000a)
β-Phenylethyl alcohol	18,000–162,000	14,000 (F)	1.3–11	Baumes et al. (1986); Guth (1997b); Ferreira et al. (2000a)
Methionol	166–2,400	500 (G), 1,000 (F)	0.2–5	Baumes et al. (1986); Ferreira et al. (2000a)
1-Hexanol	1,000–13,200	8,000 (G)	0.1–1.7	Baumes et al. (1986); Ferreira et al. (2000a)
(Z)-3-Hexenol	7.2–850	400 (G)	0–2.1	Baumes et al. (1986); Ferreira et al. (2000a)
Acetic acid	69–400 ppm	200 ppm (G)	0.3–2	Ferreira et al. (2000a)

Butyric acid	434–4,720	173 (F)	2.5–27	Ferreira et al. (2000a)
Hexanoic acid	200–6,200	420 (F)	0.5–15	Marais and Pool (1980); Shinohara (1985); Ferreira et al. (2000a)
Octanoic acid	40–7,900	500 (F)	0.1–16	Marais and Pool (1980); Shinohara (1985)
Decanoic acid	62–3,400	1,000 (F)	0–3.4	Shinohara (1985); Ferreira et al. (2000a)
Linalool	1.7–1,500	15 (G), 25.2 (F)	0.1–99	Ribereau-Gayon et al. (1975); Ferreira et al. (2000a)
α-Terpineol	0.6–145	250 (F)	0–0.6	Riberau-Gayon et al. (1975)
Geraniol	0.9–1,059	30 (G)	0–35	Riberau-Gayon et al. (1975)
(Z)-Rose-oxide	0–21	0.2 (G)	0–105	Guth (1997b)
Guaiacol	1.1–15	10 (G)	0.1–1.5	Chatonnet and Boidron (1988); Ferreira et al. (2000a)
4-Ethylguaiacol	0–400	33 (F)	0–13	Schreier et al. (1980); Ferreira et al. (2000a)
Eugenol	0.5–30	6 (F)	0.1–5	Chattonet and Boidron (1988); Guth (1997b)
4-Vinylguaiacol	0.2–710	1,100 (F)	0–0.7	Versini and Tomasi (1983); Ferreira et al. (2000a)
4-Ethylphenol	0–6,480	440 (Ch)	0–15	Chatonnet et al. (1992)
4-Vinylphenol	3–1,241	180 (Ch)	0–6.9	Chatonnet et al. (1993)
(E)-Whiskylactone	0–140	67 (Et)	0–2.2	Simpson and Miller (1984)
(Z)-Whiskylactone	0–150	790 (Et)		Simpson and Miller (1984)
γ-Nonalactone	3–41	30 (N)	0.1–1.4	Nakamura et al. (1988)
β-Damascenone	0.3–4.7	0.05 (G)	6–94	Guth (1997b); Kotseridis et al. (1999); Ferreira et al. (2000a)
α-Ionone	0.02–0.54	2.6 (Et)	0–0.2	Kotseridis et al. (1998, 1999a); Ferreira et al. (2000a)
β-Ionone	0.03–0.24	0.09 (F)	0.3–2.7	Kotseridis et al. (1998, 1999a); Ferreira et al. (2000a)
Acetaldehyde	1–160 ppm	0.5 ppm (G)	2–140	Criddle et al. (1983); Guth (1997b)
2,3-Butanodione	200–2,722	100 (G)	2–27	Hayasaka and Bartowsky (1999); Ferreira et al. (2000a)
Acetoine	0.6–159 ppm	150 ppm (Et)	0–1	Ferreira et al. (2000a)
Furfural	0.002–8.8	14.1 (F)	0–0.5	Ho et al. (1999); Ferreira et al. (2000a)
1,1-Diethoxyethane	0–500	50 (G)	0–10	Guth (1997b)

Table 6.3. *Continued*

Odorant	Concentration range (ppb)	Olfactory threshold[a] (ppb)	OUV range	References
3,4,5-Trimethyldioxolane	0–8,000	2,000	0–4	Escudero et al. (unpubl. data)
2-Methoxy-3-isopropylpirazine	0.00–0.002	0.002 (S)	0–2	Lacey et al. (1991)
2-Methoxy-3-(2-methylpropyl)pyrazine	0.00–0.04	0.002 (S)	0–20	Lacey et al. (1991)
4-Mercapto-4-methyl-2-Pentanone	0.00–0.4	0.0006 (G), 0.0008 (Tom)	0–800	Guth (1997b); Tominaga et al. (1998b)
4-Mercapto-4-methyl-2-Pentanol	0.00–0.11	0.055 (Tom)	0–2	Tominaga et al. (1998b)
3-Mercaptohexanol	0.00–12.8	0.060 (Tom)	0–213	Tominaga et al. (1998b)
3-Mercaptohexyl acetate	0.00–0.80	0.004 (Tom)	0–200	Tominaga et al. (1998b)
3-Mercapto-2-methyl-propanol	0.00–10	3 (Bou)	0–3-3	Bouchilloux et al. (1998)
Methylthiopropanaldehyde (methional)	0.00–140	0.50 (Et)	0–280	Escudero et al. (2000b)
Dimethyl sulfide	7–53	10 (G)	0.7–5.3	Guth (1997b); Rauhut et al. (1998)
Dimethyl trisulfide	0.09–0.25	0.2 (G)	0.5–1.2	Guth (1997b)
4-Hydroxy-2,5-dimethyl-3(2H)-furanone (furaneol)	0–623 (VDN) 100–3,500 (Hyb)	10 (F) 500 (G)	0–63 (VDN) 10–350	Guedes de Pinho et al. (1997); Cutzach et al. (1999)
3-Hydroxy-4,5-dimethyl-2(5H)-furanone (sotolon)	5–207 (VDN) 120–268 (vin jaune)	5 (G) 10 (M)	1–54	Guichard et al. (1993); Cutzach et al. (1999)
c-6-Dodeceno-γ-lactone	0–3	0.1 (G)	0–3	Guth (1997b)
3a,4,5,7a-Tetrahydro-3.6-Dimethylbenzofuran2(3H)one (wine lactone)	0–0.1	0.01 (G)	10	Guth (1997b)

[a] G, Guth (1997b); F, Ferreira et al. (2000a); Ch, Chatonnet et al. (1992); Et, Etievant (1991); N, Nakamura et al. (1988); Tom, Tominaga et al. (1998); Bou, Bouchilloux et al. (1998); S. Seibert et al. (1970)

and quantitative information, and (2) because it will be of great help to discuss methodological aspects of wine olfactometry. In fact, in the subsequent sections we will use these data in the tables to raise some key methodological questions and to propose the most adequate solutions given by the different authors.

6.4 Methodological Aspects

6.4.1 Headspace or Total Extraction?

This is a tricky question, since if you are studying the aroma of a product it seems more practical to study the product headspace, just to get rid of all the least volatile compounds, which, after all, probably will not be able to reach the olfactory epithelium. Following this reasoning, headspace olfactometric analysis should also give more reliable information since it does not depend on extraction recovery, the process can be automated and sample handling kept to a minimum. In addition, if one properly selects the conditions in which the headspace is formed (temperature, presence of saliva, dilution, emulsion with gas, etc.), it would be possible to imitate not only the process of orthonasal olfaction but also that of retronasal olfaction (Roberts and Acree 1995, 1996). However, all this reasoning being correct from a formal point of view, the fact is that headspace analysis does not provide, yet, a complete picture of wine aroma. The reasons for this are related to the practicability of headspace analysis and to the nature of wine aroma: (1) wine is a mainly aqueous solution and the presence of water in the headspace limits the volume of headspace that can be directly transferred into the chromatograph; (2) wine headspace also contains important amounts of alcohol, which, no doubt, is going to affect the sorption ability of the trap, lowering most of the breakthrough volumes of the different aroma compounds; (3) thermal desorption does not ensure a complete transfer of the least volatile compounds into the chromatograph. We agree with Mistry et al. (1997) that the introduction of a trap introduces a bias.

These three reasons can explain why almost all the headspace chromatograms of wine do not contain any peak eluted after β-phenylethanol in a polar column, regardless of whether it is static or dynamic (Noble et al. 1980; Shimoda et al. 1993; Salinas et al. 1994; García-Jares et al. 1995; Salinas and Alonso 1997). In contrast, Table 6.3 clearly shows that heavier compounds can have strong sensory relevance. Of course, headspace analysis make it easier to study the most volatile odorants, but a critical study of the results of the researchers working on headspace analysis shows that there is little benefit in the sense that almost all the odorants analyzed by headspace can be also directly determined. The results of Guth (1997a,b) are very illustrative. He complements AEDA with a static headspace olfactometric analysis (SHA/o) (Guth and Grosch 1993) and only 4 out of 45 compounds were detected only in the headspace. Furthermore, the headspace technique failed in the hierarchy of odorants, as was later demonstrated by the estimation of OUV and by suppression and reconstitution tests (Guth 1997b).

In conclusion, we do not recommend, today, headspace analysis as the basis of an olfactometric study (things could change after a thorough study of wine headspace analysis). However, it should be kept in mind that the most volatile compounds can be lost during the concentration of the extract, and that one should overcome this inconvenience by complementing AEDA with the olfactometric analysis of a direct injection of wine, or of a wine liquid microextract, or of a direct injection of a small volume of headspace.

6.4.2 Obtaining an Extract

The first question that one must ask when intending to carry out an olfactometric experiment is what one really wants to do and what will be done with the results of the research. At the beginning of a research, what one usually wants is to have an idea of the aroma profile of a wine, that is to say, to be able to establish a hierarchy of all the odorants present in the wine. In that case, representativity means that the extract must contain all the odorants of wine and in the same proportion. However, if one wants to focus on a single odor note, for instance an off-odor, the only requisite that the extract must fulfill is to have a high recovery of that off-odor. In other cases, one may be interested in the differences between a series of wines. In that case, representativity means that the extracts of wines perceived as different also differ. Therefore, representativity is not an absolute, but depends on the purpose of the research, and the researcher should be able to control the different extraction systems in order to select the most appropriate for each case.

In order to solve this question, we have studied the liquid–liquid and liquid–solid distribution coefficients of a selected group of odorants in different solvent and adsorbent systems (Ferreira et al. 2000c). Some results are given in Table 6.4. Data in the table clearly show that if we want to select a non-selective solvent we must choose dichloromethane. On the contrary, if we want to extract non-polar compounds, we should select Freon 113, for instance. The addition of salt in all the cases improves the recovery of the more polar analytes. The use of continuous extraction systems allows improvement of the results and reduces the amount of solvent used. However, the extraction of non-volatile material also increases, which limits the use of the most polar solvents. Selection of a high wine/solvent phase ratio, which is called microextraction, provides a high concentration but does not ensure complete recovery of all the volatiles (Ferreira et al. 1993a, 1998c).

A second and key result from the table is that most of the solid sorbents have higher distribution coefficients than liquid systems. Silica-based sorbents are not very efficient in extracting aroma compounds from wine; organic polymers are far better. Porapak Q provides a very good way to obtain extracts enriched with non-polar compounds, while Amberlite XAD-4 is the best to achieve non-selective

Table 6.4. Liquid–liquid and liquid–solid distribution coefficients between wine and some extracting systems of some selected wine odorants

Liquid solvent/solid sorbent	Ethyl decanoate	β-Ionone	Eugenol	Guaiacol	Isoamylalcohol	Sotolon
Diethyl ether/pentane 1:9	16	20	10	1	0.5	0.02
Diethyl ether/pentane 1:1	12	12	9	1	0.8	0.05
Ethyl acetate/pentane 1:3	18	19	11	1	0.7	0.02
Freon 113	21	57	23	2	0.7	0.04
Dichloromethane	24	44	161	30	5	2
C18	305	1,050	19	124	4	0.00
Amberlite XAD-4	1,018	1,020	420	114	33	14
Porapak Q	1,200	1,050	1,719	69	23	7
Lichrolut EN	>30,000	14,494	9,508	498	88	62

extractions (Ortega 1999; Escudero et al. 2000b; Ferreira et al. 2000c). Recent research with last-generation organic polymers has shown that it is possible to use a very small amount of solid sorbent to recover aroma compounds from wine, which limits the solvent evaporation steps.

Some researchers have used wine demixture by salting out to produce highly representative extracts (Ferreira et al. 1996; Priser et al. 1997). Although it is true that the ethanolic extract got by salting out wine contains almost all the wine-important odorants, it also contains high amounts of polyphenols and other non-volatile material, which hampers the use of this strategy in olfactometry. The solution devised by some authors (Moio and Etievant 1995; Priser et al. 1997) is the distillation of the extract, but this complicates sample handling and the final extract does not have the necessary concentration to study the most diluted compounds.

6.4.3 Evaluation of the Representativity of the Extract

As explained earlier, representativity is a relative concept that depends on the goals of the research. Where we need a total extract from the wine, the evaluation of representativity is not obvious. There are just a couple of works in the scientific literature that deal in depth with this question (Etievant et al. 1994; Priser et al. 1997). In these works, the authors carry out a quantitative descriptive analysis of both wines and extracts, and study the differences in the sensory descriptors between the wines and their extracts. Apart from the fact that this task is too long and tedious, the authors find that the tasters need some additional training. Otherwise, they would not be able to give similar sensory scores to two things that are clearly different: a wine and its extract. In a wine, there is a particular liquid–gas distribution coefficient for each aroma compound that depends on its boiling point and on its solubility in wine. In an extract, the liquid–gas distribution coefficients are absolutely different, the most polar compounds being more favored towards the gas phase. This makes a wine and its extract smell different. In another paper, Moio and Etievant (1995) used olfactometry to try to circumvent this problem, but they did not obtain definitive results.

A more pragmatical approach was proposed by the same group of researchers (Abbot et al. 1993b) and we have used it advantageously (Ferreira et al. 2000b; Aznar et al. 2001). In this approach, the tasters are asked not to describe the wine or the extract, but to measure if extracts coming from different wines can also be recognized as different by the tasting panel. In a second test, the tasters are asked to associate the extracts with wines, in order to ascertain if they are able to correctly match the different pairs wine/extract. These tests are not easy due to the aforementioned reasons, and will require some expertise by the tasters and some additional trials to select the optimum presentation conditions. It is particularly important to ensure that the liquid–gas distribution coefficients are as similar as possible, and to avoid the pernicious odor of the solvent. Practical details have been published (Aznar et al. 2001). It is very important to correctly select the wines for the experiment. We usually try to choose one or several samples as similar in structure as possible to the wine to be studied, but with a neutral aromatic and flavor profile.

6.4.4 Concentration of the Extracts

The concentration of extracts is one of the least studied steps of flavor analysis. However, important losses of volatile compounds take place during this stage. If the analytes are well solved by the solvent, the behavior of a concentration system is, at first, quite close to what one would expect from ideal systems: slight losses of compounds, and the higher the volatility, the higher the losses (Ferreira et al. 1995a). However, this ideal behavior is interrupted in the final stages of the concentration, i.e. when reflux stops and there only remains a thin film of evaporating liquid. In that moment, all the aroma compounds can be lost in very important amounts regardless of their boiling point. The practical conclusions of the study are that one should not let the concentration progress until that point, i.e. in a microKuderna-Danish, extracts can be conveniently concentrated from 30 or 40 to 0.3–0.4 mL, but not less. If additional concentration is needed, then the partially concentrated extract should be transferred to a second concentration system. In other study it was possible to demonstrate that polar analytes are blown off during the concentration if the solvent of the extract is non-polar, which explains the abnormally low recoveries of some fusel alcohols during the concentration of pentane, hexane and even Freon 11 extracts. The addition of a small amount of silica improved the recovery (Ferreira et al. 1998c).

Even with the best technique, the evaporation of solvent involves the loss of significant amounts of the most volatile compounds. In the best conditions, the concentration of a 2-ml extract to a final volume of 0.1 ml will mean the loss of

50% of an analyte with a boiling point 50 °C higher than that of the solvent. This figure becomes 90% if the analyte boiling point is just 25 °C higher than that of the solvent. Recoveries are satisfactory only for analytes with boiling points 70–80 °C higher than that of the solvent. In wine this means that all the aroma compounds with a Retention Index in a non-polar phase below 800 will be lost in high proportion, even if low boiling solvents such as pentane, diethylether of dichloromethane are used. The following directions can be useful to reduce the impact of this step:

1. Use those extraction strategies that reduce the use of solvents, such as solid phase extraction (SPE).
2. Try not to use non-polar solvents such as pentane and hexane. Where you do, consider the concentration protocol described in Ferreira et al. (1998c).
3. Include the crude, unconcentrated extract in the olfactometric experiments. Instead of producing a concentrated extract and later diluting it, begin with the crude extract and later concentrate it to form the corresponding solutions. Correct the results obtained for the most volatile compounds to take into account the losses during the evaporation.
4. Complement AEDA analysis with an olfactometry of a small volume of the headspace, as proposed in Section 4.1.

6.4.5 The Chromatographic System for Olfactometry

The chromatographic system must fulfill some requisites to ensure a reproducible and accurate operation. There is no single optimum system, but there are some criteria that must be properly handled in the system design.

Injector and Pneumatics. If the reader is interested in a deeper study of this question, we recommend the textbooks written by Grob (1987, 1988), or the authoritative works by Poole and Poole (1991) and Jennings et al. (1997). We will only make some general comments about this issue here. The best injector is the on-column injector because it is free from discrimination effects and because it minimizes the problems of degradation of thermal-unstable compounds. It has a drawback that can advise against its use. This is the analysis of dirty samples. A dirty sample in GC is a sample that contains important amounts of non-volatile material. Although on-column injection can tolerate a reasonable amount of dirtiness if an uncoated precolumn is used and replaced frequently (every time the chromatographic performance falls), there is a limit that it is not difficult to reach in the AEDA of wines. The authors found in the AEDA of extracts from hydrolysates of grape and wine glycosidic flavor precursors that one injection was enough to destroy the system performance. In this case it was found that the best (although not perfect) solution was splitless injection.

Splitless injection can provide very satisfactory results if it is properly done (Ferreira et al. 1993b). The critical parameter in splitless injection is the efficiency in the transfer of analytes from the hot injector to the column, which critically depends on the carrier gas velocity through the injector. If it is possible to guarantee a high linear carrier gas velocity (not in the column, but in the injector line), then both risks, that of analyte discrimination and of thermal degradation, decrease. This velocity can be roughly estimated as 0.6–1 cm/s. With conventional liners (about 0.8 ml in volume, 3.2 mm in diameter and 10 cm long) this means that the carrier gas velocity in the column must be 60–100 cm/s in a 0.32-mm-diameter column and 106–177 cm/s in a 0.25-mm column. This carrier gas velocity will ruin chromatographic efficiency unless hydrogen is used as carrier. If helium must be used, then the liner should be thinner (a 300-μL volume glass liner is a good choice). Finally, split injection should not be used unless the study is focused on a narrow range of odorants with close boiling points (as in headspace analysis). The reason for this is that the rate of evaporation of odorants in the chromatographic inlet depends on the analyte boiling point. Therefore, in the analysis of complex mixtures very different evaporation rates should be expected, which can cause important discrimination effects, more frequent if the extract is dirty (Ferreira et al. 1993b).

Column. It is highly advisable to use two columns, a polar as the main column and a non-polar as the complementary column. The most frequent choice is a cross-linked Carbowax 20 M as the polar one and a 5% phenylpolimethylsiloxane as the non-polar. It is advisable to use the polar column as the main column because of the poor elution peaks of fatty acids in non-polar phases, which greatly distort the chromatograms, and because of the higher selectivity of polar phases. The use of polar columns requires the use of highly purified gases and a frequent control of column performance through the Grob test or similar (Grob et al. 1978).

Column Outlet and Sniffing Port. The sniffing port should make the sniffing process comfortable. It is important to use make-up air, which a recent study finds optimum at 50 ml/min (Hanaoka et al. 2000). In that study, the authors also conclude that the usual practice of humidifying the air by bubbling it in water has almost no influence. The reason for this is that, on average, the flow of gas from the GC (1–100 ml/min) system is far less than the flow breathed by our lungs (1.5 L/min). An additional and very important detail not well resolved by the commercial systems is the need for an additional source of heating at the column outlet. In our laboratory we place a small resistance between the column outlet and the sniffing port. With this simple approach a very good improvement in the detection of the least volatile odorants has been observed.

6.5 Techniques for Processing the Olfactometric Signal

The olfactometric signal can take different forms, each one having its advantages and disadvantages, which we have tried to summarize in Table 6.5. In our opinion,

Table 6.5. Summary of the different olfactometric techniques

Olfactometric technique	Technique and references	Summary
Dilution techniques	AEDA (Ullrich and Grosch 1987; Grosch 1993)	Provides a hierarchical list of odorants Very simple and does not require special instrumentation Does not provide quantitative data, but is very repetitive Must be followed by quantitative analysis and by some sensory tests
	Charm analysis (Acree et al. 1984; Acree and Cotrell 1985)	Provides quantitative data of a list of odorants and allows further statistical processing Allows for meaningful comparison between samples Seems to be less repetitive than AEDA Requires specific instrumentation
Time intensity techniques	Osme (McDaniel et al. 1990; Miranda-López et al. 1992; Da Silva et al. 1994)	Provides a measurement of intensities of odors through continuous evaluation of the column effluent Requires intensive training and a panel of sniffers Its quantitative performance has been demonstrated with synthetic solutions Up to this date, its applicability to analysis of complex samples has not been proved
	Finger-span method (Etievant et al. 1999)	Provides a measurement of the intensities of odors Requires intensive training and a panel of sniffers Its quantitative performance has been demonstrated with synthetic solutions Up to this date, its applicability to analysis of real samples has not been proved, but it seems promising
Frequency of impact techniques	NIF (van Ruth and Roozen 1994; van Ruth et al. 1995, 1996; Pollien et al. 1997, 1999)	Provides a rough quantitative estimation of amount of odorant present in an extract Does not require special instrumentation Requires working with a panel Its quantitative performance has been demonstrated with synthetic solutions and with real samples

AEDA should be the method of choice if the purpose of the research is to produce a hierarchical list of odorants ordered according to their potential importance. Signal processing in AEDA is very simple, since you only record a flavor dilution factor, that is the maximum dilution at which a compound can be detected. The training of sniffers for AEDA is not very demanding and after several weeks results are very repetitive. In addition, AEDA does not require a panel and the results are really meaningful because it allows us to gather information about odorants at different concentrations. However, AEDA is not very useful for establishing comparisons between different samples because AEDA results are not very adequate for statistical processing. This task can be done far better with Charm analysis, but if one wants to use three trained sniffers, then the number of experiments becomes too high (Guichard et al. 1995).

Dilution techniques have been fiercely criticized because they are not consistent with psychophysical laws (Fritjers 1978), but they do provide an insight into the aroma of a product that cannot be obtained with time-intensity methods or with the frequency of impact techniques. It is true that neither flavor detection nor a Charm value is a direct measurement of the effective role of that compound in the product. Furthermore, the quantitative determination of the odorants and the estimation of the number of odor units (or odor activity value) is not a good measurement of that role either; it is just a further refinement of the AEDA list. In spite of this refinement, it is possible to find in the same product compounds with the same OUV but showing different olfactive intensities, or vice versa; compounds with the same intensity, but with different OUV. The only way in which it is possible to obtain actual information about the effective role of an odorant in a product is by means of the sensory evaluation of different samples, some of which can be reconstructed extracts in which some compound can have been suppressed (Buttery et al. 1990; Guth 1997b, 1998).

On the other hand, time-intensity techniques and frequency of impact techniques are claimed to be consistent with psychophysical laws, since they allow for a direct measurement of the intensity of the odorants on a previously agreed scale. Some researchers have been able to demonstrate the quantitative ability of these techniques (Da Silva et al. 1994; Pollien et al. 1999). Assuming this statement to be true, it should be taken into account that, however, this measurement of intensity is not a direct measurement of the effective sensory role played by an odorant in the product. First because it is not possible to guarantee that the concentration of product reaching the nose of the sniffer in the olfactometry is equal to, or very similar to, that reaching the olfactory epithelium of a consumer when smelling or eating that product. Second, because the final sensory role of an odorant can only be determined through the sensory evaluation of samples containing all the significant odor-active compounds and a matrix very close to that of the original product. In conclusion, these groups of techniques are really promising if the purpose of the research is the comparison between different samples, or, even, as quantitative methods of analysis.

6.6 Final Remarks

At the beginning of an experiment, an AEDA or Charm analysis should be conducted first, and special emphasis should be placed on the identification of the odor-active compounds. Once the potentially most important products are identified, sensory experiments to verify their role in the product should be conducted. Afterwards, time-intensity or frequency of impact techniques can be developed and applied to study the differences in the most important odorants between samples, or even to obtain some quantitative information about particular groups of odorants. The present knowledge about wine aroma makes it possible, for the first time, to guarantee a successful research if the aforementioned steps are followed.

References

Abbot N, Etievant P, Issanchou S, Langlois D (1993a) Critical evaluation of two commonly used techniques for the treatment of data from extract dilution sniffing analysis. J Agric Food Chem 41:1698–1703

Abbot N, Etievant P, Langlois D, Lesschaeve I, Issanchou S (1993b) Evaluation of the representativness of the odor of beer extracts prior to analysis by GC eluate sniffing. J Agric Food Chem 41:777–780

Acree TE (1997) GC/olfactometry. Anal Chem 69:170A–175A

Acree TE, Barnard J (1994) Gas chromatography-olfactometry and Charm analysis. In: Maarse JH, van der Heij DG (eds) Trends in flavour research. Elsevier, Amsterdam, pp 211–220

Acree TA, Cotrell THE (1985) Chemical indices of wine quality. In: Birch GG, Lindley MG (eds) Alcoholic beverages. Elsevier, New York, pp 145–159

Acree TE, Lavin EH (1990) o-Amino acetophenone, the "foxy" smelling component of Labruscana grapes. In: Bessière Y, Thomas AF (eds) Flavor science and technology. Wiley, New York, pp 49–52

Acree TE, Barnard J, Cunningham DG (1984) A procedure for the sensory analysis of gas chromatographic effluents. Food Chem 14:273–286

Aubry V, Etievant PX, Giniès C, Henry R (1997) Quantitative determination of potent flavor compounds in Burgundy Pinot Noir wines using a stable isotope dilution assay. J Agric Food Chem 45:2120–2123

Augustyn OPH, Rapp A, Van Wyk CJ (1982) Some volatile aroma components of *Vitis vinifera* L. cv. Sauvignon Blanc. S Afr J Enol Vitic 3:53–60

Aznar M, López R, Cacho J, Ferreira V (2001) Identification and quantitation of impact odorants of aged red wines from Rioja. A research conducted through GC-olfactometry, quantitative GC-MS, and odor evaluation of HPLC fractions. J Agric Food Chem (in press)

Baek HH, Cadwallader KR, Marroquin E, Silva JL (1997) Identification of predominant aroma compounds in muscadine grape juice. J Food Sci 62:249–252

Baumes R, Cordonnier R, Nitz S, Drawert F (1986) Identification and determination of volatile constituents in wines from different vine cultivars. J Sci Food Agric 37:927–943

Blank I (1997) Gas chromatography-olfactometry in food aroma analysis. In: Marsily R (ed) Techniques for analyzing food aroma. Dekker, New York, pp 293–330

Bouchilloux P, Darriet P, Henry R, Lavigne-Cruége V, Dubourdieu D (1998) Identification of volatile and powerful odorous thiols in Bordeaux red wine varieties. J Agric Food Chem 46:3095–3099

Buttery R, Teranishi R, Ling L, Turnbaugh J (1990) Quantitative and sensory studies on tomato paste volatiles. J Agric Food Chem 38:336–340

Charles M, Martin B, Ginies C, Etievant P, Coste G, Guichard E (2000) Potent aroma compounds of two red wine vinegars. J Agric Food Chem 48:70–77

Chatonnet P, Boidron JN (1988) Dosage de phenols volatils dans les vins par chromatographie en phaxe gazeuse. Sci Aliments 8:479–488

Chatonnet P, Dubourdieu D, Boidron JN, Pons M (1992) The origin of ethylphenols in wines. J Sci Food Agric 60(2):165–178

Chatonnet P, Barbe C, Boidron JN, Dubourdieu D (1993) Origines et incidences organoleptiques de phenols volatils dans les vins. Application à la maîtrise de la vinification et de l'élevage. In: Bayonove C, Crouzet J, Flanzy C, Martin JC, Sapis JC (eds) Connaissance aromatique des cepages et qualité des vins. Revue Frangaise Oenologie, Montpellier, pp 279–287

Chisholm MG, Samuels JM (1992) Determination of the impact of the metabolites of sorbic acid on the odor of a spoiled red wine. J Agric Food Chem 40:630–633

Chisholm MG, Guiher LA, Vonah TM, Beaumont JL (1994) Comparison of some French-American hybrid wines with white Riesling using gas chromatography-olfactometry. Am J Enol Vitic 45:201–212

Chisholm MG, Guiher LA, Zaczkiewicz SM (1995) Aroma characteristics of aged Vidal Blanc wine. Am J Enol Vitic 46:56–62

Criddle WJ, Goswell RW, Williams MA (1983) The chemistry of Sherry maturation. II. An investigation of the volatile components present in standard Sherry base wine. Am J Enol Vitic 34:61–71

Cunningham DG, Acree TE, Barnard J, Butts RM, Braell PA (1986) Charm analysis of apple volatiles. Food Chem 19:137–147

Cutzach I, Chatonnet P, Henry R, Dubourdieu D (1997) Identification of volatile compounds with a toasty aroma in heated oak used in barrelmaking. J Agric Food Chem 45:2217–2224

Cutzach I, Chatonnet P, Dubourdieu D (1999) Study of the formation mechanisms of some volatile compounds during aging of sweet fortified wines. J Agric Food Chem 47:2837–2846

Da Silva AA.P, Lundhal DS, McDaniel MR (1994) The capability and psychophysics of Osme: a new GC-olfactometry technique. In: Maarse H, van der Heij DG (eds) Trends in flavour research. Elsevier, Amsterdam, pp 191–209

Darriet P, Lavigne V, Boidron JN, Dubourdieu D (1991) Caractérisation de l'arome variétal des vins de Sauvignon par couplage chromatographie en phase gazeuse-odométrie. J Int Sci Vigne Vin 25:167–174

Darriet P, Tominaga T, Lavigne V, Boidron JN, Dubourdieu D (1995) Identification of a powerful aromatic component of *Vitis vinifera* L. var. Sauvignon wines: 4-mercapto-4-methylpentan-2-one. Flavour Fragrance J 10:385–392

Escudero A, Etievant P (1999) Effect of antioxidants on the flavor characteristics and the GC-O profiles of champagne extracts. J Agric Food Chem 47:3303–3308

Escudero A, Charpentier M, Etievant P (2000a) Characterization of aged champagne wine aroma by GC-O and descriptive profile analysis. Sci Aliment 20:331–346

Escudero A, Cacho J, Ferreira V (2000b) Clues about the role of methional as character impact odorant of some oxidized wines. J Agric Food Chem 47:4268–4272

Escudero A, Cacho J, Ferreira V (2000c) Isolation and identification of odorants generated in wine during its oxidation: a gas chromatography-olfactometric study. Eur Food Res Technol 211(2):105–110

Etievant PX (1991) Wine. In: Maarse H (ed) Volatile compounds of food and beverages. Dekker, New York, pp 483–546

Etievant PX (1996) Artifacts and contaminants in the analysis of food. CRC Food Sci Nutr 36:733–745

Etievant PX, Issanchou SN, Bayonove CL (1975) The flavour of muscat wine: the sensory contribution of some volatile compounds. J Sci Food Agric 34:497–504

Etievant PX, Issanchou SN, Bayonove CL (1983) The flavour of Muscat wine: the sensory contribution of some volatile compounds. J Sci Food Agric 34(5):497–504

Etievant PX, Moio L, Guichard E, Langlois D, Lesschaeve I, Schlich P, Chambellant E (1994) Aroma extract dilution analysis (AEDA) and the representativeness of the odour of food extracts. In: Maarse H, van der Heij DG (eds) Trends in flavour research. Elsevier, Amsterdam, pp 179–192

Etievant PX, Callement G, Langlois D, Issanchou S, Coquibus N (1999) Odor intensity evaluation in gas chromatography-olfactometry by finger span method. J Agric Food Chem 47:1673–1680

Ferreira V, Rapp A, Cacho J, Hastrich H, Yavas I (1993a) Fast and quantitative determination of wine flavor compounds using microextraction with Freon 113. J Agric Food Chem 41:1412–1419

Ferreira V, Escudero A, Salafranca J, Fernández P, Cacho J (1993b) Matrix effects and solute discrimination when injecting dirty samples in capillary columns. A comparative study between classical split and splitless injections. J Chromatogr A 655:257–266

Ferreira V, Fernández P, Meléndez J, Cacho J (1995a) Analytical characteristics of sample evaporation with the microKuderna-Danish concentrator. J Chromatogr A 695:41–55

Ferreira V, Fernández P, Peña C, Escudero A, Cacho JF (1995b) Investigation on the role played by fermentation esters in the aroma of young Spanish wines by multivariate analysis. J Sci Food Agric 67:381–392

Ferreira V, Sharman M, Cacho J, Dennis J (1996) New and efficient microextraction, solid-phase fractionation method for the gas chromatographic analysis of wine volatiles. J Chromatogr A 731:247–259

Ferreira V, López R, Escudero A, Cacho J (1998a) The aroma of Grenache red wine: hierarchy and nature of its main odorants. J Sci Food Agric 77:259–267

Ferreira V, Ardanuy M, López R, Cacho J (1998b) Relationship between flavor dilution values and odor unit values in hydroalcoholic solutions: the role of volatility and a practical rule for its estimation. J Agric Food Chem 46:434–440

Ferreira V, Peña C, López R, Crespo C, Cacho J (1998c) Concentration of small volumes of apolar solutions containing trace volatile compounds. J Chromatogr A 824:195–203

Ferreira V, Hernández-Orte P, Escudero A, López R, Cacho J (1999) Semipreparative reversed-phase liquid chromatographic fractionation of aroma extracts from wine and other alcoholic beverages. J Chromatogr A 864:77–88

Ferreira V, López R, Cacho JF (2000a) Quantitative determination of the odorants of young red wines from different grape varieties. J Sci Food Agric 80:1659–1667

Ferreira V, Aznar M, Escudero A, López R, Cacho J (2000b) Aroma composition of the "classic" old red wine from Rioja (Spain). A research through AEDA, odor evaluation of RP-HPLC flavor fractions and quantitative GC-MS. Comparison of techniques. In: Proc 6th Wartburg Aroma Symp, Eisenach, 10–13 Apr

Ferreira V, Ortega L, Escudero A, Cacho JF (2000c) A comparative study of the ability of different solvents and adsorbents to extract aroma compounds from alcoholic beverages. J Chromatogr Sci 38:469–476

Fritjers JE.R (1978) A critical analysis of the Odour Unit number and its use. Chem Senses Flavour 3:227–223

Fuller GH, Steltenkamp R, Tisserand GA (1964) The gas chromatograph with human sensor: perfumer model. Ann NY Acad Sci 116:711

García-Jares C, García-Martín S, Cela-Torrijos R (1995) Analysis of some highly volatile compounds of wine by means of purge and cold trapping injector capillary gas chromatography. Application to the differentiation of Rias Baixas Spanish white wines. J Agric Food Chem 43:764–768

Grob J (1987) On-column injection in capillary gas chromatography. Huethig, Heidelberg

Grob K (1988) Classical split and splitless injection in capillary GC, 2nd edn. Huethig, Heidelberg

Grob K Jr, Grob G, Grob K (1978) Comprehensive, standardized quality test for glass capillary columns. J Chromatogr 156:1–13

Grosch W (1993) Detection of potent odorants in foods by aroma extract dilution analysis. Trends Food Sci Technol 41:68–73

Grosch W (1994) Determination of potent odourants in foods by aroma extract dilution analysis (AEDA) and calculation of odour activity values (OAVs). Flavour Fragrance J 9:147–153

Guedes de Pinho P, Bertrand A (1995) Analytical determination of furaneol (2,5-dimethyl-4-hydroxy-3(2H)-furanone). Application to differentiation of white wines from hybrid and various *Vitis vinifera* cultivars. Am J Enol Vitic 46:181–186

Guedes de Pinho P, Anocibar Beloqui A, Bertrand A (1997) Détection d'un composé soufré responsable de l'arôme typique de vins issus de certains cépages non-*Vitis vinifera*. Sci Aliment 17:341–348

Guichard E, Pham TT, Etievant P (1993) Quantitative determination of sotolon in wines by high-performance liquid chromatography. Chromatographia 37:539–542

Guichard H, Guichard E, Langlois D, Issanchou S, Abbot N (1995) GC sniffing analysis: olfactive intensity measurement by two methods. Z Lebensmitt Unters Forsch 201:344–350

Guth H (1997a) Identification of character impact odorants of different white wine varieties. J Agric Food Chem 45(8):3022–3026

Guth H (1997b) Quantitation and sensory studies of character impact odorants of different white wine varieties. J Agric Food Chem 45:3027–3032

Guth H (1998) Comparison of different white wine varieties in odor profiles by instrumental analysis and sensory studies. In: Waterhouse AL, Ebeler SE (eds) Chemistry of wine flavor. ACS Symp Ser 714. ACS, Washington, DC, pp 39–52

Guth H, Grosch W (1993) Identification of potent odourants in static headspace samples of green and black tea powders on the basis of aroma extract dilution analysis (AEDA). Flavour Fragrance J 8:173–178

Hanaoka K, Sieffermann J-M, Giampaoli P (2000) Effects of the sniffing port air makeup in gas chromatography-olfactometry. J Agric Food Chem 48:2368–2371

Hayasaka Y, Bartowsky E (1999) Analysis of diacetyl in wine using solid-phase microextraction combined with gas chromatography-mass spectrometry. J Agric Food Chem 47:612–617

Heimann W, Rapp A, Völter J, Knipser W (1983) Beitrag zur Entstehung des Korktons im Wein. Dtsch Lebensmitt Rundsch 79:103–107

Ho P, Hogg TA, Silva CM (1999) Application of a liquid chromatographic method for the determination of phenolic compounds and furans in fortified wines. Food Chem 64:115–122

Jennings W, Mittlefehldt E, Stremple P (1997) Analytical gas chromatography, 2nd edn. Academic Press, San Diego

Kotseridis Y, Baumes R (2000) Identification of impact odorants in Bordeaux red grape juice, in the commercial yeast used for its fermentation, and in the produced wine. J Agric Food Chem 48:400–406

Kotseridis Y, Anocibar Beloqui A, Bertrand A, Doazan JP (1998) An analytical method for studying the volatile compounds of Merlot Noir clone wines. Am J Enol Vitic 48:44–48

Kotseridis Y, Baumes RL, Skouroumounis GK (1999) Quantitative determination of free and hydrolytically liberated β-damascenone in red grapes and wines using a stable isotope dilution assay. J Chromatogr A 849:245–254

Lacey MJ, Allen MS, Harris RLN, Brown WV (1991) Methoxypyrazines in Sauvignon grapes and wines. Am J Enol Vitic 42:103–108

Lavigne V, Henry R, Dubourdieu D (1998) Identification et dosage de composés soufrés intervenant dans l'arôme grillé des vins. Sci Aliment 18:175–191

Le Fur Y, Etievant P (1998) Données relatives à l'étude de quatre composés volatils suspectés d'intervenir sur la qualité d'odeur et d'arôme des vins de Chardonnay de Bourgogne: l'exemple d'une démarche analytique. Rev Oenolog 88:13–16

López R, Ferreira V, Hernández P, Cacho J (1999) Identification of impact odorants of young red wines made with Merlot, Cabernet Sauvignon and Grenache grape varieties: a comparative study. J Sci Food Agric 79:1461–1467

Maarse H, Visscher CA (1989) Volatile compounds in food, alcoholic beverages. TNO-CIVO Food Analysis Institute, Aj Zeist

Marais J, Pool HJ (1980) Effect of storage time and temperature on the volatile composition and quality of dry white table wines. Vitis 19:151–164

Marin AB, Acree TE, Barnard J (1988) Variation in odor detection thresholds determined by Charm analysis. Chem Senses 13:435–444

Martin B, Etievant PS, Le Quéré JL, Schlich PJ (1992) More clues about the impact of sotolon in Sherry wines. J Agric Food Chem 40:475–478

McDaniel MR, Miranda-López R, Walson BT, Micheals NJ, Libbey LM (1990) Pinot Noir aroma: a sensory/gas chromatographic approach. In: Charalambous G (ed) Flavors and off-flavors. Elsevier, Amsterdam, pp 23–36

Miranda-López R, Libbey LM, Watson BT, McDaniel MR (1992) Odor analysis of Pinot Noir wines from grapes of different maturities by a gas-chromatography-olfactometry technique (Osme). J Food Sci 57:985–993, 1019

Mistry BS, Reineccius T, Olson LK (1997) Gas chromatography-olfactometry for the determination of key odorants in foods. In: Marsily R (ed) Techniques for analyzing food aroma. Dekker, New York, pp 265–292

Moio L, Chambellant E, Lesschaeve I, Issanchou S, Schlich P, Etievant PX (1995) Production of representative wine extracts for chemical and olfactory analysis. Ital J Food Sci 3:265–278

Moio L, Etievant PX (1995) Ethyl anthranilate, ethyl cinnamate, 2,3-dihydrocinnamate and methyl anthranilate: four important odorants identified in Pinot Noir wines of Burgundy. Am J Enol Vitic 46:392–398

Moio L, Schlich P, Issanchou S, Etievant PX, Feuillat M (1993) Description de la typicité aromatique de vins de Bourgogne issus du cépage Chardonnay. J Int Sci Vigne Vin 27:179–189
Moio L, Schlich P, Etievant P (1994) Acquisition et analysis d'aromagrammes de vins de Bourgogne issus du cépage Chardonnay. Sci Aliment 14:601–608
Nakamura S, Crowell EA, Ough CS, Totsuka A (1988) Quantitative analysis of g-nonalactone in wines and its threshold determination. J Food Sci 53:1243–1244
Nelson RR, Acree TE (1978) Concord wine composition as affected by maturity and processing technique. Am J Enol Vitic 29:83–86
Noble AC, Flath RA, Forrey RR (1980) Wine headspace analysis. Reproducibility and application to varietal classification. J Agric Food Chem 28:346–353
Ong PKC, Acree TE (1999) Similarities in the aroma chemistry of Gewürztraminer variety wines and lychee fruit. J Agric Food Chem 47:665–670
Ong PKC, Acree TE, Lavin EH (1998) Characterization of volatiles in rambutan fruit (*Nephelium lappaceum*, L). J Agric Food Chem 46:611–615
Ortega L (1999) Uso de coeficientes de reparto para la predicción del comportamiento analítico de sistemas de extracción en fase sólida. Optimización de distintas técnicas de aislamiento de los volátiles del vino. Postgraduate Research Project (Tesina), University of Zaragoza
Peña C (1997) Nuevos métodos analíticos para la caracterización química y sensorial del aroma de bebidas alcohólicas. Aplicación a la caracterización y génesis del aroma del vino de Chardonnay de la D.O. Somontano. PhD Thesis, University of Zaragoza
Pollien P, Fay LB, Baumgartner M, Chaintreau A (1999) First attempt of odorant quantitation using gas chromatography. Anal Chem 71:5391–5397
Pollien P, Ott A, Montigon F, Baumgartner M, Muñoz-Box R, Chaintreau A (1997) Hyphenated headspace–gas chromatography–sniffing technique: screening of impact odorants and quantitative aromagram comparisons. J Agric Food Chem 45:2630–2637
Poole CF, Poole SK (1991) Chromatography today. Elsevier, Amsterdam
Priser C, Etievant PX, Nicklaus S, Brun O (1997) Representative champagne wine extracts for gas chromatography olfactometry analysis. J Agric Food Chem 45:3511–3514
Rapp A, Pretorius PJ (1990) Foreign and undesirable flavours in wine. In: Charalambous G (ed) Flavors and off-flavors '89. Elsevier, Amsterdam, pp 1–21
Rapp A, Knipser W, Engel L, Ullemeyer H, Heimann W (1980) Fremdkomponenten im Aroma von Trauben und Weinen interspezifischer Rebsorten I. Die Erdbeernote. Vitis 19:13–23
Rauhut D, Kürbel H, MacNamara K, Grossman M (1998) Headspace GC-SCD monitoring of low volatile sulfur compounds during fermentation and in wine. Analusis 26:142–145
Reglero G (1985) Estudio sobre los distintos tipos de columnas para cromatografía de gases. Aportación al conocimiento de la relación naturaleza química/actividad sensorial de los componentes del aroma del vino. PhD Thesis, University Autonoma of Madrid
Riberau-Gayon P, Boidron JN, Terrier A (1975) Aroma of muscat grape varieties. J Agric Food Chem 23:1042–1047
Roberts D, Acree TE (1995) Simulation of retronasal aroma using a modified headspace technique: investigating the effects of saliva, temperature, shearing and oil on flavor release. J Agric Food Chem 43:2179–2186
Roberts DD, Acree TE (1996) Effects of heating and cream addition of fresh raspberry aroma using a retronasal aroma simulator and gas chromatography olfactometry. J Agric Food Chem 44:3919–3925
Salinas MR, Alonso GL (1997) Adsorption–thermal desorption–gas chromatography applied to the determination of wine aromas. In: Linskens HF, Jackson JF (eds) Modern methods of plant analysis, vol 19. Plant volatile analysis. Springer, Berlin Heidelberg New York, pp 175–192
Salinas MR, Alonso GL, Esteban-Infantes FJ (1994) Adsorption–thermal desorption–gas chromatography applied to the determination of wine aromas. J Agric Food Chem 42:1328–1331
Schlich P, Moio L (1994) Corrélation entre les profils aromatiques et les arògrammes de vins de Bourgogne issus de cépage Chardonnay. Sci Aliment 14:609–615
Schneider R, Baumes R, Bayonove C, Razungles A (1998) Volatile compounds involved in the aroma of sweet fortified wines (Vins Doux Naturels) from Grenache Noir. J Agric Food Chem 46:3230–3237

Schreier JP, Drawert P, Abraham KO (1980) Identification and determination of volatile constituents in Burgundy Pinot Noir. Lebensmitt Wiss Technol 13:318–321

Seibert RM, Buttery RG, Guadagni DG, Black DR, Harris JG (1970) Synthesis of some 2-methoxy-3-alkylpyrazines with strong bell pepper-like odors. J Agric Food Chem 18:246–249

Shimoda M, Shibamoto T, Noble AC (1993) Evaluation of headspace volatiles of Cabernet Sauvignon wines sampled by an on-column method. J Agric Food Chem 41:1664–1668

Shinohara T (1985) Gas chromatographic analysis of volatile fatty acids in wines. Agric Biol Chem 49:2211–2212

Simpson RF, Miller GC (1984) Aroma composition of Chardonnay wine. Vitis 23:143–158

Slingsby RW, Kepner RE, Muller CJ, Webb AD (1980) Some volatile components of *Vitis vinifera* variety Cabernet Sauvignon wine. Am J Enol Vitic 31:360–363

Stern DJ, Guadagni D, Stevens KL (1975) Aging of wine: qualitative changes in the volatiles of zinfandel wine during two years. Am J Enol Vitic 26:208–213

Tominaga T, Darriet P, Dubourdieu D (1996) Identification of 3-mercaptohexylacetate in Sauvignon wines, a powerful aromatic compound exhibiting box-tree odor. Vitis 35:207–210

Tominaga T, Furrer A, Henry R, Dubourdieu D (1998a) Identification of new volatile thiols in the aroma of *Vitis vinifera* L. var. Sauvignon Blanc wines. Flavour Fragrance J 13:159–162

Tominaga T, Murat ML, Dubourdieu D (1998b) Development of a method for analyzing the volatile thiols involved in the characteristic aroma of wines made from *Vitis vinifera* L. cv. Sauvignon Blanc. J Agric Food Chem 46:1044–1048

Tominaga T, Blanchard L, Darriet P, Dubourdieu D (2000) A powerful aromatic volatile thiol, 2-furanmethanethiol, exhibiting roast coffee aroma in wines made from several *Vitis vinifera* grape varieties. J Agric Food Chem 48(5):1799–1802

Ullrich F, Grosch W (1987) Identification of the most intense volatile flavour compounds formed during auto-oxidation of linoleic acid. Z Lebensmitt Untersch Forsch 184:277–282

van Ruth SM, Roozen JP (1994) Gas chromatography/sniffing port analysis and sensory evaluation of commercially dried bell peppers (*Capsicum annuum*) after rehydration. Food Chem 51:165–170

van Ruth SM, Roozen JP, Cozijnsen JL (1995) Changes in flavour release from rehydrated diced bell peppers (*Capsicum annuum*) by artificial saliva components in three mouth model systems. J Sci Food Agric 67:189–196

van Ruth SM, Roozen JP, Cozijnsen JL (1996) Gas chromatography/sniffing port analysis evaluated for aroma release from rehydrated French beans (*Phaseolus vulgaris*). Food Chem 56: 343–346

van Wyk CJ, Augustyn OPH, de Wet P, Joubert WA (1978) Isoamyl acetate – a key fermentation volatile of wines of *Vitis vinifera* cv. Pinotage. Am J Enol Vitic 30:167–173

Versini G (1985) Sull'aroma del vino Traminer aromatico or Gewürtztraminer. Vignevini 5:57–65

Versini G, Tomasi T (1983) Confronto tra i componenti volatili dei vini rossi ottenuti con macerazione tradizionale e macerazione carbonica. Enotecnico 19:595–600

Wildenradt HL, Christensen EN, Stackler B, Caputi A, Slinkard K, Scutt K (1975) Volatile constituents of grape leaves. I. *Vitis vinifera* variety Chenin Blanc. Am J Enol Vitic 26:148–153

Williams AA, Mottram DS (1981) Gas chromatographic analysis of furaneol. High Resol Chromatog Capillary Chromatogr 4:421–422

7 Analysis of Volatile Components of *Citrus* Fruit Essential Oils

G. Ruberto

7.1 Introduction

The geographical origin of *Citrus* fruits is localised in a large region of Southeast Asia comprising southern China, India, Indo-China, Indonesia and the Malay Archipelago. Very slowly, and thanks mainly to nomads, merchants, conquest wars and explorers, *Citrus* fruits became known in Africa and Europe. Several witnesses to the presence of *Citrus* fruits during Alexander the Great's reign and the Greek and Roman periods may be found in literature, sculptures, mosaics and so on. However, it was under Arab rule throughout most of the Mediterranean basin, in a long period between the end of the first and the beginning of the second millennium, that the diffusion as well as the first cultivation of *Citrus* plants occurred (Webber 1967). Indeed, in this period an Arab botanist, Al-Beithâr, reported, in his treatise "Dictionary of the Simple Remedies" (1200), the first technical description of essential oil extraction from citron fruits (Calabrese 1990). However, for many centuries the cultivation of *Citrus* plants was limited to their ornamental use. Their industrial exploitation started in Sicily between 1500 and 1600 precisely with the extraction of essential oil from lemon fruits. The process entailed the removal of the peel from the pulp, which was considered a waste product, and only successively was the pulp used industrially to prepare the juice, called "agro", exclusively for the preparation of calcium citrate (Safina 1984).

Today, *Citrus* are the most widespread arboreal plants in the world and represent one of the most important crops. They are cultivated in over 130 countries between 40°N and 40°S, extending over 4 million ha (Spina and Di Martino 1991; Starrantino 1994). Data from the Food and Agricultural Organisation (2000) report a world *Citrus* production of near to 100,000 metric tons in 1999, with an increasing trend in the last decade. Brazil, the USA and China are, in that order, the main producer countries, accounting for about 47% of the world production; Spain, Mexico, the Islamic Republic of Iran, India, Nigeria and Italy are the other most important producers.

Orange is the most cultivated fruit, followed by the mandarin, tangerine, clementine and satsuma group, then by lemon and lime, and by grapefruit

Istituto del C.N.R. per lo Studio delle Sostanze Naturali di Interesse Alimentare e Chimico-Farmaceutico, Via del Santuario 110, 95028 Valverde, CT, Italy and pomelo. A further group is represented by particular *Citrus* fruits such as bergamot, citron, bigarade and kumquat.

For a long time the main commercial outlet of *Citrus* fruits has been the fresh market. However, in more recent years an even higher proportion of fruits has been processed, mainly for juice production (Food and Agriculture Organisation 2000). In some countries, such as Brazil and the USA more than 80% of the entire production is industrially transformed into different kinds of juices. Another important product of *Citrus* fruits is the essential oil, which is mainly obtained from *Citrus* peels, while other *Citrus* essential oils are produced from leaves, twigs and blossoms, though their amounts are very much lower than peel essential oils. From an industrial point of view, juice and peel essential oils production necessarily proceed together. In fact the various industrial plants, even if adopting different technologies, provide the contemporary production of juice and essential oils. Therefore the increase in juice production, especially that of orange, has caused a similar increase in essential oil. Data on the real amount of the *Citrus* essential oils produced are not available or up to date, and in some cases are also contrasting. In 1989 Thomas and Bessière reported that the output of (+)-limonene, the main component of *Citrus* essential oils, was ca. 50,000 tons, whereas Lawrence (1995a) reports a worldwide orange oil production of ca. 26,000 tons in 1990. However, given the increase in the supply of *Citrus* and its transformation products in these last 10 years, it is presumable that the actual total *Citrus* essential oil production may be near to 75,000 tons.

Orange essential oil is, as previously mentioned, the main *Citrus* oil, and at the same time given its large production it is also the cheapest, while all other *Citrus* essential oils are much more expensive. The main end-uses of these oils are for flavouring in the food industry, especially in the production of several kinds of beverages. Furthermore, most of them, but particularly the bergamot, bitter orange, petitgrain and neroli oils, find an end-use in soap, cologne and perfume.

An important feature that differentiates *Citrus* essential oils from similar oils obtained from common aromatic plants is the extraction methodology. In fact, whereas plant essential oils are prepared by steam distillation, *Citrus* peel essential oils are obtained by a mechanical process which involves the rupture of the oleic glands (utricles) containing the essential oil present in the *Citrus* peel. Only lime oil is obtained by steam distillation of the whole fruit. Depending on the particular industrial process ["pelatrice", "sfumatrice", "in-line F.M.C." (Food Machinery Corporation)], breaking of the glands can be carried out before or after drawing the juice. Whatever the process the oil flowing out from the broken glands is emulsified with water, which also acts as a vehicle agent, and successively throughout filtration and centrifugation, it is purified by residues and separated from the water. The oil obtained in such a way is the so-called cold-pressed oil. The water used in this procedure is continuously recycled, so that at the end of the process it is saturated with essential oils. Normally, this water is subjected to distillation either to obtain mainly limonene or in particular cases (i.e. bergamot and lemon) to distillate a second-choice essential oil.

7.2 Chemical Composition of *Citrus* Peel Essential Oils

The volatile components present in the *Citrus* peel essential oils largely belong to terpene compounds. Mono and sesquiterpene hydrocarbons together with several oxygenated derivatives, comprising alcohols, aldehydes, ketones, esters and epoxides, are the main compounds. Depending on the particular species and/or cultivar a more or less significant amount of non-isoprenoid compounds can be present, the main component being alcohols, esters and aldehydes of aliphatic origin. Table 7.1 lists the components detected in the essential oils of some of the most important *Citrus* species, namely blond and blood sweet orange (*Citrus sinensis* L. Osbeck), bitter orange (*C. aurantium* L.), lemon (*C. limon* L. Burman), mandarin (*C. reticulata* Blanco), clementine (*C. clementina* Hort ex Tan.), grapefruit (*C. paradisi* MacFayden), bergamot (*C. bergamia* Risso), lime (*C. aurantifolia* Christm. Swing) and the fruit of the hybrid citrange Carrizo (*C. sinensis* × *Poncirus trifoliata*). The last plant is one of the main rootstocks for *Citrus* propagation, although its fruit is not edible and it is cultivated only for seed production.

The monoterpene hydrocarbons are the principal components of all *Citrus* oils, their amount ranges between 98 and 60% of the total oil, limonene being the main compound. Notwithstanding the high proportion of these components, the fragrance of each of the *Citrus* oils is well differentiated from each other, given the different amount of oxygenated derivatives, which are mainly responsible for the particular aroma of a particular *Citrus* oil (Shaw 1979; Bauer and Garbe 1985; Di Giacomo and Mincione 1994; Tateo and Bonomi 1995; Caccioni et al. 1998; Dugo et al. 1999a).

Sweet orange oil (*C. sinensis*): the limonene content is always more than 90%, whatever the cultivar. The oxygenated components such as linalool, the octanal and decanal aldehydes, the neral/geranial and α-/β-sinensal couples, as well as some esters (octyl, neryl and geranyl acetate) are the compounds that strongly affect the orange fragrance. The amount of oxygenated components is higher in blood orange cultivars (Tarocco, Moro, Sanguinello) than in blond orange cvv. (Washington, Valencia) (Caccioni et al. 1998), and this accounts for the stronger aroma of the blood oranges. The main fields of application of this fragrance are the food industry (beverages and sweets), and for perfumes, soap and household products.

Bitter orange (*C. aurantium*): the essential composition is very similar to the previous sweet orange, but with less aldehydes and more terpenoid esters. This apparently "small" difference makes this oil particularly appreciated in the production of perfumes and liqueurs.

Lemon oil (*C. limon*): the average content of limonene in this oil is about 65%, and is associated with two other monoterpene hydrocarbons, β-pinene and γ-terpinene (10–12% each). The typical lemon fragrance is due to the high content of neral and geranial. A further contribution is also given by the corresponding esters neryl and geranyl acetate. Lemon, unlike other *Citrus* species, can fructify several times a year, and chemical compositional differences are

Table 7.1. Chemical components of *Citrus* peel essential oils. Compounds are listed according to the elution order on apolar capillary column (5% phenyl-methylpolysiloxane). All samples are cold-pressed oils, except for Carrizo Citrange oil which was obtained by peel hydrodistillation. (Shaw 1979; Caccioni et al. 1998; Dugo et al. 1999a)

Compound	Blond sweet orange	Blood sweet orange	Bitter orange	Lemon	Mandarin	Clementine	Grapefruit	Bergamot	Lime	Carrizo Citrange
Hexanol			✓							
Tricyclene				✓				✓	✓	
α-Thujene	✓	✓	✓	✓	✓	✓	✓	✓	✓	✓
α-Pinene	✓	✓	✓	✓	✓	✓	✓	✓	✓	✓
Camphene	✓	✓	✓	✓	✓	✓	✓	✓	✓	✓
Thuja-2,4(10)-diene									✓	
Heptanol					✓					
Sabinene	✓	✓	✓	✓	✓	✓	✓	✓	✓	✓
β-Pinene	✓	✓	✓	✓	✓	✓	✓	✓	✓	✓
6-Methyl-5-hepten-2-one				✓	✓			✓	✓	
Myrcene	✓	✓	✓	✓	✓	✓	✓	✓	✓	✓
α-Phellandrene	✓	✓	✓	✓	✓	✓	✓	✓	✓	✓
Octanal	✓	✓	✓	✓	✓	✓	✓	✓	✓	✓
Decane									✓	
Hexyl acetate								✓		
δ-3-Carene	✓	✓	✓	✓	✓	✓		✓	✓	✓
α-Terpinene	✓	✓	✓	✓	✓	✓	✓	✓	✓	✓
p-Cymene	✓	✓	✓	✓	✓	✓	✓	✓	✓	✓
Limonene	✓	✓	✓	✓	✓	✓	✓	✓	✓	✓
β-Phellandrene	✓	✓	✓	✓	✓	✓	✓	✓	✓	
1,8-Cineole				✓	✓			✓	✓	
(*Z*)-β-Ocimene	✓	✓	✓	✓	✓	✓	✓	✓	✓	✓
(*E*)-β-Ocimene	✓	✓	✓	✓	✓	✓	✓	✓	✓	✓
γ-Terpinene	✓	✓	✓	✓	✓	✓	✓	✓	✓	✓
cis-Sabinene hydrate	✓	✓	✓	✓	✓	✓	✓	✓	✓	✓
cis-Linalool oxide			✓					✓		
Octanol	✓	✓	✓	✓	✓	✓	✓	✓		✓

Terpinoolene	✔	✔	✔	✔	✔	✔	✔	✔	✔	✔
trans-Linanool oxide			✔				✔	✔		
trans-Sabinene hydrate		✔	✔	✔	✔	✔		✔		
Linalool	✔	✔	✔	✔	✔	✔	✔	✔	✔	✔
Nonanal	✔	✔	✔	✔	✔	✔	✔	✔	✔	
1,3,8-*p*-Menthatriene					✔					
endo-Fenchol									✔	
cis-*p*-Menth-2-en-1-ol									✔	
Heptyl acetate								✔		
cis-Limonene oxide	✔	✔	✔	✔	✔	✔	✔	✔	✔	
allo-Ocimene[a]				✔				✔	✔	
trans-Pinocarveol									✔	
trans-Limonene oxide	✔	✔	✔	✔	✔	✔	✔	✔	✔	
Camphor				✔	✔			✔	✔	
trans-*p*-Menth-2-en-1-ol									✔	
Isopulegol							✔	✔		
(*E*)-Miroxide			✔							
Citronellal	✔	✔	✔	✔	✔	✔	✔	✔	✔	✔
Borneol			✔	✔					✔	
cis-Pinocamphone									✔	
Terpinen-4-ol	✔	✔	✔	✔	✔	✔	✔	✔	✔	✔
Nonanol			✔				✔			
p-Cymen-8-ol					✔					
α-Terpineol	✔	✔	✔	✔	✔	✔	✔	✔	✔	✔
cis-Piperitol									✔	
Decanal	✔	✔	✔	✔	✔	✔	✔	✔	✔	
Dodecane								✔	✔	
trans-Carveol	✔					✔	✔			
Nerol	✔	✔	✔	✔	✔		✔	✔	✔	✔
Citronellol			✔	✔	✔					
Cytronellyl formate	✔	✔								
Carvone	✔	✔			✔	✔	✔	✔		
Neral	✔	✔	✔	✔	✔	✔	✔	✔	✔	✔
Piperitone			✔	✔	✔				✔	
Geraniol	✔	✔	✔	✔	✔		✔	✔	✔	✔

Table 7.1. *Continued*

Compound	Blond sweet orange	Blood sweet orange	Bitter orange	Lemon	Mandarin	Clementine	Grapefruit	Bergamot	Lime	Carrizo Citrange
Linalyl acetate			✔		✔	✔	✔	✔		
(*E*)-2-Decenal	✔	✔			✔	✔		✔		
Perillaldehyde	✔	✔	✔	✔	✔	✔	✔	✔	✔	
Geranial	✔	✔	✔	✔	✔	✔	✔	✔	✔	
Decanol			✔				✔			
Bornyl acetate				✔				✔	✔	
Perillalcohol					✔	✔				
trans-Pinocarvyl acetate									✔	
Thymol					✔	✔				
Carvacrol									✔	
Tridecane									✔	
Undecanal	✔	✔	✔	✔	✔	✔	✔	✔	✔	
(*E*,*E*)-2,4-Decadienal	✔	✔	✔		✔	✔				
Nonyl acetate	✔	✔	✔	✔	✔			✔		
Methyl geranoate				✔				✔		
δ-Elemene			✔					✔	✔	
Linalyl propanoate								✔		
α-Terpenyl acetate	✔	✔	✔			✔	✔	✔	✔	✔
trans-Carvyl acetate							✔			
Citronellyl acetate	✔	✔	✔	✔	✔		✔	✔	✔	
Neryl acetate	✔	✔	✔	✔	✔	✔	✔	✔	✔	
α-Copaene	✔	✔			✔	✔	✔			✔
Geranyl acetate	✔	✔	✔	✔	✔	✔	✔	✔	✔	
β-Elemene	✔	✔			✔	✔	✔		✔	
β-Cubebene	✔	✔	✔			✔				✔
Methyl-*N*-methylanthranilate					✔	✔				
Dodecanal	✔	✔	✔		✔	✔	✔	✔	✔	
β-Caryophyllene	✔	✔	✔	✔	✔	✔	✔	✔	✔	✔
cis-α-Bergamotene				✔				✔	✔	

Decyl acetate	✔	✔	✔	✔	✔	✔	✔	✔	✔	
α-Santalene				✔				✔	✔	
α-Cadinene	✔	✔				✔				
α-Elemene									✔	✔
trans-α-Bergamotene			✔	✔	✔		✔	✔	✔	✔
α-Humulene	✔	✔	✔	✔	✔	✔	✔	✔	✔	✔
β-Santalene				✔				✔	✔	
(*Z*)-β-Farnesene	✔	✔	✔	✔		✔	✔	✔	✔	✔
(*E*)-2-Dodecenal			✔		✔	✔	✔			
Dodecanol								✔		
Germacrene-D	✔	✔	✔		✔	✔	✔	✔		
γ-Muurolene		✔								
(*E*)-β-Farnesene				✔					✔	✔
α-Selinene					✔				✔	
Valencene	✔	✔		✔	✔	✔				✔
Bicyclogermacrene	✔	✔	✔	✔	✔	✔	✔	✔		
γ-Cadinene		✔					✔			✔
7-*epi*-α-Selinene	✔									
δ-Guaiene		✔								
(Z)-α-Bisabolene				✔				✔	✔	
β-Bisabolene			✔	✔	✔		✔	✔	✔	
Aristolene										✔
(*E*,*E*)-α-Farnesene	✔	✔			✔	✔	✔	✔	✔	✔
(*Z*)-γ-Bisabolene				✔				✔	✔	
Tridecanal	✔	✔								
Undecyl acetate								✔		
δ-Cadinene	✔	✔	✔		✔	✔	✔			
(*E*)-γ-Bisabolene									✔	
(*E*)-α-Bisabolene				✔						
Elemol	✔	✔				✔	✔			
Germacrene-B								✔	✔	
(*E*)-Nerolidol	✔	✔	✔		✔	✔	✔	✔		✔
Germacrene-D-4-ol	✔	✔		✔			✔			
Caryophyllene oxide	✔	✔					✔		✔	
cis-Sesquisabinene hydrate									✔	

Table 7.1. *Continued*

Compound	Blond sweet orange	Blood sweet orange	Bitter orange	Lemon	Mandarin	Clementine	Grapefruit	Bergamot	Lime	Carrizo Citrange
Tetradecanal	✔	✔	✔		✔	✔	✔	✔	✔	
Selin-11-en-4-ol	✔	✔		✔					✔	
Campherenol				✔				✔	✔	
β-Bisabolol				✔				✔	✔	
α-Bisabolol				✔				✔	✔	
Cadinol[a]										✔
α-Asarone										✔
β-Sinensal	✔	✔				✔	✔			✔
Herniarin									✔	
Pentadecanal									✔	
(*E*,*E*)-Farnesol							✔			
(*E*,*Z*)-Farnesol							✔		✔	
α-Sinensal	✔	✔	✔		✔	✔				
Nootkatone	✔	✔	✔	✔			✔	✔		
Hexadecanal	✔	✔			✔		✔		✔	

[a] Correct isomer not identified

observed in the lemon fruits harvested in different periods (Di Giacomo and Mincione 1994; Caccioni et al. 1998). Lemon oil is largely used in the food sector.

Lime oil (*C. aurantifolia*) (Mexican lime): unlike other *Citrus* oils, lime oil is obtained by distillation of the whole fruit. The oil is characterised by 50–65% of limonene and about 10% each of β-pinene and γ-terpinene. Among the oxygenated components the content of 1,8-cineole, terpinen-4-ol and α-terpineol is particularly high. These are probably enhanced during the distillation process, contributing to the particular lime aroma. The main use of lime oil is in the preparation of soft drinks.

Mandarin oil (*C. reticulata*): limonene is the main component (65–75%) followed by γ-terpinene (16–23%). Among the oxygenated compounds the aldehydes (octanal, decanal, α-sinensal) and alcohols (linalool, α-terpineol) are the principal. However, the characteristic aromatic component of mandarin oil is methyl-*N*-methyl anthranilate, which is present at trace levels only in the clementine oil, not in other *Citrus* species. Mandarin oil is often used to reinforce sweet orange oil aroma. The main applications are in the formulation of liqueurs and perfumes.

Grapefruit oil (*C. paradisi*): limonene exceeds 90%. Octanal and decanal, linalool and a small presence of esters (mainly geranyl acetate) represent the oxygenated components. However, the compound that confers the particular aroma is the oxygenated sesquiterpene nootkatone. The grapefruit oil is applied as flavouring in beverages.

Bergamot oil (*C. bergamia*): the production of this species and its essential oil is limited almost exclusively to the Italian region of Calabria. The Ivory Coast, Guinea and Brazil are other producer countries. This is the sole *Citrus* peel essential oil in which the amount of monoterpene hydrocarbons reaches, at the most, 60%, the main compounds being limonene, γ-terpinene and β-pinene in that order. Extraordinarily high, with respect to the other peel *Citrus* oils, is the content of oxygenated components, linalyl acetate (ca. 30%) and the corresponding alcohol linalool (ca. 10%) being the main ones. On account of this particular chemical composition bergamot oil is the most appreciated and expensive *Citrus* oil. The main end-use is in the perfume industry, i.e. it is the fundamental ingredient of Eaux-de-Cologne (Forlot 2000).

Petitgrain oils: these oils are obtained by steam distillation of leaves and twigs of nearly all *Citrus* species. However, the most appreciated is that from bitter orange (*C. aurantium*), which is in fact rich in oxygenated components such as linalyl acetate (>50%) and linalool (>27%) (Mondello et al. 1996) and is mainly used in luxury soap and cosmetics. Other petitgrain oils deriving from sweet orange (Mondello et al. 1997b), mandarin (Mondello et al. 1997a), lemon (Mondello et al. 1997c) and combava (*C. hystrix* DC) are commercially available. A large compilation of the chemical composition of these oils has been reviewed by Lawrence (1995b).

Neroli oil: this is obtained by distillation of the blossom of the bitter orange blossom. Its composition is characterised by a high amount of oxygenated components, namely linalool, linalyl acetate, geranyl acetate and nerolidol (Tateo and Bonomi 1995; Dugo et al. 1999b). It is a particularly expensive oil and finds its main application in the formulation of perfumes.

The chemical composition of the cold-pressed *Citrus* peel essential oils, given the particular mechanical extraction procedure, is not exhausted with the volatile components. In fact, together with these, a non-volatile residue, ranging from 1 to 15% depending on the *Citrus* species or cultivar, is always present. The chemical composition of this non-volatile fraction is quite varied and different kinds of hydrocarbons, fatty acids, sterols, carotenoids, waxes and a group of oxygen heterocyclic compounds, namely coumarins, psolarens and polymethoxylated flavones, can be present. Of course all these components are completely absent in the distilled oils.

Oxygen heterocyclic compounds are those detected in higher amounts and more than 50 different compounds have been identified in the *Citrus* oils (McHale and Sheridan 1989; Ziegler and Spiteller 1992; Stremple 1998; Bonaccorsi et al. 1999; Dugo et al. 1999c). Many of these components are specific to a single or a few species; therefore their analysis represents a further tool to establish the quality and the genuineness of a given *Citrus* oil. Furthermore, these natural products, considered for a long time as non-functional components in a living system, are now receiving growing attention since many of them, mainly coumarins and flavonoids, have shown interesting biological activities, such as antiplatelet aggregating, and antimicrobial, antimutagenic, as well as antitumor and antioxidant properties (Kawai et at al. 1999a,b; Miyake et al. 1999; Merken and Beecher 2000; Ogawa et al. 2000). The re-evaluation of their biological role in living systems is testified by several in vitro and in vivo assays, as well as by epidemiological studies, which have positively correlated the intake of these components in the diet and the development of particular pathologies such as coronary heart disease and cancer, to quote the most devastating (Gordon 1996). These and many other secondary metabolites are now called nutraceuticals to underline their nutritional and pharmaceutical role (Bracke et al. 1996; Widmer and Montanari 1996). On the contrary, the interest towards these components is also due to their toxicity in living systems. This is the case of with psolarens or furocoumarins and, regarding the *Citrus* oils, in particular of 5-methoxypsoralen or bergapten (Fig. 7.1). These components are suspected of having a phototoxic action on human skin; therefore international bodies have defined quantitative limits of these components in products to be used on the skin (Poiana et al. 1999).

In Fig. 7.1 some of the most common oxygen heterocyclic compounds isolated from *Citrus* essential oils are shown.

Finally, in order to complete the picture of the aromatic components of *Citrus* fruit it is necessary to mention the juice aroma. This topic has been exhaustively reviewed by Nagy and Shaw (1990), and also subjected to more recent studies (Buettner and Schieberle 1999; Hinterholzer and Schieberle 1999), and therefore it will not be examined thoroughly on this occasion. However, a fundamental aspect emerging from these studies is that if, on the one hand, it is true that the peel oil

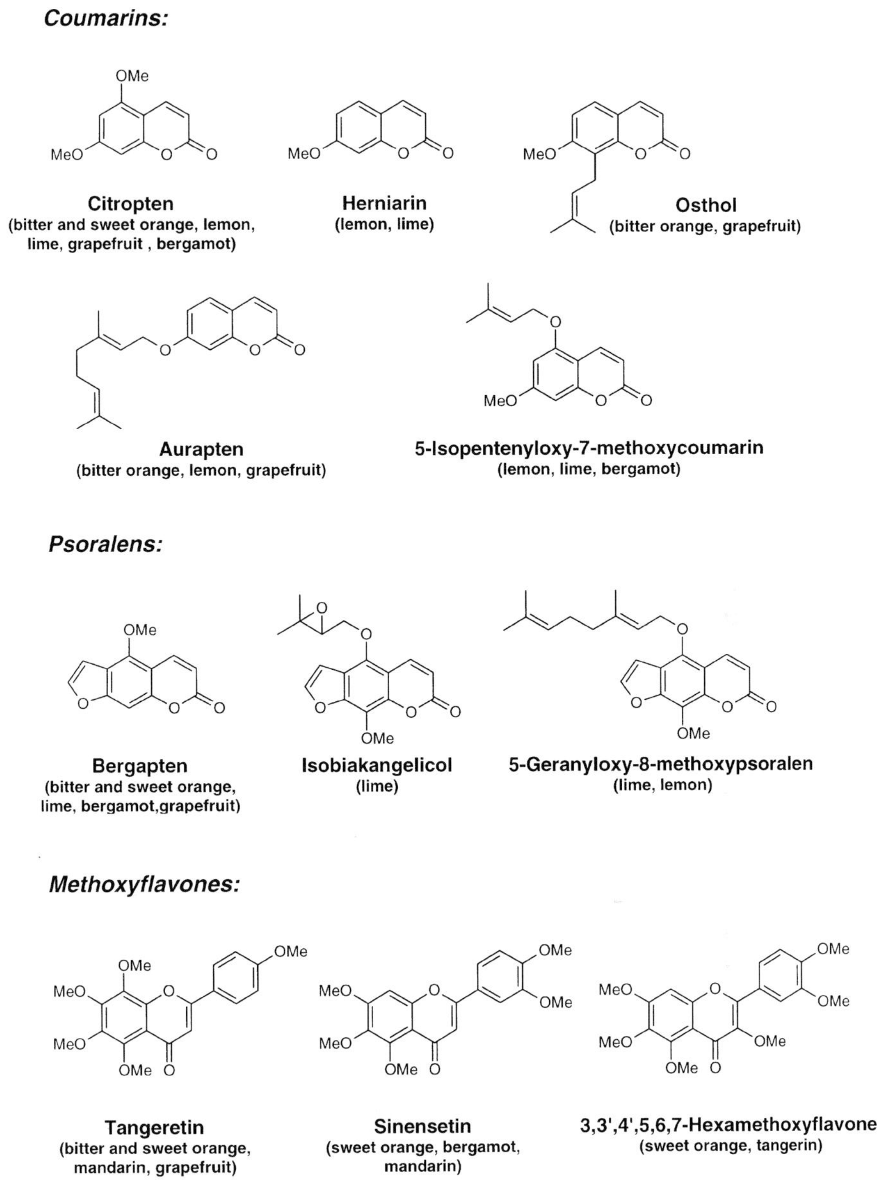

Fig. 7.1. Selected oxygen heterocyclic compounds present in *Citrus* peel essential oils

gives the characteristic aroma of any *Citrus* fruit, it is also true, on the other hand, that the peel oil is not influential to the juice aroma, unless some peel oil dissolves in the juice during the juice preparation. The juice-oil is responsible for both the pulp and juice aroma. It is contained in juice sacs of the pulp and from a chemical point of view it is quite different from the peel oil – normally it is much richer in ester derivatives (Nagy and Shaw 1990), which confer the particular fruity-note.

This diversity mainly influences the juice obtained from concentrate. The procedure for concentrate causes almost the total loss of its aromatic components; therefore during the juice reconstruction procedure it is necessary to add an aliquot of peel oil, which even if deterpenated will not confer the characteristic fruity-note of freshly squeezed juice.

7.3 Analysis of *Citrus* Peel Essential Oils

7.3.1 High Resolution Gas Chromatography (HRGC)

The *Citrus* essential oils as well as every other natural volatile component have been the main protagonists of the extraordinary development of gas chromatography, the separation techniques that more than all others have contributed to the increase in the knowledge of this field. Martin and Synge were the first to conceive of gas chromatography in 1941, but only in the 1950s did James and Martin build the first gas chromatograph. The development of this technique was very impressive, so much so as to slow down the other chromatographic techniques. In the 1960s gas chromatography was already a mature system, contributing decidedly to the recognition of chromatographic analysis, in general, as well established analytical technique (Scott 1987).

Gas chromatography is essentially a separation and not an identification technique, unless it is combined with an identification instrument such as a mass spectrometer. However, a gas chromatograph can be used to some extent to contribute to the identification of compounds, even if no gas chromatographic method alone can give an unambiguous determination of the components, but its use can narrow the number of possibilities. Retention time is a characteristic of a component in well-established chromatographic conditions, but unfortunately many compounds possess identical retention times and this makes this measurement of nearly no use. A much more useful tool is the relative retention time, namely with respect to a standard mixture. The more commonly adopted system is that developed by Kovats (1958), in which the standard is a mixture of *n*-paraffins. It is advisable to use two different chromatographic columns, one apolar and the other polar, in order to obtain two relative retention indexes, which are quite characteristic of a given component. Kovats developed his model in isothermal conditions, and this in some cases might represent a limit. Other authors, such as McReynolds (1966), Jennings and Shibamoto (1980) and Adams (1995), developed analogous systems using a temperature-programmed analysis. Today several lists of these values, calculated on apolar and polar columns, are available for a great number of components. However, even if these values are specific for a given component, they are not sufficient to establish the identity of a component, being used mainly as secondary proof (Mussinan 1993).

One of the peculiar features which, more than others, has contributed in these last years to the success of gas chromatography is the high number of possible detectors, some of them specific for certain classes of compounds, others endowed with a general applicability. The flame ionisation detector (FID) is probably one of the most used, both for its high degree of universality as well as its very low detectability limit, which can reach values around 10^{-12} g/s (Uden 1987). Therefore because of these characteristics the FID can be used for quantitative measurements in a great number of analyses, even if it is important to underline that any compound possesses its own response factor, and thus a correction factor must be introduced for fine quantitative measurements.

7.3.2 High Resolution Gas Chromatography–Mass Spectrometry (HRGC-MS)

Concerning the identification of the complex mixtures normally injected in a gas chromatograph, the mass spectrometer (MS) detector is one of the most effective instruments. The success of the GC-MS combination has contributed very consistently to the development of mass spectrometry technology. The miniaturisation of a complex and large mass spectrometer, thanks also to the introduction of quadrupolar and/or ion-trap analysers, has allowed the easy assembly of two instruments. Furthermore, the very high number of mass spectra in the literature has favoured the compilation of several libraries, some of which regard only particular classes of components; and finally the contemporary development of several data processing systems have contributed to make the GC-MS technique an irreplaceable tool in the study of any unknown mixture of volatile components. Early on, the main difficulty of this combination was the interface between GC and MS, which as is well known work in a contrasting way, namely under pressure in gas chromatography and under high vacuum in mass spectrometry. Therefore it was initially necessary to create opportune transfer devices – jet separator or open-split (Uden 1987) – to overcome the impasse. However, with the introduction of capillary column it was no longer necessary to install an interface, since the 1- to 2-ml gas flows applied to these columns are perfectly compatible with the vacuum systems

Electron impact (EI) is the most common ionisation system used in mass spectrometry when combined with gas chromatography. A typical GC-MS run is carried out in automatic scanning mode, during which the selected mass range (a 40- to 400-a.m.u. (atomic mass unit) range is typical for an essential oil analysis) is continuously scanned by the instrument, apart from the column eluant. Therefore at the end of the analysis the results will be a total ion chromatogram (TIC) trace, similar in aspect to a FID analysis, and some thousands of mass spectra stored in the data processing in the bar-graph format. The successive step will be the matching operation between the spectra collected in the computer and those present in the libraries, obtaining a list of compounds (5–10 or more) whose mass spectra have a decreasing percentage of likeness with the component under investigation.

Figure 7.2 shows the TIC of a lemon essential oil together with the result of the matching procedure for two components. These two examples allow us to evidence advantages and limits of this procedure. In fact, in the first case (γ-terpinene) the matching with the spectra of libraries gives an immediate identity of the compound, which is also confirmed by the value of the retention index. In the second case (α-bergamotene) the matching procedure shows some ambiguity. This is due to different factors: library mass spectra obtained with different systems, presence of different mass spectra for the same compound obtained with different spectrometers, presence of isomers with similar mass spectra (in the essential oils this is quite common). The result is that the first choice given by the data processing is dubious. In most cases this limit is easily overcome by using the relative retention indexes (calculated on apolar and polar columns) as secondary confirmatory evidence of a component. Therefore through a combination of mass spectra and retention indexes most of the doubtful cases can be solved. A further advantage is the use of home-made libraries which annul or drastically reduce some of the limits of the commercial libraries (different operative systems, presence of many duplicates, etc.) making the matching procedure more reliable.

Mass spectrometry offers a wide number of different techniques which can be used in combination with gas chromatography. One of these is called selected ion detection, by which it is possible to detect the presence of the characteristic ions of a component if these are, of course, present in the analysed mixture. With this technique the sensitivity of the mass spectrometer increases enormously, allowing the detection of a suspected component also at very low trace level (Mussinan 1993; Ragunathan et al. 1999). A further mass spectrometry technique is chemical ionisation (CI), in which a gas, such as methane, isobutane or ammonia, is introduced to the ion source together with the sample. The gas ions, produced by electron impact, transfer the charge to the compound, forming a molecular ion (with a positive or negative charge depending on the kind of gas used), which has a slight tendency to fragment later. Therefore a CI spectrum will always show fewer fragments than the corresponding EI spectrum, the molecular ion peak being easily detectable. This technique is particularly useful for compounds, such as alcohols or esters, whose molecular ions in the EI spectra are weak or completely absent (Mussinan 1993; Ragunathan et al. 1999).

Further and more sophisticated improvements in mass spectrometry techniques are represented by the MS-MS or tandem MS, and the time-of-flight MS (TOF-MS), both already applied as detectors of gas chromatographic analyses (Ragunathan et al. 1999).

The MS-MS, also known as the sequential mass measurement technique, is as a matter of fact a methodology that can be used without any sample preparation, such as extraction, purification or separation. The crude sample, such as a complex mixture, can be directly analysed since the MS-MS system is able to identify individual particular ions which testify the presence of given a component with great accuracy (Mussinan 1993). Nevertheless, the MS-MS system finds several applications as a GC and also as a high performance liquid chromatography (HPLC) detector, mainly in clinical and forensic diagnosis, as well as in environmental

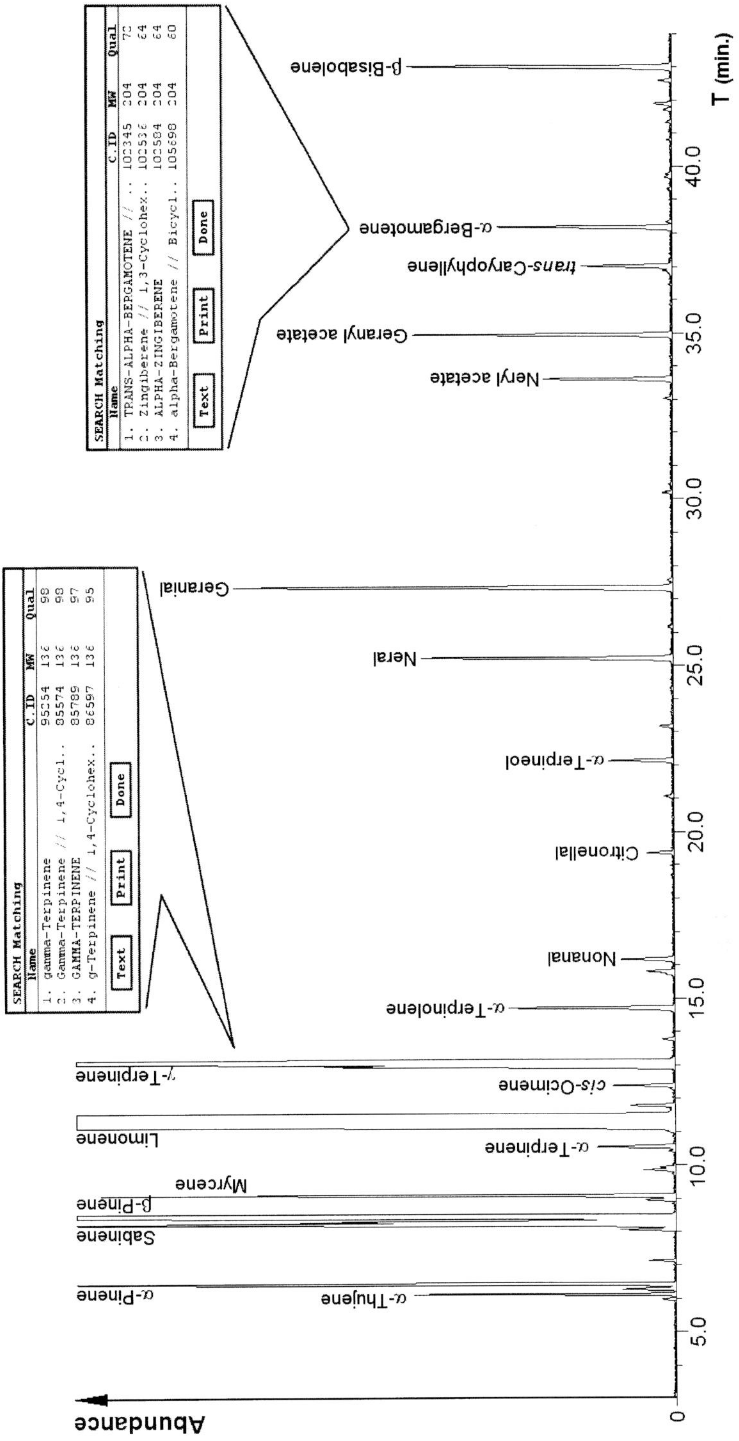

Fig. 7.2. Total ion chromatogram (TIC) of a cold-pressed lemon oil with two windows reporting the matching procedure of two components

analyses owing to its intrinsic high accuracy response (Ragunathan et al. 1999). Until now few applications have been reported in essential oil studies, though one of these is the determination of nootkatone among the volatile components released from a whole orange (Mussinan 1993). Instead, the TOF-MS technique has wider application possibilities as a GC detector (Vreuls et al. 1999). The peculiar characteristic of this mass analyser is its very high-speed detection, which is well suited to the high-speed chromatographic separations. One of the main developments of the GC-MS technique is the drastic shortening of time analyses, keeping the actual sensitivity and reproducibility. These high-speed systems, recently reviewed by Blumberg (1999), Cramers et al. (1999), Ragunathan et al. (1999) and McNair and Reed (2000), are not compatible with the actual scanning mass spectrometers; therefore "high-speed detector, such as TOF mass spectrometer, is necessary to successfully acquire mass spectral data" (Ragunathan et al. 1999).

7.3.3 High Resolution Gas Chromatography–Fourier Transform IR Spectroscopy (HRGC-FTIR)

High resolution gas chromatography, mass spectrometry and retention indexes allow the determination of most compositional details of a *Citrus* essential oil. However, despite the abundance of instrumental tools described above a limit is still present: the characterisation of isomer components. No MS technique is, in fact, able to unambiguously distinguish a mixture of isomers. This is a concern for essential oil analysis, in which the presence of isomers is not a rarity. Most of the classes of components show several isomers, many of which can be fortunately distinguished throughout their retention indexes, but with the sesquiterpenes, both hydrocarbons and oxygenated, the identification is often impossible, owing to the very high number of isomers, the overlapping of retention indexes and the lack of standard compounds.

In these cases a complementary method to MS, such as HRGC detector, is represented by Fourier transform IR spectroscopy (FT-IR). An FT-IR spectrum can be run in solid, liquid and gaseous phase, and gives important information on the molecular structure with particular regard to chemical functionality, being able to distinguish between different isomers. It is not by chance that IR spectroscopy has been for a long time the main spectroscopic technique in organic chemistry, before being superseded by nuclear magnetic resonance (NMR) in the structural elucidation.

In the 1970s the development of FT-IR spectroscopy was propaedeutic for the first on-line capillary GC-FT-IR in 1981 (Smith 1987). Instruments started to spread and, as for the MS, their technical development is still in progress (Ragunathan et al. 1999). The IR spectrum during an HRGC-FT-IR is measured in gaseous phase, but in these conditions it is rather different from the spectrum of the same product obtained in solid or liquid phase, making an opportune comparison impossible, since most of the IR spectra compilations are in the latter format. It would be necessary to build libraries with gaseous IR spectra (Mussinan

1993). However, adopting the cryofocusing GC-FT-IR the aforesaid limit can be overcome. With this technique the individual peaks emerging from the GC column are frozen at very low temperature (–165 °C), and the IR spectrum is run. The spectra obtained in such a way are very similar to the literature spectra measured in solvent or as KBr pellets, and thus easily comparable. Recently this technique has been applied to identify some sesquiterpenes in *Citrus* oils, in particular γ-elemene, bicyclogermacrene and germacrene-B show mass spectra almost superimposable with each other (Fig. 7.3). The corresponding IR spectra of the three components (Fig. 7.4) are, instead, quite different and easily comparable with the reference spectra, making the unambiguous identification of the three sesquiterpenes possible (Chamblee et al. 1997).

In analysing the essential oil literature of these last years, the applications of the HRGC-FT-IR are not so widespread, the MS being the leading detector. This may be due to some existing technical hesitation, the lack of essential-oil-specific IR libraries, and lower detectability limit with respect, for example, to MS. This a pity because IR detection offers enormous potential. Moreover, it is not a destructive technique, and therefore the sample can be used for further analysis. This allows us to link IR and MS detectors to obtain GC-FT-IR-MS instruments already

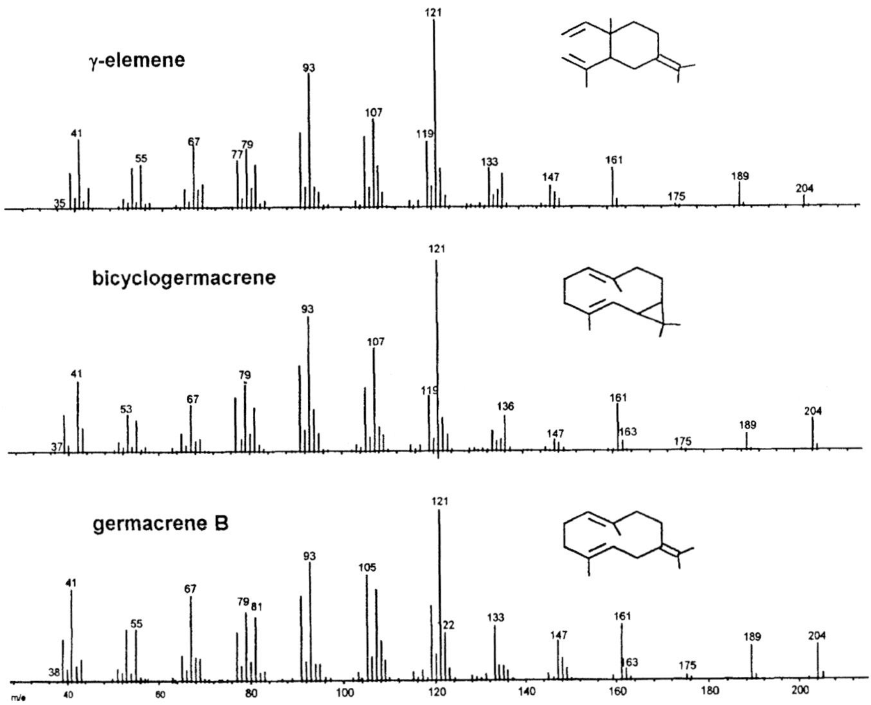

Fig. 7.3. Comparison of mass spectra of γ-elemene, bicyclogermacrene and germacrene B. (From Chamblee et al. 1997, reproduced with permission of Allured Publishing Corp.)

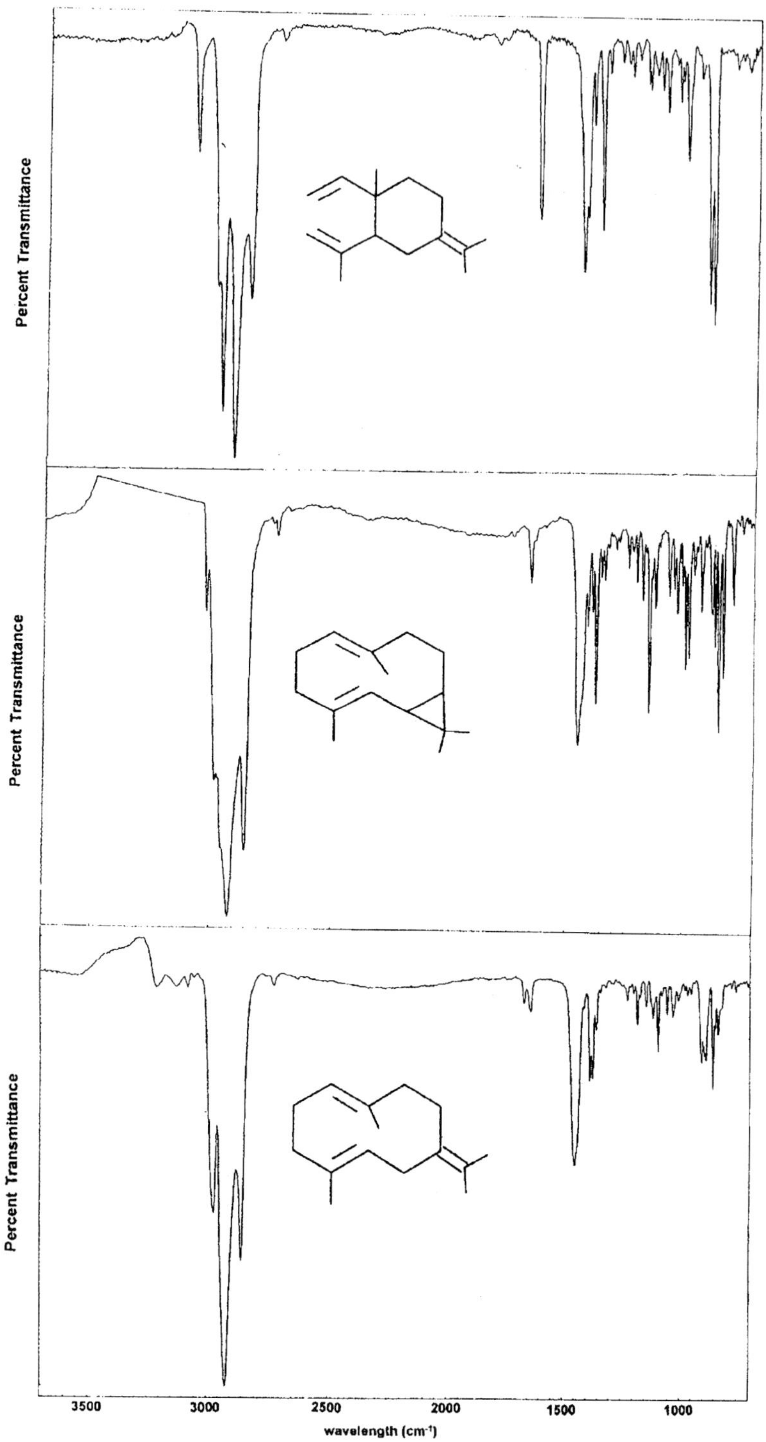
Percent Transmittance
Percent Transmittance
Percent Transmittance
3500
3000
2500
2000
1500
1000
wavelength (cm-1)

commercially available (Ragunatahn et al. 1999). This last kind of sophisticated technique has recently been applied to detect the alteration of some *Citrus* essential oils (Oberhofer et al. 1999).

7.3.4 Liquid Chromatography–High Resolution Gas Chromatography–Mass Spectrometry (LC-HRGC-MS)

The definitive goal of every chromatographic process is to obtain the best separation of a mixture of components. Gas chromatography is probably the best separation technique for many reasons: high separation efficiency (more than ten-fold with respect to HPLC); wide choice of specific and/or universal detectors; large number of stationary phases suitable for several classes of compounds; detectability limit ≤1 ng of product (Vreuls et al. 1994). However, given the complexity of many essential oils, as is the case with *Citrus* oils, gas chromatographic profiles usually present quite considerable overlapping of components, even using different stationary phases. Furthermore, in these cases it will be almost impossible to get information on trace level components eluting in crowded portions of the chromatogram, or to establish the enantiomeric ratio of many components.

New analytical methods have been developed over the last years, precisely to solve these kinds of problems. These methods consist of the combination of liquid chromatography, of the off-line mode as open column or in-line mode as HPLC, with the HRGC; or the coupling of two gas chromatographs, obtaining the so-called multidimensional gas chromatography (MDGC) (Vreuls et al. 1994; Phillips and Beens 1999; Ragunathan et al. 1999; Bertsch 1999, 2000). Many of these methods have been successfully applied in the studies of *Citrus* essential oils (Cotroneo et al. 1986; Bernreuther and Schreier 1991; Mondello et al. 1994, 1995, 1998a–c, 1999; Dugo et al. 1999b).

Off-line liquid chromatography combined with gas chromatography is the simplest hyphenation technique. The essential oil is fractionated on an open allumina or silica gel column eluting with an increasing amount of ethyl ether in hexane. Exploiting the capacity of the liquid chromatography to separate the different classes of components present in the essential oils, the end result will be represented by several fractions each containing selected components; namely hydrocarbons, esters, carbonyl compounds and alcohol, which can be injected into the GC-MS system for their identification. In Fig. 7.5 the result of the application of this methodology for a lemon oil is reported (Cotroneo et al. 1986). The simplification of the total essential oil profile is quite clear, and thus the easier identification of components.

The combination of liquid and gas chromatography has received a significant improvement with the assembly of the on-line instruments, which have allowed the use of HPLC, with complete automated control. As in the off-line system, the oil is previously analysed by HPLC normally using a UV or a photodiode array

◀

Fig. 7.4. GC-FT-IR spectra of γ-elemene (*above*), bicyclogermacrene (*middle*) and germacrene B (*below*). (Chamblee et al. 1997, reproduced with permission of Allured Publishing Corp.)

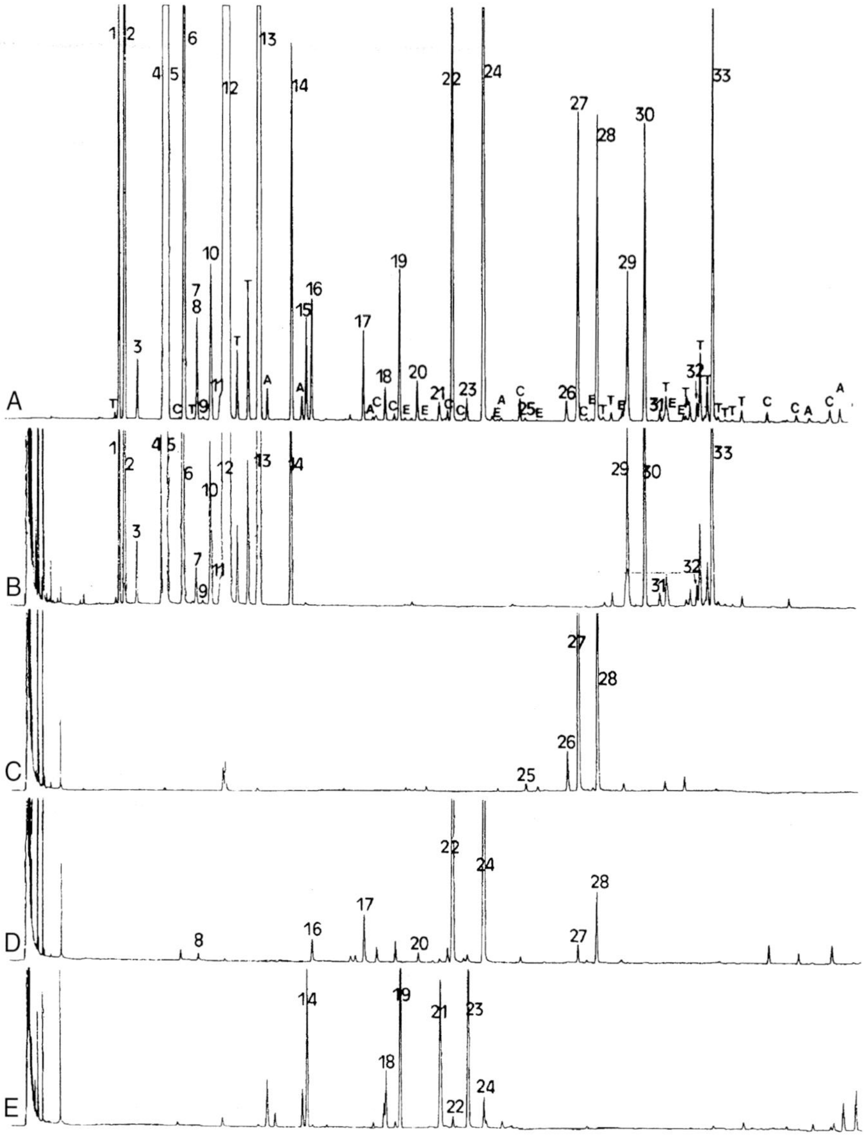

Fig. 7.5. Off-line LC-HRGC of lemon oil. **A** GC-FID of total essential oil; **B** terpene fraction; **C** esters; **D** carbonyl compounds; **E** alcohols. List of components: *1* α-thujene; *2* α-pinene; *3* camphene; *4* sabinene; *5* β-pinene; *6* myrcene; *7* α-phellandrene; *8* octanal; *9* Δ^3-carene; *10* α-terpinene; *11* *p*-cymene; *12* limonene; *13* γ-terpinene; *14* terpinolene; *15* linalool; *16* nonanal; *17* citronellal; *18* terpinen-4-ol; *19* α-terpineol; *20* decanal; *21* nerol + citronellol; *22* neral; *23* geraniol; *24* geranial; *25* nonyl acetate; *26* citronellyl acetate; *27* neryl acetate; *28* geranyl acetate; *29* caryophyllene; *30* bergamotene; *31* humulene; *32* valencene; *33* bisabolene; *T* unidentified mono- and sesquiterpenes; *E* unidentified esters; *C* unidentified carbonyl compounds; *A* unidentified alcohols. (Cotroneo et al. 1986, reproduced with permission of John Wiley and Sons Ltd.)

detector. On the basis of the HPLC chromatogram profile particular fractions can be selected and injected, by a transfer line, into the gas chromatograph (Mondello et al. 1994, 1995; Vreuls et al. 1994). The main problem of the HPLC-GC hyphenation has been due to the solvent which elutes from the HPLC column together with the component(s). It must be necessarily and selectively removed before introducing the component(s) into the gas chromatograph. The overcoming of this problem has been achieved by the introduction of particular transfer lines, which allow the evaporation of the solvent and the introduction of LC fractions in the GC. The most common transfer line devices are the fully concurrent solvent evaporation (FCSE) or the partially concurrent solvent evaporation (PCSE) and the retention gap system (Vreuls et al. 1994). In conclusion, with the on-line LC-GC system it is possible to carry out a more selective transfer with respect to the off-line system, particularly by exploiting the higher separation efficiency of the HPLC system. In Fig. 7.6 the separation of the three kinds of aldehydes (aliphatic, monoterpene and sesquiterpene) present in a sweet orange oil is shown (Mondello et al. 1994).

7.3.5 Multidimensional Gas Chromatography (MDGC)

Another method addressed at the improvement of the resolving capacity of a single gas chromatographic column is multidimensional gas chromatography (MDGC), also named GC-GC, in which two columns are connected to each other through an opportune transfer line (David and Sandra 1987; Beens et al. 1998; Bertsch 1999, 2000; Phillips and Beens 1999; Ragunathan et al. 1999). This assembly allows the transfer of fractions and/or singular components from the first to the second column. Depending on the kind of analysis, the first column, also called the pre-column, can be a packed or wide-bore column, but also a high-resolution capillary column; the second column, or analytical column, is always a high-resolution column, with a different polarity than the first column, or containing a chiral stationary phase for enatiomer resolution (Juvancz and Petersson 1996). Figure 7.7 shows a scheme of a typical GC-GC apparatus developed by Mondello et al. (1998b), in which the central switching valve, the heart of the system, is evidenced. Furthermore, from the figure it is clear that the two GC systems can also be used singularly.

The basic principle of GC-GC or MDGC is the so-called heart-cutting, by which selected fractions of the first column are transferred to the second one, in a completely automated way. The main applications of this technique are to enhance the separation efficiency by using two columns with different polarity, to evidence trace components by repeated cuts of several injections and to establish the presence of unwanted solvents in essential oil mixtures (Mussinan 1993). However, one of the most interesting applications of GC-GC is the determination of the enantiomeric distribution of essential oil constituents, owing to the impossibility of carrying out such analysis directly on the complex essential oil. *Citrus* essential oils have been examined extensively for the enatiomeric distribution of several

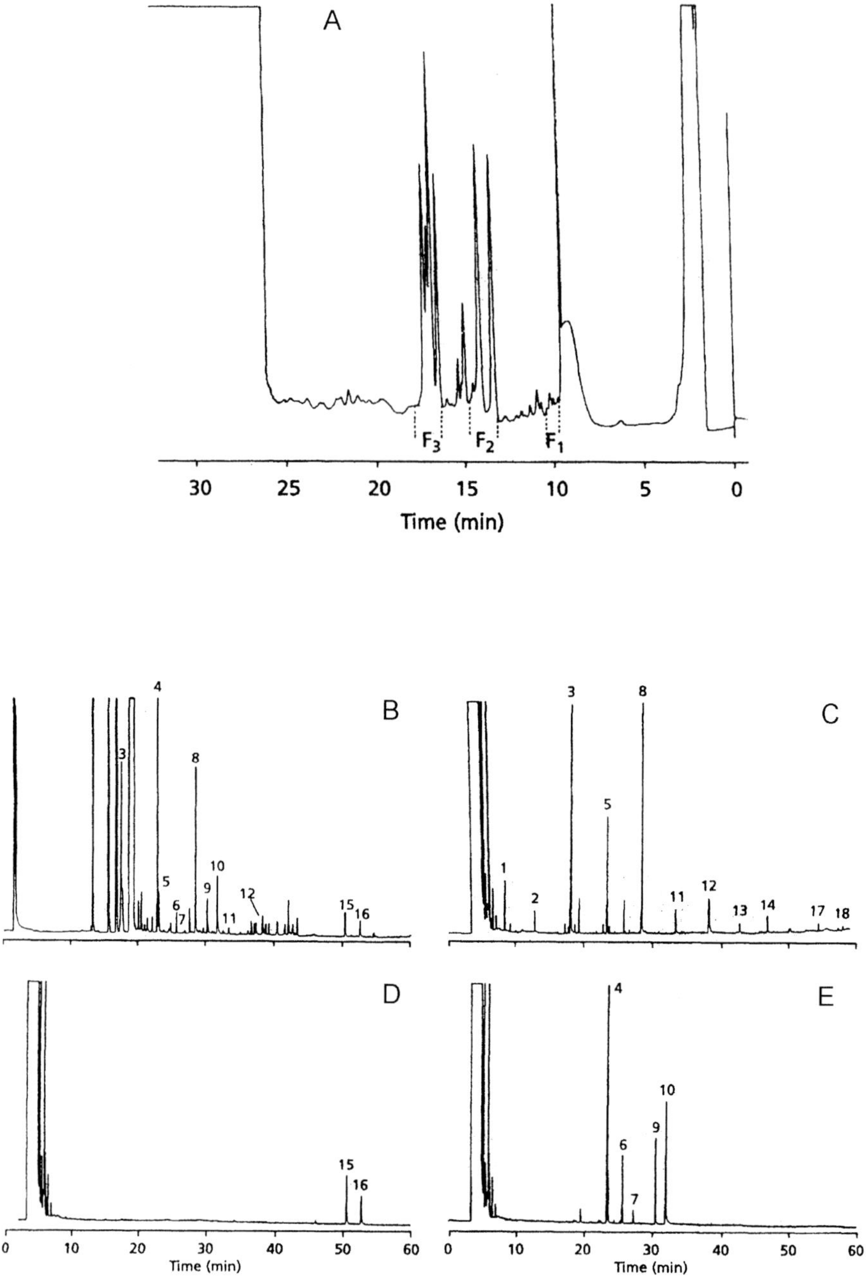

Fig. 7.6. On-line HPLC-HRGC of sweet orange oil. **A** HPLC chromatogram of sweet orange oil in which the fractions (*F1*, *F2*, *F3*) to be transferred to GC analyses are evidenced; **B** GC chromatogram of total sweet orange oil; **C** GC profile of aliphatic aldehydes fraction (F1); **D** GC profile of sesquiterpene aldehydes fraction (F2); **E** GC profile of monoterpene aldehydes (F3). List of components: *1* hexanal; *2* heptanal; *3* octanal; *4* linalool; *5* nonanal; *6* citronellal; *7* terpinen-4-ol; *8* decanal; *9* neral; *10* geranial; *11* undecanal; *12* dodecanal; *13* tridecanal; *14* tetradecanal; *15* β-sinensal; *16* α-sinensal; *17* hexadecanal; *18* heptadecanal. (Mondello et al. 1994, reproduced with permission of Wiley-VCH Verlag GmbH)

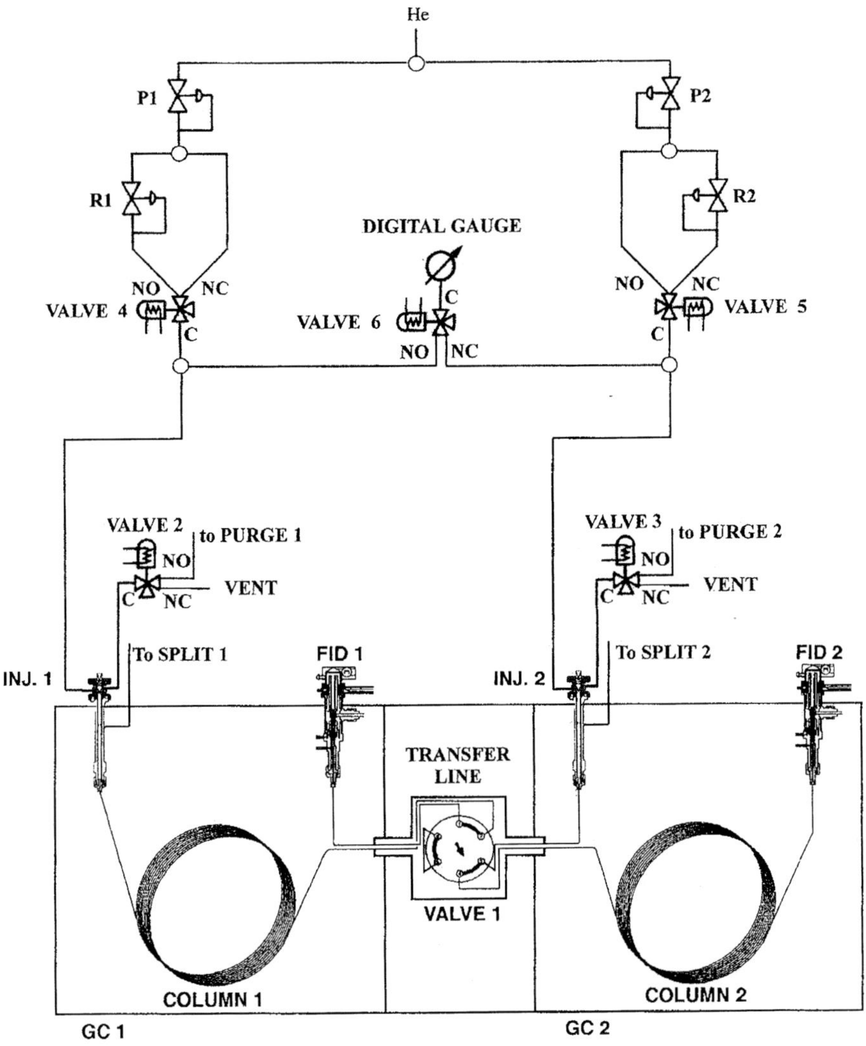

Fig. 7.7. Scheme of a GC–GC system. (Mondello et al. 1998b, with permission of the American Chemical Society)

components in order to provide information on adulterations and in general to establish their quality (Mondello et al. 1998a–c). Figure 7.8 shows the application of GC-GC analysis for the enantiomeric distribution of some components of cold-pressed bergamot oil. In this case the enantiomeric ratios of β-pinene, sabinene, limonene, linalool, terpinen-4-ol, α-terpineol and linalyl acetate have been determined by a single analysis.

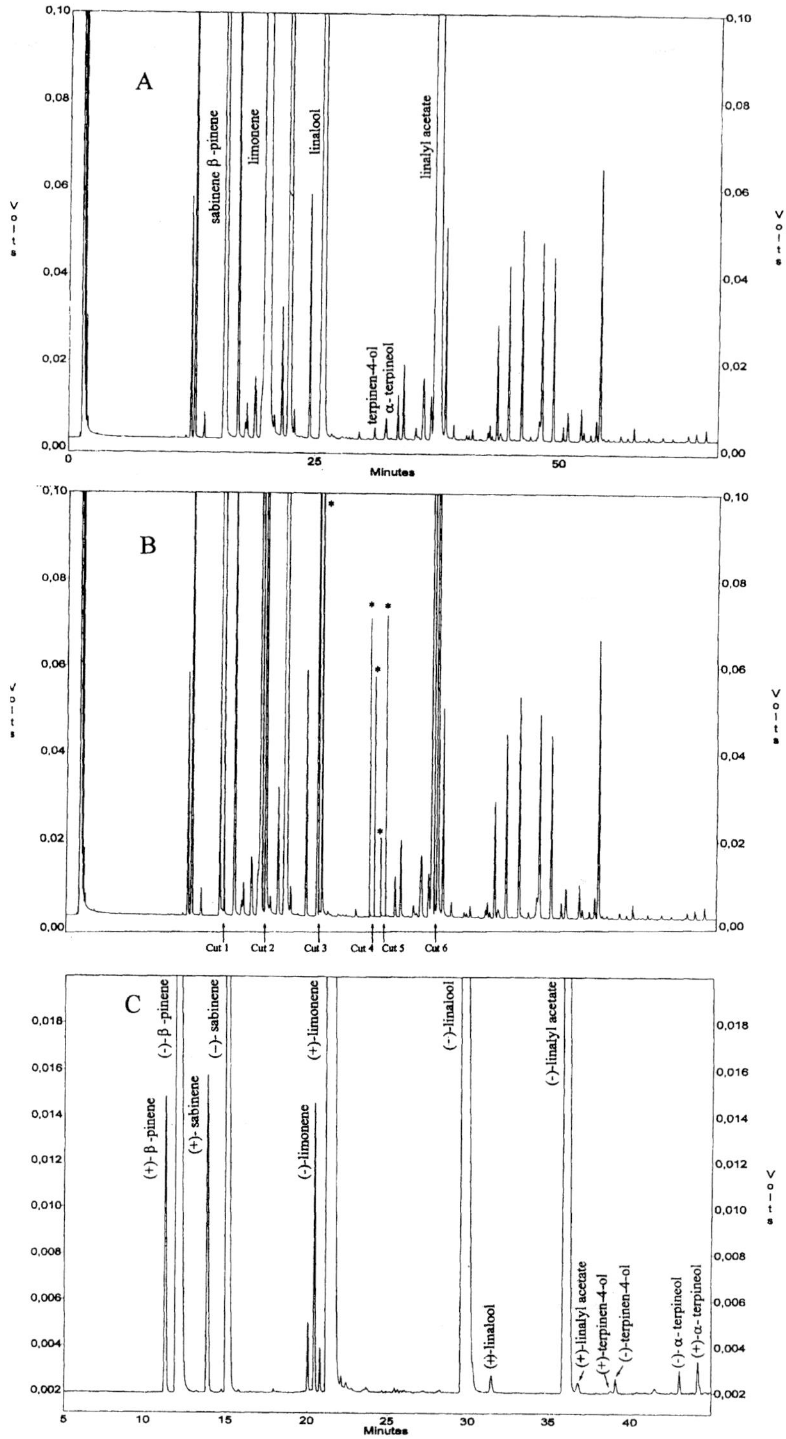

Fig. 7.8. Gas chromatogram of a cold-pressed bergamot oil (**A**), GC with the six heart cuts (**B**), and GC–GC chiral detection of the transferred components (**C**). *Stars* in **B** show the electric spike coming from the valve switching. (Mondello et al. 1998b, with permission of the American Chemical Society)

7.4 Deterpenation of *Citrus* Essential Oils

As recently reported, the flavours and fragrances market is experiencing a continuing expansion (Shamel and Udis-Kessler 2000). Their main end-uses are in beverages, concerning the food sector, and in soaps, detergents and perfumes, concerning the cosmetic industry. *Citrus* essential oils finds an optimal and in many cases a pre-eminent placing in both fields. Normally a *Citrus* oil is used without any further purification; however, over the last years a growing request for "deterpenated" *Citrus* oils has arisen in the most important application sectors. Deterpenation, also known as folding, means the selective removal of terpene hydrocarbons to obtain a cold-pressed oil rich in oxygenated components, which, as previously mentioned, are mainly responsible for the typical aroma of the different *Citrus* species and cultivars. Nevertheless, the terpene hydrocarbons, present in large amount in *Citrus* oils (60–98%), scarcely contribute to the *Citrus* oil aromas; furthermore, among the different chemical classes of components present in the oils, they are the more unstable ones, being subjected to various degradation phenomena with formation of several off-flavours. Therefore the deterpenation procedure, beyond enhancing the *Citrus* flavour, improves the oil stability, and makes the oil more soluble in water and alcohols. Actually the deterpenation is carried out by vacuum distillation, chromatographic separation and solvent extraction. Despite the care spent in these procedures, poor results are often obtained owing to scarce yield, formation of artefacts and presence of residual solvents.

An alternative procedure is thus desirable, which may be represented by supercritical fluid extraction (SFE) using carbon dioxide as supercritical fluid (Modey et al. 1996). Manifold advantages are offered by supercritical carbon dioxide extraction: the critical point (T = 31.05 °C, P = 73.3 bar) is easy to reach and maintain with respect to the critical points of many other substances; furthermore, CO_2 is a GRAS (generally regarded as safe) substance, it is neither combustible nor explosive, and is easily available and cheap. Also, being a gas in normal conditions it escapes from the extract without leaving any trace, is less toxic than most of the traditional solvent and, finally, since commercial CO_2 is a by-product of the fermentation process its use does not affect the amount present in the atmosphere, thus it does not contribute to the "greenhouse effect" (Moyler 1993).

From an operative point of view the supercritical CO_2 extraction is an extremely modulable procedure; in fact acting on temperature and pressure the solvent power can be opportunely modified, therefore selecting the extract composition. Moreover, the solvent power of supercritical CO_2 can be enhanced by the use of traditional co-solvents (or entrainer), even if this partially annuls one of the advantages of SFE, that is the absolute absence in the extract of any solvent trace. Figure 7.9 shows the pressure-temperature (PT) phase diagram for carbon dioxide with the operative zones.

The most important and historic application of supercritical CO_2 extraction is the decaffeination of coffee, and, successively, of tea. However, many other

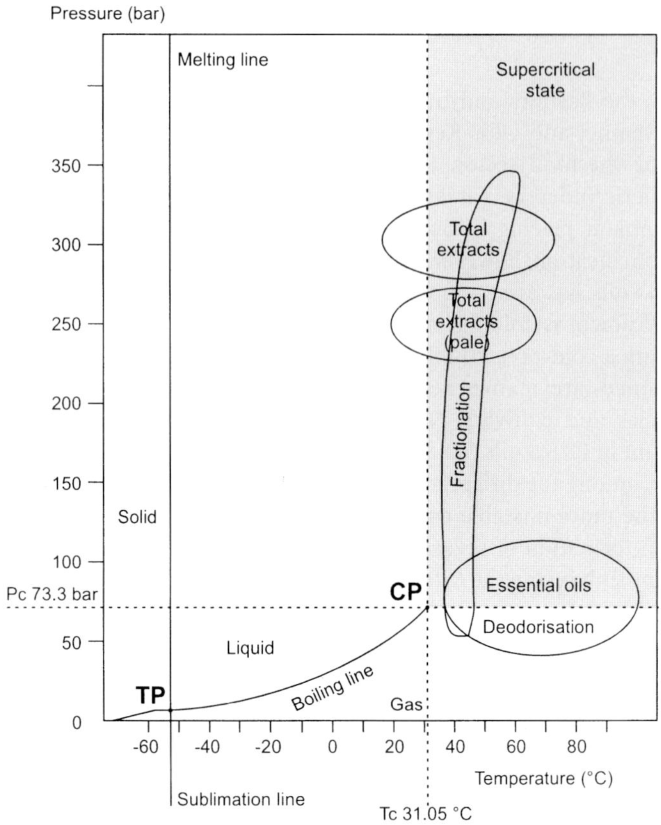

Fig. 7.9. Pressure–temperature (PT) phase diagram for carbon dioxide with the extraction and fractionation zones

natural extracts are becoming commercially available, especially the essential oils, despite the unique but significant limit of this procedure: the high cost of the industrial plants.

The success of this extraction procedure is mainly due to the particularly soft conditions, which give extracts endowed with the most naturalness. This is especially appreciated in the essential oils obtained, normally, by steam distillation which causes various degrees of degradation (Moyler 1993; Modey et al. 1996).

Concerning *Citrus* peel essential oils the SFE does not set itself as an alternative procedure to the actual industrial processes, which are quite cheap and soft (not denaturing). Some studies show that the SFE of *Citrus* peels gives higher yields of essential oils, whose contents are slightly different from cold-pressed-oils, namely a lower amount of monoterpene hydrocarbons (mainly less limonene) and a higher presence of oxygenated components (Poiana et al. 1994, 1998a,b, 1999).

Table 7.2. Amount (in %) of components of cold-pressed bigarade and lemon oils and their fraction produced by desorption with supercritical carbon dioxide

Class of components/ Components	Bigarade oil[a]				Lemon oil[b]		
	Cold-pressed oil	F1[c]	F2[d]	Residue	Cold-pressed oil	F1[c]	F2[d]
Terpene hydrocarbons	**91.74**	**99.39**	**49.65**	**4.20**	**93.23**	**99.34**	**19.94**
Limonene	82.12	88.79	32.04	1.16	60.46	65.60	13.16
Oxygenated terpenes	**5.73**	**0.48**	**47.63**	**27.91**	**5.06**	**0.02**	**62.56**
Linalool	0.52	0.10	4.70	1.40	0.14	0.01	2.62
Terpinen-4-ol	0.15	0.02	1.29	0.40	0.06	0.01	3.54
α-Terpineol	0.11	–	3.08	0.85	0.22	–	0.80
trans-Carveol	0.62	–	6.92	2.99			
cis-Carveol	0.34	–	5.46	2.21			
Carvone	1.21	–	2.96	2.24			
Linalyl acetate	0.68	0.15	6.09	2.06			
Neral					1.20	–	10.05
Geranial					1.95	–	22.21
Neryl acetate	–	–	0.33	–	0.61	–	10.12
Geranyl acetate	0.13	–	2.75	1.09	0.66	–	11.96
Aliphatic aldehydes	**0.22**	**0.05**	**1.21**	**0.10**	**0.30**	–	**8.02**
Octanal	–	–	0.19	–	0.10	–	4.35
Nonanal	0.07	0.05	0.14	0.10	0.13	–	2.30
Non-volatile constituents	–	–	–	**64.90**			
Bergapten	–	–	–	2.16			
Suberosin	–	–	–	5.20			
Auraptenol	–	–	–	2.81			
Isomeranzin	–	–	–	23.47			
Osthol	–	–	–	4.24			
Meranzin hydrate	–	–	–	6.42			
Meranzin	–	–	–	14.27			
Tangeretin	–	–	–	2.75			

[a] Adapted from Chouci et al. (1996)
[b] Adapted from Dugo et al. (1995)
[c] Terpenic fraction
[d] Deterpenated fraction

The deterpenation or folding of *Citrus* peel oils could be, instead, carried out by using supercritical CO_2 selective extraction. Studies performed in this field indicate that the best procedure is the initial adsorption of the cold-pressed oil on an inert substrate (silica-gel seems to give the best results), and successive desorption of the components by flushing supercritical CO_2. The temperature of the entire process does not exceed 40 °C, while the pressure involved can be between 75 and 120 bar. A further parameter is represented by the desorption times. Normally, varying opportunely pressure values and desorption times, three fractions are obtained, a terpene fraction, a deterpenated fraction and a residue (Dugo et al. 1995; Chouchi et al. 1996; Budich et al. 1999). Table 7.2 lists some selected values of the aforesaid fractions. The content of oxygenated constituents in the deterpenated fraction is very high, if compared to the similar oil obtained by distillation. This entails a very high qualitative level of these oils, as well as an unmistakable naturalness.

As has been mentioned many times, cold-pressed *Citrus* peel oils contain a non-volatile portion (1–15%), mainly constituting oxygen heterocyclic compounds. During the fractionation with supercritical CO_2, these components, owing to their molecular features, are not extracted and form the residue of the extraction process. This represents a further advantage of the SFE procedure since it allows us to selectively remove the potentially toxic components from a *Citrus* essential oil, e.g. the bergapten content in the bergamot oil obtained by SFE is six times lower than in the cold-pressed oil (Poiana et al. 1994). Moreover, the almost exclusive presence of the non-volatile components in the residue makes their analysis easier, especially for those present at trace levels which cannot be detected in the crude oils.

7.5 Novel *Citrus* Fruits

I would like to conclude this review on the analytical studies of *Citrus* peel essential oils with some considerations regarding the exploitation of these oils in the *Citrus* genetic improvement field.

Citrus fruits represent one of the most important crops, influencing, similarly to the other common crops, the social-economic aspects of large regions of the main producer countries. This also promotes a widespread research aimed at the improvement of extant species and/or development of new fruits. The main targets of such studies, which involve geneticists, biologists and chemists, concern the increase of quality and quantity of production, best adaptability to different environmental conditions and improved resistance to several diseases, as well as the production of seedless fruits with optimal size and shape, easily removable peel, new and original organoleptic characteristics, and possibly precocious or late ripening.

From a biological point of view *Citrus* plants are highly heterozygous species and present a wide range of variation and different modalities of vegetative reproduction (Scora and Kumamoto 1983; Ruberto et al. 1999a). This wide biological diversification makes the taxonomy of the *Citrus* genus quite complex; in fact more than a dozen different classifications are reported in the literature (Scora et al. 1969). However, the most widely accepted classification is the one proposed by Swingle and Reece (1967), with some successively introduced modifications, which lists 16 species in the genus *Citrus* (Scora and Kumamoto 1983). At the same time the aforesaid biological differentiation gives the breeder many opportunities to carry out programs aimed at genetic improvement. The latter can be developed according to traditional methodologies such as the selection of natural or induced mutations, nucellar selection, hybridisation, or by recent biotechnological methods such as in vitro tissue culture, somaclonal mutations, somatic hybridisation via protoplast fusion, and gene introduction (Rousseff et al. 1994; Ruberto et al. 1999a).

This kind of research on *Citrus* plants, as well as on other fruit-bearing species, is rather complex since the results, positive or negative, can require several years, unlike similar studies on herbaceous plants. Notwithstanding this limit, the progress achieved in this field offers growers a large number of cultivable varieties, which allow us to lengthen their harvest-period, to expand and to diversify production range. This, of course, positively affects the consumer's choice, who now finds a more variegated number of high-quality fruits in the market.

A *Citrus* peel essential oil is specific for an individual *Citrus* taxa representing a sort of fingerprint for a given plant species. With this in mind we are carrying out a systematic study of the essential oils of new *Citrus* fruits whose aim is the characterisation of the essential oils of the fruits as a qualitative parameter of their quality and the evaluation of using these essential oil profiles as "fingerprints" in the prosecution of the hybridisation project and the identification of new flavour components or new fragrances for industrial uses.

In co-operation with the Istituto Sperimentale per l'Agrumicoltura (ISA, Acireale) we are evaluating the essential oil composition of new triploid hybrids. The procedure entails the cross-breeding between a diploid female parent and a tetraploid male parent whose result is the production of seedless triploid hybrids ($2n = 3x = 27$), this being due to the unbalanced chromosomic order which makes these fruits totally sterile (Esen and Soost 1973; Starrantino 1992).

From a general point of view the essential oil composition of a hybrid is the superimposition of that of its parents. The principal differences between hybrid–parent oils are quantitative, namely the different amounts of one or more components from one or both parents. Furthermore, the absence of some components, present in one or both parents, or, in turn, the presence in the hybrid oil of some components not detected in both parents, occasionally occur (Ruberto et al. 1999b). All these differences are to be ascribed to the minor components, monoterpene hydrocarbons (i.e. limonene) being the main ones. However, as has been previously mentioned, precisely these minor components, and mainly the oxygenated ones, give the particular aroma to each *Citrus* oil, and therefore the hybrids often have original organoleptic characteristics, in some cases better than both parents (Soost and Cameron 1980, 1985; Moshonas et al. 1991).

Most of the new *Citrus* fruits have been produced by the aforesaid procedure and the chemical composition of their essential oils has been analysed (Ruberto et al. 1993, 1994, 1997a,b, 1999a,b). Surely the best new fruits are those coming from a blood orange (*C. sinensis* cvv. Tarocco, Moro and Sanguinello) used as tetraploid male parent, and clementine (*C. clementina*) or mandarin (*C. reticulata*) as diploid female parent. Two of the new hybrids, named Tacle and Clara, coming from the cross between the Monreal clementine and Tarocco orange have optimal characteristics, the first one already being cultivated. In Fig. 7.10, as an example of the comparative analysis of the essential oils of new fruits and parents, partial

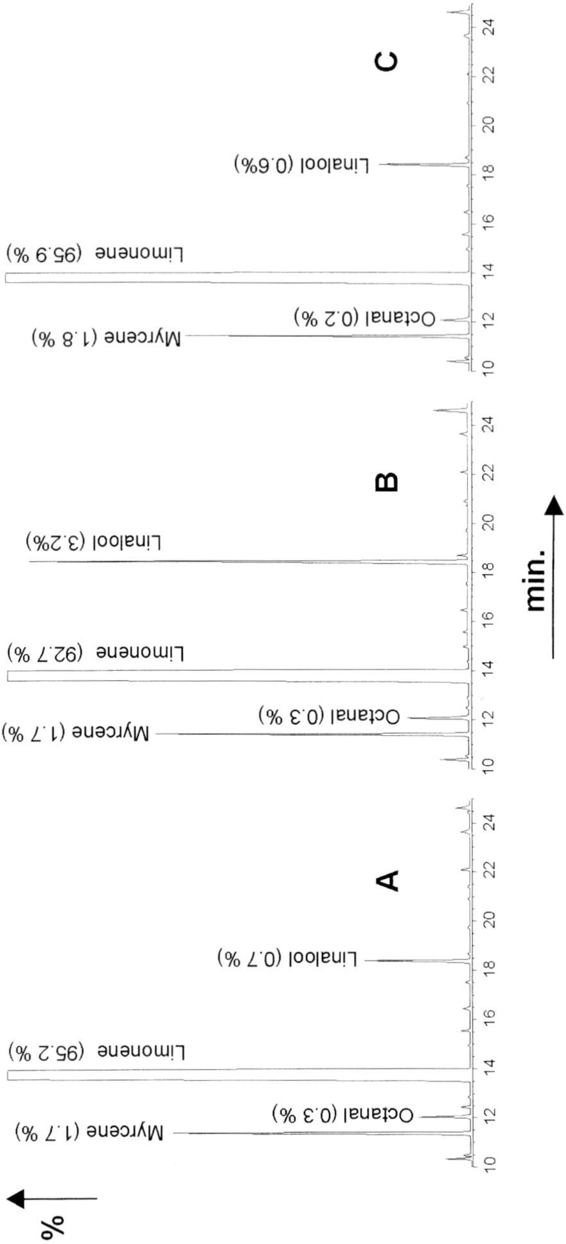

Fig. 7.10. Selected GC-FID profiles of Monreal clementine (A), Clara hybrid (B) and Tarocco orange (C) essential oils

portions of the gas chromatographic profiles of Clara, clementine and orange essential oils are reported. The content of linalool in the Clara oil is much higher than both parents and this is one of the main reasons for the valuable and appreciated characteristics of the new fruit.

Finally, recent results on the studies of the essential oil composition of new *Citrus* hybrids, obtained by somatic hybridisation (Fatta del Bosco et al. 1998; Alonzo et al. 2000), further confirm the undeniable role that a *Citrus* essential oil analysis can assume in disclosing some still unknown aspects of *Citrus* gene expression.

Acknowledgements. The author wishes to express his gratitude to colleagues involved in *Citrus* new hybrid essential oils research cited in the reference list. Special thanks are due to Mr. A. Renda for graphical artworks and to Dr. P. Rapisarda for critical revision of the manuscript and for his suggestions. Support from the Consiglio Nazionale delle Ricerche (CNR, Rome) is gratefully acknowledged.

References

Adams RP (1995) Identification of essential oil components by gas chromatography mass spectrometry. Allured Publishing Corporation, Carol Stream, Illinois

Alonzo G, Fatta Del Bosco S, Palazzolo E, Saiano F, Tusa N (2000) *Citrus* hybrid leaf essential oil. Flavour Fragr J 15:91–95

Bauer K, Garbe D (1985) Common fragrance and flavor materials. Preparation, properties and uses. VCH Verlagsgesellschaft, Weinheim

Beens J, Boelens H, Tijssen R, Blomberg J (1998) Quantitative aspects of comprehensive two-dimensional gas chromatography (GC × GC). J High Resol Chromatogr 21:47–54

Bernreuther A, Schreier P (1991) Multidimensional gas chromatography/mass spectrometry: a powerful tool for the direct chiral evaluation of aroma compounds in plant tissues. II. Linalool in essential oils and fruits. Phytochem Anal 2:167–170

Bertsch W (1999) Two-dimensional gas chromatography. Concepts, instrumentation, and applications, part 1. Fundamentals, conventional two-dimensional gas chromatography, selected applications. J High Resol Chromatogr 22:647–665

Bertsch W (2000) Two-dimensional gas chromatography. Concepts, instrumentation, and applications, part 2: comprehensive two-dimensional gas chromatography. J High Resol Chromatogr 23:167–181

Blumberg LM (1999) Theory of fast capillary gas chromatography, part 4: column performance vs. liquid film thickness. J High Resol Chromatogr 22:501:508

Bonaccorsi IL, McNair HM, Brunner LA, Dugo P, Dugo G (1999) Fast HPLC for the analysis of oxygen heterocyclic compounds of *Citrus* essential oils. J Agric Food Chem 47:4237–4239

Bracke ME, Vermeulen SJ, Bruyneel EA, Vennekens K, Mareel MM (1996) The *Citrus* flavonoid tangeretin inhibits invasion of human breast cancer cells. In: Finley JW, Armstrong DJ, Nagy S, Robinson SF (eds) Hypernutritious foods. Agscience, Auburndale, FL, pp 205–217

Budich M, Heilig S, Wesse T, Leibküchler V, Brunner G (1999) Countercurrent deterpenation of *Citrus* oils with supercritical CO_2. J Supercrit Fluids 14:104–114

Buettner A, Schieberle P (1999) Characterisation of the most odor-active volatiles in fresh, hand-squeezed juice of grapefruit (*Citrus paradisi* MacFayden). J Agric Food Chem 47: 5189–5193

Caccioni RL, Guizzardi M, Biondi DM, Renda A, Ruberto G (1998) Relationship between volatile components of *Citrus* fruit essential oils and antimicrobial action on *Penicillium digitatum* and *Penicillium italicum*. Int J Food Microbiol 43:73–79

Calabrese F (1990) La favolosa storia degli agrumi. Agricoltura 208:81–128

Chamblee TS, Karelitz RL, Radford T, Clark BC Jr (1997) Identification of sesquiterpenes in *Citrus* essential oils by cryofocusing GC/FT-IR. J Essential Oil Res 9:127–132

Chouchi D, Barth D, Reverchon E, Della Porta G (1996) Bigarade peel oil fractionation by supercritical carbon dioxide desorption. J Agric Food Chem 44:1100–1104

Cotroneo A, Verzera A, Lamonica G, Dugo G (1986) On the genuineness of *Citrus* essential oils, part 10: research on the composition of essential oils produced from Sicilian lemons using "pelatrice" and "sfumatrice" extraction during the entire 1983/84 production season. Flavour Fragr J 1:69–86

Cramers CA, Janssen H-G, van Deursen MM, Leclercq PA (1999) High-speed gas chromatography: an overview of various concepts. J Chromatogr A 856:315–329

David F, Sandra P (1987) Possibilities of multidimensional gas chromatography in essential oil analysis. In: Sandra P, Bicchi C (eds) Capillary gas chromatography in essential oil analysis. Huethig, Heidelberg, pp 387–429

Di Giacomo A, Mincione B (1994) Gli olii essenziali agrumari in Italia. Laruffa Editore, Reggio Calabria

Dugo G, Bartle KD, Bonaccorsi I, Catalfamo M, Cotroneo A, Dugo P, Lamonica G, McNair H, Mondello L, Previti P, Stagno d'Alcontres I, Trozzi A, Verzera A (1999a) Advanced analytical techniques for the analysis of *Citrus* essential oils, part 1: volatile fraction: HRGC/MS Analysis. Essenze Deriv Agrum 69:79–111

Dugo G, Bartle KD, Bonaccorsi I, Catalfamo M, Cotroneo A, Dugo P, Lamonica G, McNair H, Mondello L, Previti P, Stagno d'Alcontres I, Trozzi A, Verzera A (1999b) Advanced analytical techniques for the analysis of *Citrus* essential oils, part 2: volatile fraction: LC-HRGC and MDGC. Essenze Deriv Agrum 69:159–217

Dugo G, Bartle KD, Bonaccorsi I, Catalfamo M, Cotroneo A, Dugo P, Lamonica G, McNair H, Mondello L, Previti P, Stagno d'Alcontres I, Trozzi A, Verzera A (1999c) Advanced analytical techniques for the analysis of *Citrus* essential oils, part 3. Oxygen heterocyclic compounds: HPLC, HPLC/MS, OPLC, SFC, Fast HPLC analysis. Essenze Deriv Agrum 69:251–283

Dugo P, Mondello L, Bartle KD, Clifford A, Breen DGPA, Dugo G (1995) Deterpenation of sweet orange and lemon essential oils with supercritical carbon dioxide using silica gel as an adsorbent. Flavour Fragr J 10:51–58

Esen A, Soost RK (1973) Seed development in *Citrus* with special reference to 2X × 4X crosses. Am J Bot 60:448–462

Fatta Del Bosco S, Palazzolo E, Scarano MT, Germanà MA, Tusa N (1998) Comparison between essential oil yield and constituents of an allotetraploid somatic hybrid of *Citrus* and its parents. Adv Hort Sci 12:72–77

Food and Agriculture Organisation (2000) *Citrus* fruit. Fresh and processed – annual statistics. http://www.fao.org

Forlot P (2000) Cosmetic and cosmeceutic uses of bergamot oil. Essenze Deriv Agrum 70:2–7

Gordon MA (1996) Dietary antioxidants in disease prevention. Nat Prod Rep 13:265–273

Hinterholzer A, Schieberle P (1999) Identification of the most odour-active volatiles in fresh, hand-extracted juice of grapefruit of Valencia Late orange by odour dilution techniques. Flavour Fragr J 13:49–55

Jennings W, Shibamoto T (1980) Qualitative analysis of flavor and fragrance volatiles by glass capillary gas chromatography. Academic Press, New York

Juvancz Z, Petersson P (1996) Enantioselective gas chromatography. J Microcolumn Sep 8:99–114

Kawaii S, Tomono Y, Katase E, Ogawa K, Yano M (1999a) Quantitation of flavonoid constituents in *Citrus* fruits. J Agric Food Chem 47:3565–3571

Kawaii S, Tomono Y, Katase E, Ogawa K, Yano M (1999b) Isolation of furocoumarins from fruits as HL-60 differentiation-inducing compounds. J Agric Food Chem 47:4073–4078

Kovats E (1958) Gas chromatographic characterisation of organic compounds. I. Retention indexes of aliphatic halides, alcohols, aldehydes and ketones. Helv Chim Acta 41:1915–1932

Lawrence BM (1995a) Essential oils 1992–1994. Allured Publishing Corporation, Carol Stream, Illinois, pp 187–204

Lawrence BM (1995b) Essential oils 1992–1994. Allured Publishing Corporation, Carol Stream, Illinois, pp 107–126

McHale D, Sheridan JB (1989) The oxygen heterocyclic compounds of *Citrus* peel oils. J Essential Oil Res 1:139–149

McNair HM, Reed GL (2000) Fast gas chromatography: the effect of temperature programming. J Microcolumn Sep 12:351–355

McReynolds WO (1966) Gas chromatographic retention data. Preston Technical Abstracts Company, Evanston, Illinois

Merken HM, Beecher GR (2000) Measurement of food flavonoids by high-performance liquid chromatography: a review. J Agric Food Chem 48:577–599

Miyake Y, Murakami A, Sugiyama Y, Isobe M, Koshimizu K, Ohigashi H (1999) Identification of coumarins from lemon fruit (*Citrus limon*) as inhibitors of in vitro tumor promotion and superoxide and nitric oxide generation. J Agric Food Chem 47:3151–3157

Modey WK, Mulholland DA, Raynor MW (1996) Supercritical fluid extraction of natural products. Phytochem Anal 7:1–15

Mondello L, Bartle KD, Dugo G, Dugo P (1994) Automated HPLC-HRGC: a powerful method for essential oil analysis, part III: aliphatic and terpene aldehydes of orange oil. J High Resol Chromatogr 17:312–314

Mondello L, Dugo P, Bartle KD, Dugo G, Cotroneo A (1995) Automated HPLC-HRGC: a powerful method for essential oils analysis, part V: identification of terpene hydrocarbons of bergamot, lemon, mandarin, sweet orange, bitter orange, grapefruit, clementine and Mexican lime oils by coupled HPLC-HRGC-MS (ITD). Flavour Fragr J 10:33–42

Mondello L, Dugo G, Dugo P, Bartle KD (1996) Italian *Citrus* petitgrain oils. Part I. Composition of bitter orange petitgrain oil. J Essent Oil Res 8:597–609

Mondello L, Basile A, Previti P, Dugo G (1997a) Italian *Citrus* petitgrain oils, part II: composition of mandarin petitgrain oil. J Essential Oil Res 9:255–266

Mondello L, Cotroneo A, Stagno d'Alcontres I, Dugo G (1997b) Italian *Citrus* petitgrain oils, part III: composition of sweet orange petitgrain oil. J Essential Oil Res 9:379–392

Mondello L, Cotroneo A, Dugo G, Dugo P (1997c) Italian *Citrus* petitgrain oils, part IV: composition of lemon petitgrain oil. J Essential Oil Res 9:495–508

Mondello L, Catalfamo M, Proteggente AR, Bonaccorsi I, Dugo G (1998a) Multidimensional capillary GC-GC system for the analysis of real complex samples. 3. Enantiomeric distribution of monoterpene hydrocarbons and monoterpene alcohols of mandarin oils. J Agric Food Chem 46:54–61

Mondello L, Verzera A, Previti P, Cripso F, Dugo G (1998b) Multidimensional capillary GC-GC system for the analysis of complex samples. 5. Enantiomeric distribution of monoterpene hydrocarbons, monoterpene alcohols, and linalyl acetate of bergamot (*Citrus bergamia* Risso et Poiteau) oils. J Agric Food Chem 46:4275–4282

Mondello L, Catalfamo M, Dugo G, Dugo P (1998c) Multidimensional capillary GC-GC system for the analysis of real complex samples, part II: enantiomeric distribution of monoterpene hydrocarbons and monoterpene alcohols of cold-pressed and distilled lime oils. J Microcolumn Sep 10:203–212

Mondello L, Catalfamo M, Cotroneo A, Dugo G, Dugo G, McNair H (1999) Multidimensional capillary GC-GC system for the analysis of real complex samples, part IV: enantiomeric distribution of monoterpene hydrocarbons and monoterpene alcohols of lemon oils. J High Resol Chromatogr 22:350–356

Moshonas MG, Shaw PE, Carter RD (1991) Ambersweet orange hybrid: compositional evidence for variety classification. J Agric Food Chem 39:1415–1421

Moyler DA (1993) Extraction of flavours and fragrances with compressed CO_2. In: King MB, Bott TR (eds) Extraction of natural products using near-critical solvents. Blackie Academic and Professional, Chapman and Hall, Glasgow

Mussinan CJ (1993) Instrumental analysis in the flavor industry. In: Acree TE, Teranishi R (eds) Flavor science. Sensible principles and techniques. ACS Professional Reference Book, Washington, DC, pp 169–224

Nagy S, Shaw PE (1990) Factors affecting the flavour of *Citrus* fruits. In: Morton ID, MacLeod AJ (eds) Food flavours, part C: the flavour of fruits. Elsevier, Amsterdam, pp 93–124
Oberhofer B, Nikiforov A, Buchbauer G, Jirovetz L, Bicchi C (1999) Investigation of the alteration of the composition of various essential oils in aroma lamp applications. Flavour Fragr J 14:293–299
Ogawa K, Kawasaki A, Yoshida T, Nesumi H, Nakano M, Ikoma Y, Yano M (2000) Evaluation of aurapten content in *Citrus* fruits and their products. J Agric Food Chem 48:1763–1769
Phillips JB, Beens J (1999) Comprehensive two-dimensional gas chromatography: a hyphenated method with strong coupling between two dimensions. J Chromatogr A 856:331–347
Poiana M, Reverchon E, Sicari V, Mincione B, Crispo F (1994) Supercritical carbon dioxide extraction of bergamot oil: bergapten content in the extracts. Ital J Food Sci 4:459–466
Poiana M, Sicari V, Mincione B (1998a) A comparison between the chemical composition of the oil, solvent extract and supercritical carbon dioxide extract of *Citrus medica* cv. Diamante. J Essential Oil Res 10:142–152
Poiana M, Sicari V, Mincione B (1998b) Supercritical carbon dioxide (SC-CO_2) extraction of grapefruit flavedo. Flavour Fragr J 13:125–130
Poiana M, Fresa R, Mincione B (1999) Supercritical carbon dioxide extraction of bergamot peels. Extraction kinetics of oil and its components. Flavour Fragr J 14:358–366
Ragunathan N, Krock KA, Klawun C, Sasaki TA, Wilkins CL (1999) Gas chromatography with spectroscopic detectors. J Chromatogr A 856:349–397
Rousseff R, Gmitter F, Grosser J (1994) *Citrus* breeding and flavour. In: Piggott JR, Paterson A (eds) Understanding natural flavours. Blackie Academic and Professional, Chapman and Hall, Glasgow, pp 113–127
Ruberto G, Biondi D, Piattelli M, Rapisarda P, Starrantion A (1993) Profiles of essential oils of new *Citrus* hybrids. Flavour Fragr J 8:179–184
Ruberto G, Biondi D, Piattelli M, Rapisarda P, Starrantion A (1994) Essential oil of a new *Citrus* hybrid, *Citrus clementina* x *C. limon*. J Essential Oil Res 6:1–8
Ruberto G, Renda A, Piattelli M, Rapisarda P, Starrantion A (1997a) Essential oil of a new pigmented *Citrus* hybrid, *Citrus clementina* × *C. sinensis*. J Agric Food Chem 45:467–471
Ruberto G, Biondi D, Rapisarda P, Renda A, Starrantion A (1997b) Essential oil of Cami, a new *Citrus* hybrid. J Agric Food Chem 45:3206–3210
Ruberto G, Starrantino A, Rapisarda P (1999a) *Citrus* improvement – chemical and genetic aspects. In: Pandalai SG (ed) Recent research development in agricultural and food chemistry, vol 3. Research Signpost, Trivandrum, pp 445–470
Ruberto G, Starrantino A, Rapisarda P (1999b) Essential oils from new *Citrus* fruits. Essenze Deriv Agrum 69:15–26
Safina G (1984) Derivati agrumari. Stazione Sperimentale per l'Industria delle Conserve Alimentari, Parma
Scora RW, Kumamoto J (1983) Chemotaxonomy of genus *Citrus*. In: Waterman PG, Grundon MF (eds) Chemistry and chemical taxonomy of Rutales. Academic Press, London, pp 343–351
Scora RW, Coggins CW Jr, Knapp JFC (1969) Influence of maturation upon the essential oil components in *Citrus sinensis*. In: Proc 1st Int *Citrus* Symp vol 3, pp 1171–1175
Scott RPW (1987) Introduction to the chromatographic process. In: Katz E (ed) Quantitative analysis using chromatographic techniques. Wiley, Chichester, pp 1–29
Shamel R, Udis-Kessler A (2000) The sweet taste of success. Chem Ind, pp 27–28
Shaw PE (1979) Review of quantitative analyses of *Citrus* essential oils. J Agric Food Chem 27:246–257
Smith SL (1987) Possibilities, limitations and future development in GC-FT-IR analysis of essential oils. In: Sandra P, Bicchi C (eds) Capillary gas chromatography in essential oil analysis. Huethig, Heidelberg, pp 367–385
Soost RK, Cameron JW (1980) "Oroblanco" a triploid pummelo-grapefruit hybrid. HortScience 15:667–669
Soost RK, Cameron JW (1985) "Melagold" a triploid pummelo-grapefruit hybrid. HortScience 29:1134–1135
Spina P, Di Martino E (1991) Gli agrumi. Edagricole – Edizione agricole, Bologna
Starrantino A (1992) Use of triploids for production of seedless cultivars in *Citrus* improvement programs. In: Proc Int Soc Citriculture, Acireale, 8–13 Mar, vol 1, pp 117–121

Starrantino A (1994) Quadro attuale e orientamenti varietali per la trasformazione. In: Proc Conf on Innovazioni nell'industria dei derivati agrumari, 5 Mar, Acireale

Stremple P (1998) GC/MS of polymethoxyflavones in *Citrus* oils. J High Resol Chromatogr 21: 587–591

Swingle WT, Reece PC (1967) The botany of *Citrus* and its wild relative. In Reuther W, Webber HJ, Batchelor LD (eds) The *Citrus* industry, vol 1. University of California Press, Berkeley, pp 190–430

Tateo F, Bonomi M (1995) Chimica analitica degli aromi. GM Ricchiuto Editore, Bussolengo

Thomas AF, Bessière Y (1989) Limonene. Nat Prod Rep 6:291–309

Uden PC (1987) Detection in quantitative gas chromatography. In: Katz E (ed) Quantitative analysis using chromatographic techniques. Wiley, Chichester, pp 99–155

Vreuls JJ, de Jong GJ, Ghijsen RT, Brinkman UAT (1994) Liquid chromatography coupled on-line with gas chromatography: state of the art. J Assoc Off Anal Chem Int 77:306–327

Vreuls RJJ, Dallüge J, Brinkman UAT (1999) Gas chromatography–time-of-flight mass spectrometry for sensitive determination of organic microcontaminants. J Microcolumn Sep 11:663–675

Webber HJ (1967) History and development of the *Citrus* industry. In: Reuther W, Webber HJ, Batchelor LD (eds) The *Citrus* industry, vol 1. University of California Press, Berkeley, pp 1–39

Widmer WW, Montanari AM (1996) The potential for *Citrus* phytochemicals in hypernutritious foods. In: Finley JW, Armstrong DJ, Nagy S, Robinson SF (eds) Hypernutritious foods. Agscience, Auburndale, FL, , pp 75–89

Ziegler H, Spiteller G (1992) Coumarins and psoralens from Sicilian lemon oil (*Citrus limon* (L.) Burm. f.). Flavour Fragr J 7:129–139

8 Aroma Volatiles in Fruits in Which Ethylene Production Is Depressed by Antisense Technology

A.D. Bauchot[1], D.S. Mottram[2], and P. John[1]

8.1 Why Use Antisense Technology to Study Fruit Aroma?

During ripening, fruit undergoes striking changes in composition, texture, aroma, flavor and color. In climacteric fruits, such as tomatoes, bananas, melons, apricots and apples, these changes are associated with a respiratory burst (climacteric) and an increase in ethylene production (see Abeles et al. 1992; Tucker 1993). Also, ethylene applied to these fruits induces or accelerates ripening. Thus, ethylene plays an essential role in climacteric fruit ripening. Until now, the relationship between ethylene and fruit quality has not been fully understood. For instance, the development of a characteristic aroma is an important trait in many fruits, especially dessert fruits. Unfortunately, the role of ethylene on aroma volatile formation during ripening is not clear. Genetic modifications of the ethylene biosynthetic pathways in fruit have been achieved first in tomato (Hamilton et al. 1990; Oeller et al. 1991) and subsequently elsewhere (Ayub et al. 1996). These strategies provide useful tools to study the role of ethylene in fruit aroma development in general.

Inhibition of ethylene production by transgenic methods promises to deliver advantages in postharvest longevity and in fruit quality. However, the effects of ethylene inhibition have proved to be complex; some ripening processes being affected more strongly than others, with some appearing to be effectively ethylene-independent (Hadfield et al. 2000). Similarly, an analysis of aroma development in ACC oxidase antisensed melons has revealed that some components of the aroma are affected more strongly than others by the inhibition of ethylene production (Bauchot et al. 1998). The delivery of commercial benefits by the antisense technology depends upon the maintenance, and even enhancement, of dessert fruit quality, especially in terms of aroma. Thus, an analysis of aroma volatiles in antisensed dessert fruit is of prime importance in any program for the improvement of storability and extension of shelf life in dessert climacteric fruit.

Tomato has been the model climacteric fruit to study ripening with molecular techniques. However, tomato aroma is not as typically "fruity" as in dessert fruits. In general, the latter have high levels of volatiles, with esters being one of

[1] Department of Agricultural Botany, School of Plant Sciences, University of Reading, Reading, RG6 6AS, UK
[2] Department of Food Biosciences, University of Reading, Reading, RG6 6AS, UK

the most important classes (Schreier 1984). Among the dessert climacteric fruits, it is in cantaloupe melon that ethylene production had been inhibited through antisense RNA technology (Ayub et al. 1996). The target enzyme was ACC oxidase, the last enzyme responsible for ethylene biosynthesis. Up to 99.5% of ethylene emission was inhibited in fruit transformed with the antisense ACC oxidase gene. Our group has been assessing aroma quality in these ethylene-depleted melons. To our knowledge, we have been the first group to report on volatile aroma of fruit in which ethylene biosynthesis was genetically modified (Bauchot et al. 1998). Thus, we will describe in this chapter the methods used in studying volatile aroma of anti-ACC oxidase (anti-ACO) melon, although they might be applied to other fruits after appropriate transgenic modifications.

8.1.1 Successful Inhibition of Ethylene Biosynthesis in Fruit

The ethylene biosynthesis pathway involves three enzymatic steps, catalyzed by methionine adenosyltransferase, ACC synthase, and ACC oxidase. Both ACC synthase and ACC oxidase have been shown to control ethylene biosynthesis. Earlier work showed that inhibitors of ACC synthase or ACC oxidase slow or abolish ethylene production. Later, the cloning of ACC synthase and ACC oxidase permitted the use of molecular techniques to inhibit the expression of both enzymes. Antisense RNA technology has been applied first to tomatoes by Hamilton et al. (1990) who inhibited ACC oxidase expression. Later, Oeller et al. (1991) reported the successful inhibition of AAC synthase expression. Both transformations led to successful inhibition of ethylene biosynthesis in transgenic tomatoes, with inhibition of softening, color development, and the aroma changes normally associated with ripening.

More recently, other molecular techniques have been described, such as sense RNA technology (Hamilton et al. 1998), random mutagenesis (Tarun et al. 1998), and double-stranded RNA technology (see Fire 1999) to transform plants and to inhibit the expression of targeted proteins. These gene-silencing techniques have been, or can be, applied to reduce ethylene biosynthesis. Introduction of ACC deaminase (Klee et al. 1991) or SAM hydrolase (Good et al. 1994) diverts ethylene precursors and can also decrease ethylene production. Modifications of genes involved in the perception of ethylene and/or signal transduction (Ecker 1995; Bleecker and Schaller 1996) can also delay or inhibit ripening parameters controlled by ethylene. These techniques are beyond the scope of this chapter and will not be described here.

8.1.2 Studying Fruit Aroma in Ethylene-Depleted Fruit

By definition, aroma is the odor of a food product. It is detected when the volatiles emitted enter the nasal passage and are perceived by the olfactory system. Aroma can be characterized by a variety of sensory analysis techniques (as reviewed in

Meilgard et al. 1999). These techniques would be ideal tools to measure fruit aroma variations during ripening and the effect of antisense transformation depleting ethylene biosynthesis. However, these techniques do have limitations. They need either the training of an expert panel or the survey of many consumer groups. A great number of samples would be required, often over several weeks or even months. For consumer studies, availability of large samples of fruits is essential. Producing enough transgenic material for these purposes would not be always possible, especially owing to the inherent seasonality of fruit crops. Even if storage were possible, postharvest changes are likely to introduce other parameters to take into account. Furthermore, the assessment of transgenic food is nowadays a controversial matter, particularly where an antibiotic resistance gene marker is present. Even though the product is innocuous, contact of the transgenic material with human assessors in the early phases of work is to be avoided in order to comply with regulatory requirements. Thus, alternative methods of analyses have to be sought.

The aroma component of a fruit, sensed when sniffed through the nose, is due to a complex mixture of volatile organic compounds. Chemical approaches based on the chemical and physical properties of these volatiles allow the isolation and concentration of fruit aroma compounds prior to their separation and identification by gas chromatography-mass spectrometry (GC-MS).

8.1.3 Fruit Volatile Compound Analyses

Sample preparation techniques include direct thermal desorption, solvent extraction, steam distillation with continuous solvent extraction, headspace analyses and solid phase microextraction (SPME), and supercritical fluid extraction. The extract is then separated by GC. The chromatographic profile will vary with the extraction technique employed. None yields an aroma isolate that truly represents either qualitatively or quantitatively the original fruit aroma (Parliment 1997). Some methods can generate biases or artifacts. For example, solvents extract more effectively heavier compounds than lighter compounds, and vice versa for headspace sampling. Some compounds are degraded with excessive heat, or enzymatic reactions can occur with some extraction procedures (see Reineccius 1993).

Nevertheless, comparison between transgenic and non-transgenic material is possible as long as the limits of the chosen method are known. Careful interpretation of the results leads to valuable information related to the quality of the aroma of the transgenic fruit compared to the non-transformed fruit. In our laboratory, flavor volatiles are routinely analyzed through headspace techniques, although solvent extraction and distillation techniques have been used on melons (Schieberle et al. 1990; Homatidou et al. 1992; Wyllie and Leach 1992; Moshonas et al. 1993), and much information on these techniques is available elsewhere (Parliment 1997). We will describe below only three types of extraction: a solvent extraction method, Tenax trapping, and SPME.

Once the volatiles are extracted, separated, identified, and quantified, results have to be interpreted. Volatile analysis in relation to aroma is valid only if it can be reunited with its first aim: when presented with the product, what does the consumer smell? How strong or pleasant is the aroma? How typical is it of the product? How different is it from the original product? If so, is it perceived by the consumer?

The most abundant volatiles may have little or no odor in a fruit. The combination of olfactometry with CG (GCO or "CG-sniffing") involves smelling the GC effluent after its separation. A human assessor may describe the odor character, the perceived intensity and/or the duration of the signal. Several techniques have been developed to relate these characteristics to the overall food aroma (see Mistry et al. 1997). The most common techniques allow searching for key odorants contributing to the aroma of a food. Aroma values or odor values (OV) quantify a compound present in an extract in relation to its perceived threshold. They are obtained by dividing the concentration of a compound by its known odor threshold value in water (Guadani et al. 1966). Thus, compounds with OVs above 1 are likely to contribute to the overall aroma of the product. The method has some limitations (Mistry et al. 1997). For instance, aroma compounds are considered individually, without taking into account interactions between them or with the food matrix; also, the estimation of sensory thresholds can be difficult. Nevertheless, the method gives an indication of the relative importance of the compounds to the overall aroma. Based on data available in the literature, OVs of the main volatile esters usually present in fruits can be estimated.

8.2 Methods

We will present methods implemented to study volatile aroma composition of anti-ACO melons. Adjustments for other plant material may be necessary, although principles described here will be similar.

8.2.1 Inhibition of Ethylene Biosynthesis: Fruit Transformation

This step itself presupposes that a transformation method and a regeneration method have been optimized. Our study involved the transformation of Charentais melons (*Cucumis melo* var. *cantalupensis,* Naud. cv. Védrandais) with an antisense construct of ACC oxidase. We worked on fruit derived from the work of Ayub et al. (1996). The authors co-cultivated cotyledon explants with disarmed *Agrobacterium tumefaciens* strain LB4404 harboring an antisense construct of ACC oxidase and the neomycin phosphotranspherase gene (NPTII, kanamycin resistance gene). They regenerated transformed plants and selected those with lower levels of ethylene synthesis.

A high frequency of tetraploids in the regenerated plants was reported. Unfortunately, tetraploid plants are not marketable, and cannot be used in breeding programs. Recently, Guis et al. (2000) established a protocol leading to a majority of diploid transformed plants from leaf explants of *Cucumis melo* L. var. *cantalupensis*. For future studies on cantaloupe melon, we would recommend the use of their protocol as described below.

8.2.1.1 Tissue Regeneration

The coats of mature seeds of *C. melo* cv. Védrandais are removed, and the seeds soaked for 2 h in water. The seeds are then washed in 70% ethanol for 1 min, rinsed in distilled water for 5 min, sterilized in 2.5% calcium chloride containing 0.1% (v/v) Tween 20 for 20 min and washed three times. To obtain leaf explants, seeds are germinated on MS medium (Murashige and Skoog 1962) supplemented with 1% sucrose and solidified with 0.8% agar. The pH of all media is adjusted to 5.7 before autoclaving for 20 min at 120°C. The first leaf (0.3–0.5 mm diameter) is excised from 10-day-old seedlings. The petiole and some basal parts of the leaf are removed, and the leaf cut transversally along the vein. The leaf explants are placed on organogenic induction medium. The induction medium is based on MS medium with 1 μM BAP, 1 μM 6-(γ,γ-dimethylallylamino)purine (2iP), 3% sucrose, and 0.8% agar. Shoots are excised from the explant and transferred to a development medium (MS medium with 1 μM BAP, 3% sucrose, and 0.8% agar; GA_3 sterilized by filtration is added after autoclaving to the concentration of 0.29 μM). After 3 weeks, the elongated shoots are separated and placed on a rooting medium (MS medium, 3% sucrose and 0.8% agar). Once rooted, the plantlets are rinsed free of medium, transplanted in a mix of vermiculite and peat-moss and incubated in a saturated humidity atmosphere. After a week, acclimatized plants are transferred in plastic pots and grown in a culture room for a month (25°C, light flux of 100 $\mu mol\,m^{-2}\,s^{-1}$ from OSRAM L36 W/36 Nature tube, 16-h light/8-h dark photoperiod). Plants are transferred to a greenhouse, trained, and cultivated according to standard practices for fertilization and pesticide treatment. Plants are self-pollinated and two fruits are kept on each vine. Under the conditions described by Guis et al. (2000), less than 15% of the plants regenerated by organogenesis on leaf explants were tetraploids.

8.2.1.2 *Agrobacterium* Transformation

The *Agrobacterium tumefaciens* LBA4404 strain used is disarmed. It contains the binary vector PGA643 (An et al. 1988) with a NOS/NPTII chimeric gene, a selectable marker for kanamycin resistance and a CaMV35 S/antisense cDNA MEL1 encoding for melon ACC oxidase (Balagué et al. 1993). Young leaf explants are prepared as above and incubated with an *Agrobacterium* suspension culture (10^6 to 10^8 cells ml^{-1}) for 30 min under gentle agitation. Leaf explants are then blotted on Whatman filter paper, allowed to dry for 5 to 10 min under the hood, placed abaxially on the induction medium for 3 days and transferred to the selective media. The induction medium is supplemented with 100 $mg\,l^{-1}$ kanamycin and 500 $mg\,l^{-1}$ carbenicilin; the development and rooting media contain 100 and 50 $mg\,l^{-1}$ kanamycin respectively. The high concentration of kanamycin leads to very few plants being able to root: Guis et al. (2000) recovered only 13 rooted plantlets.

Flow cytometry is necessary to confirm diploidy. Genomic DNA of the putative transformed plants has to be subjected to preliminary screenings by PCR (polymerase chain reaction) analysis using two primers in the NPTII gene region. A PCR amplification band indicates a stable integration of the NPTII gene in the transformed plants. To show that the amplification does not result from remaining bacteria, no amplification should be observed when using Vir primers. A Southern blot confirms stable integration of the transgene.

8.2.1.3 Generation of Transformed Plants

R0 plants are self-pollinated and R1 seeds are allowed to germinate aseptically on MS medium with 2% sucrose, 0.8% agar, and 75 $mg\,l^{-1}$ kanamycin. After 3 weeks, seedlings able to develop secondary roots are considered kanamycin resistant. Plants transformed with the anti-ACO gene generally present the same phenotype as non-transformed plants (Ayub et al. 1996; Guis et al. 2000). Measuring fruit ethylene production reveals expression of the transgene. Ethylene levels can range from 0.5 to 20% that of the non-transformed fruit.

8.2.1.4 Production of Hybrids

For melon, F1 hybrids are considered of superior agronomic quality over monozygote lines. In our study, the R4 progeny of one of the transformed line obtained by Ayub et al. (1996) was used as the antisense parental line (AS). Non-transformed melons (nT) were bred using the equivalent line without the ACO antisense gene. Parental lines conferring important traits (such as resistance to pathogens or improved agronomic performances) were crossed with the nT line and with the AS line. Transformed and non-transformed hybrid plants were obtained. Antisense fruits were assessed against non-transformed fruits from the corresponding hybrid obtained with the original non-transformed line. Plants were grown in a greenhouse in Spain, according to standard cultural practices.

One of the difficulties of this type of study is to determine the stage of "ripening" of the transformed fruit. In reality, it is nearly impossible to truly compare them with non-transformed fruit, especially if maturity is ethylene-dependent. Non-transformed fruits were harvested when the rind turned yellow. Antisense fruits were harvested when the first leaf next to the fruit yellowed, as in commercial practice.

Fruit quality at harvest and postharvest conditions can affect dramatically fruit physiology and subsequently fruit quality. Storage temperatures, ventilation, humidity levels, and spoilage are all to be taken into consideration. To simulate transport and distribution from production sites to distribution sites, fruits were air-freighted to our laboratory, stored for 3 days at 12°C and held overnight at 20°C before analysis. To test the effect of ethylene on volatile biosynthesis recovery, some antisense fruit were enclosed with 400 ppm ethylene for 3 days at 20°C.

8.2.2 Volatile Analyses

8.2.2.1 Solvent Extraction

Extraction is based on the partitioning of volatile compounds between the fruit tissue and a solvent. Generally, the fruit tissue is mashed in the presence of calcium chloride, which inhibits some enzyme activities and reduces the solubility of some volatile compounds in the aqueous phase. The resulting slurry is extracted repetitively with the solvent. The final extract is dehydrated in the presence of sodium sulfate and concentrated. Diethyl ether, diethyl ether/pentane mixtures, methylene chloride, hexane, and dichloromethane are common solvents. The main drawbacks of these methods are the amount of fruit necessary, the potential hazards due to the solvents, solvent contamination, and the time required for the extraction.

For melon tissue, 50 to 80 g of pericarp was cut into cubes in a 1-l conical flask, 5 g of $CaCl_2$ added and the whole submersed in 50 ml of dichloromethane (DCM). The tissue was blended, filtered and extracted with 2×50 ml DCM. Filtrates were collected, dried with anhydrous sodium sulfate, and 100 µl internal standard added (in our case, 1,2-dichlorobenzene at 100 $\mu g\,ml^{-1}$ in ether). After 30 min at least, the DCM extract was concentrated to 0.3 ml in a Lickens-Nickerson apparatus at 45 °C and 1 µl was injected into the GC.

8.2.2.2 Headspace Sampling

To be effective as a contributor to the aroma, the components must leave the fruit matrix and travel through air to reach the nose, where they are perceived. Headspace techniques take advantage of this volatility, with the atmosphere adjacent to the matrix being sampled. These techniques have the following advantages: no solvents and no solvent impurities are involved, sampling is completed in a short time, and there is no intermediate processing. Purging and trapping are accomplished by venting a carrier gas of the headspace through a collection trap. The organic compounds are concentrated on the trap while the carrier gas passes through. Solid phase microextraction (SPME) is an adsorption/desorption technique. It involves the partitioning of organic compounds between a vapor phase and a thin polymeric film coated onto a fused silica fiber in the SPME apparatus (Sides et al. 2000).

For both techniques, the actual volatiles detected in the headspace are dependent on the temperature, the nature and geometry of the matrix, the volumes of the headspace, and the sampling time. Thus, it is clear that care must be taken to keep these parameters constant to allow consistency in the results.

All equipment should preferably be glass and oven-baked (120 °C) overnight before use; septa should be Teflon-coated. This is because plastic polymers release compounds that can contaminate the analyses.

8.2.2.2.1 Tenax Trapping

Many volatile organic compounds can be removed from a stream of gas by passing the stream through a tube packed with a sorbent (Fig. 8.1). Because of its high surface area, the organic vapor may be adsorbed onto it. By heating the trap carefully, the adsorbed compounds are volatilized again without thermal degradation. Most of the sorbent materials are porous polymers similar to the materials filling GC columns. Tenax (poly-2,6-diphenyl-*p*-phenyl oxide) is most widely used. It is capable of trapping a wide range of organic volatiles, especially aromatics, it may be heated to relatively high temperature for desorption, and it is long lasting. It is not suitable for very light volatiles (such as methanol, ethanol, and hydrocarbons lighter than hexane). This is an advantage in the case of fruit volatile analyses where emission of ethylene, ethanol, and ethanal can be high.

Fruit preparation. To analyze the volatiles emitted by the whole fruit, the fruit is enclosed in a glass container. A 6- or 9-l desiccator is adequate. A known volume of air purified on charcoal (about 10 l) is then passed over the fruit before collection (Fig. 8.1). For melons, 100 to 200 ml of air is generally enough, since fruit emit high levels of volatiles.

Alternatively, volatiles present in flesh tissue can be analyzed destructively. Five plugs (1 cm diameter, 2 cm long, about 10 g of tissue) are excised from the equator of a fruit, weighed, and dropped into a 250-ml flask previously filled with nitrogen gas. Because of the inherent variability between fruit, we would recommend that the sampling be repeated on at least four to six fruits. The flask is placed in a water bath at 37 °C to simulate volatile release in the mouth. A flow of nitrogen sweeps the tissue for 15 min at 100 ml min^{-1}. Care should be taken to have as much turbulence in the flask as possible. This can be achieved if the nitrogen is

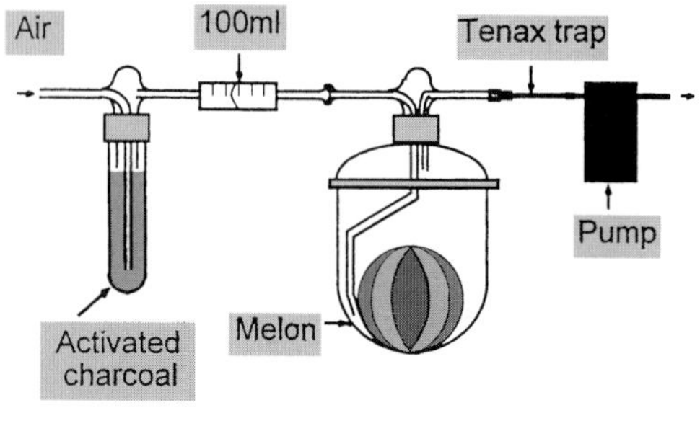

Fig. 8.1. Experimental set-up to collect flavor volatiles of fruit in a non-destructive manner

flushed as close as possible to the tissue, through a fine tubing (about 1 mm i.d.). A 1-ml disposable pipette tip can be fitted to the end of the tubing.

Trapping. The flow of gas sweeping the vessel containing the sample (fruit or tissue) is collected onto a glass-lined, stainless-steel trap (105 × 3 mm i.d.) containing 85 mg Tenax TA (Scientific Glass Engineering Ltd., Melbourne, Australia); 1 μl of internal standard (100 μg ml^{-1} 1,2-dicholorobenzene diluted in ether or methanol) is added to the trap just before collection. Methanol is preferred because it is not as volatile as ether and it is not retained on Tenax. Excess solvent and any water retained on the trap are removed by purging the trap with nitrogen at 100 ml min^{-1} for 5 min after collection. The trap is then screwed onto the CHIS injection port (Scientific Glass Engineering Ltd.) of the chromatograph. If a CHIS injection port is not fitted on the chromatograph, the trap can be desorbed with freshly distilled diethyl ether (Buttery et al. 1982). The extract is then concentrated to a small volume by using a warm water bath and a distillation column of the Lickens-Nickerson type, and injected onto the GC.

8.2.2.2.2 SPME

Headspace analysis by SPME is based on the partitioning of compounds between a vapor phase (headspace above the sample) and an organic phase coating a fiber exposed temporarily to the headspace. The fiber is a fused silica fiber coated with a thin layer of polymer. Volatile compounds will adsorb onto the coating exposed to the headspace. They will be desorbed when the fiber is exposed in the GC injection port held at high temperature. As with solvent extractions, the properties of the coatings will determine the type of compounds preferably isolated.

For fruit volatiles where the levels of the esters are high, we prefer fibers coated with 75 μm carboxen/polydimethylsiloxane (Supelco, Bellefonte, PA, USA). As for Tenax trapping, volatiles can be analyzed from whole fruit or tissue. Equilibrium does not necessarily need to be attained before sampling is realized, as long as everything is kept strictly consistent throughout the experiment.

To use the SPME unit (Supelco), the fiber is drawn into the needle. The needle is passed through a Teflon-coated septum that seals the sample vial. The plunger is depressed, to expose the fiber to the headspace above the fruit sample for 5 min. Internal standard (0.1 μl 1,2-dicholorobenzene at 100 μg ml^{-1} in ether) is injected into the injection port of the GC before the SPME syringe is placed through a pre-drilled thermogreen LB-2 septum (Supelco), and the fiber exposed for 5 min.

8.2.2.3 Gas Chromatography-Mass Spectrometry

We perform all analyses on a Hewlett Packard 5890 Series II gas chromatograph fitted with a HP5972 mass spectrometer. Volatiles (trapped on Tenax or SPME fiber) are thermally desorbed at 250 °C in the injection port onto the front of a capillary column. BPX-5 fused silica capillary column (50 m × 0.32 mm i.d., 0.5-μm film thickness, Scientific Glass Engineering Ltd.) or CP-sil 8 CB capillary column (60 m × 0.25 mm i.d., 0.5-μm film thickness, Chrompack) are suitable. The temperature of the oven is maintained at 0°C for 8 min with liquid nitrogen. This allows for the focusing of the desorbed volatiles at the front end of the column, and results in sharper peaks and enhanced chromatographic separation. After desorption, the oven temperature is raised to 50°C over 1 min and held for 2 min. The oven temperature is increased to 250 °C at a rate of 2.5 °C min^{-1}. Alternatively, a loop at the beginning of the capillary column is kept in a beaker full of dry-ice for 4.5 min while the oven temperature is maintained at 40 °C. After 5 min, the oven temperature is increased to 250 °C as above. Depending on the type of volatiles studied, the temperature ramp can be adjusted. The carrier gas is helium (8 psi), at a flow rate of 1.75 ml min^{-1} at 40 °C. Mass spectra are recorded in the electron impact mode (ionization voltage: 70 eV, source temperature: 200 °C). A scan range of 29–400 *m/z* with a scan time of 0.69 s is employed. The data are controlled and stored by the HP G1034C ChemStation system.

Volatiles are identified by comparison of each mass spectrum with spectra from authentic compounds analyzed in the laboratory, or with spectra in reference collections (NIST/EPA/NIH Mass Spectral database). A linear retention index (LRI) is calculated for each compound using the retention times of a homologous series of C_6-C_{22} *n*-alkanes loaded onto a Tenax trap analyzed under the conditions described above. Whenever possible, mass spectral identifications are confirmed by comparing their LRI with those of authentic compounds analyzed under the same conditions.

Quantification of the volatiles is based on the relation between their peak areas and that of the 1,2-dichlorobenzene internal standard, obtained from the total ion chromatograms, using a response factor of 1. No correction for extraction efficiency of recovery is achieved using this technique; it is rather a means of quantifying the levels of components present on the adsorbent traps.

8.3 Illustration: Our Results

The results presented below are mostly adapted from Bauchot et al. (1998). They illustrate some of the information obtained on aroma when comparing control fruit to fruit from the same cultivar that had been antisensed to reduce ethylene biosynthesis. In anti-ACO cantaloupe melons, ethylene production was successfully inhibited, with less than 2% that of the ethylene emitted by non-transformed fruit (data not shown).

The bulk of the volatiles emitted by cantaloupe melon pericarp are straight-chain and branched-chain esters (Table 8.1). Total volatile compounds were higher in hybrids 1 than in hybrids 2. In non-transformed hybrids, nT1 and nT2, acetates represented 77 and 85% of the total volatiles respectively. In the corresponding anti-ACO hybrids, AS1 and AS2, total volatile emission was inhibited by 70 and

Table 8.1. Volatiles identified in the headspace of melon hybrids (ng g^{-1} tissue). A flow of nitrogen swept plugs of melon pericarp, and 1.5 L was collected on a Tenax trap. Average values of four replicates

Class of compounds	Hybrids 1			Hybrids 2		
	nT1	AS1	Inhibition (%)	nT2	AS2	Inhibition (%)
Acetates	1,065	369	65	723	86	88
Propanoates	68	16	76	25	8	68
Methylpropanoates	40	8	80	5	2	60
Butanoates	136	26	80	75	14	81
Methylbutanoates	41	1	98	14	1	93
Total volatiles	1,384	420	70	848	111	87

Table 8.2. Estimated odor values of the major volatile compounds (concentration/threshold) in the headspace of melon tissue swept by nitrogen. Average values of four replicates. *na* Not available; ratios given by amounts of propyl acetate present in fruit tissue

	Odor threshold in water[a] (ng ml^{-1})	Hybrids 1			Hybrids 2		
		nT1	AS1	Ratio	nT2	AS2	Ratio
Ethyl acetate	5	64	8	8	37	8	5
Ethyl propanoate	10	6	<1	15	2	<1	>20
Propyl acetate	na	–	–	3	–	–	3
Ethyl butanoate	1	108	6	18	67	5	13
Butyl acetate	66	2	<1	2	1	<1	2
Ethyl 2-methylpropanoate	0.1	361	12	30	37	3	12
Methylpropyl acetate	65	3	2	1	2	1	2
Ethyl 2-methylbutanoate	0.3	136	2	68	47	2	23
Methylbutyl acetate	11	9	3	3	12	6	2
Ethyl hexanoate	1	27	<1	>27	4	<1	>4
Hexyl acetate	2	42	16	3	20	13	2

[a] From Takeoka et al. (1989, 1991, 1992)

87% respectively. However, different classes of volatile compounds were not affected at the same rate. For instance, methylbutanoates were much more affected by the transformation than straight-chain esters.

These observations led us to think in terms of aroma potency and odor values. Generally, straight-chain esters have lower odor thresholds than branched-chain esters. The same is true for acetates compared to ethyl esters (Table 8.2). As a consequence, odor values of ethyl esters with branched-chain acyl groups were higher than the odor values of the most common acetates identified in the headspace of melon tissue. Ethyl 2-methylbutanoate and ethyl butanoate have been identified previously among the most intense odorants in muskmelons (Yabumoto et al. 1977; Schieberle et al. 1990). Together with ethyl 2-methylpropanoate, they had the higher odor values in non-transformed hybrids. The anti-ACO transformation preferentially affected these volatile compounds. These results indicate clearly that ethylene is essential to the biosynthesis of potent odorants during ripening in cantaloupe melons.

In a different experiment, volatile biosynthesis recovered after giving an appropriate ethylene treatment (400 ppm for 3 days) to the antisensed fruit (data not shown). It showed that inhibition of aroma compound biosynthesis due to anti-ACO is reversible.

8.4 Conclusions

The techniques presented above allowed successful extraction of volatile compounds of melon fruits that were transformed with cDNA in the antisense orientation. Ethylene biosynthesis was effectively depleted, although not fully. The transformation modified dramatically volatile composition of the fruits. The use of odor values emphasized the role of potent odorants (such as ethyl esters versus acetates). Following the anti-ACO transformation, their biosynthesis was strongly inhibited. The inhibition was reversible because exogenous ethylene restored volatile levels to that of non-transformed fruits. This work shows that ester pathways are activated by ethylene during ripening. Moreover, in the pathways leading to ethyl esters, potent odorants are more strongly regulated by ethylene than the pathways leading to acetates.

The use of the anti-ACO gene in melon helped identify those steps in volatile aroma biosynthesis that were ethylene-controlled or at least those steps that required relatively high levels of ethylene. The extension of the technique to other climacteric fruits should reveal the extent to which ethylene controls the production of key components of flavor generally. In non-climacteric fruits, it is conventional to view ethylene as being just one of an array of plant regulators that control ripening and flavor development. However, the effect of the ethylene binding inhibitor, 1-methylcyclopropene (1-MCP) on ripening in non-climacteric fruits (Ku et al. 1999; Porat et al. 1999; Mullins et al. 2000) suggests that an extension of the antisense work to non-climacteric fruit might prove to be equally revealing.

Acknowledgements. We would like to thank our partners, Dr. F. Romojaro (CEBAS, Murcia, Spain), Tezier SA (Valence, France), and Tezier Iberica (Almeria, Spain), and Prof. J.C. Pech (ENSA, Toulouse, France). The melon project was sponsored by the European Program FAIR CT96-1138.

References

Abeles FB, Morgan PW, Saltveit ME Jr (1992) Ethylene in plant biology, 2nd edn. Academic Press, San Diego

An G, Hebert PR, Mitra A, Ha SB (1988) Binary vectors. In: Stanton BG, Schilperoort RA (eds) Plant molecular biology manual. Kluwer, Dordrecht, pp 1–19

Ayub R, Guis M, Ben Amor M, Gillot L, Roustan JP, Latché A, Bouzayen M, Pech JC (1996) Expression of ACC oxidase antisense gene inhibits ripening of cantaloupe melon fruits. Nat Biotech 14:862–866

Balagué C, Watson CF, Turner AJ, Rougé P, Picton S, Pech JC, Grierson D (1993) Isolation of a ripening and wound-induced cDNA from *Cucumis melo* L., with homology to the ethylene-forming enzyme. Eur J Biochem 212:27–34

Bauchot AD, Mottram DS, Dodson AT, John P (1998) Effect of aminocyclopropane-1-carboxylic acid oxidase antisense gene on the formation of volatile esters in cantaloupe Charentais melon (cv. Védrantais). J Agric Food Chem 46:4787–4792

Bleecker AB, Schaller GE (1996) The mechanism of ethylene perception. Plant Physiol 111: 653–660

Buttery RG, Seifert RM, Ling LC, Soderstrom EL, Ogawa JM, Turnbaugh JG (1982) Additional aroma components of honeydew melon. J Agric Food Chem 30:1208–1211

Ecker JR (1995) The ethylene signal transduction pathway in plants. Science 268:667–675

Fire A (1999) RNA-triggered gene silencing. Trends Genet 15:358–363

Good X, Kellogg JA, Wagoner W, Langhoff D, Matsumara W, Bestwick RK (1994) Reduced ethylene synthesis by transgenic tomatoes expressing S-adenosylmethionine hydrolase. Plant Mol Biol 26:781–790

Guadani DG, Buttery RG, Harris J (1966) Odour intensities in hop oil components. J Sci Food Agric 17:142–144

Guis M, Ben Amor M, Latché A, Pech JC, Roustan JP (2000) A reliable system for the transformation of cantaloupe Charentais melon (*Cucumis melo* L. var. *cantalupensis*) leading to a majority of diploid regenerants. Sci Hortic 84:1–2

Hadfield KA, Dang T, Guis M, Pech JC, Bouzayen M, Bennet AB (2000) Characterization of ripening-regulated cDNAs and their expression in ethylene-suppressed Charentais melon fruit. Plant Physiol 122:977–983

Hamilton AJ, Lycett GW, Grierson D (1990) Antisense gene that inhibits synthesis of the hormone ethylene in transgenic plants. Nature 346:284–287

Hamilton AJ, Brown S, Yanhai H, Ishizuka M, Lowe A, Alpuche Solis AG, Grierson D (1998) A transgene with repeated DNA causes high frequency, post-transcription and suppression of ACC-oxidase gene expression in tomato. Plant J 15:737–746

Homatidou VI, Karvouni SS, Dourtoglou VG, Poulos CN (1992) Determination of total volatile components of *Cucumis melo* L. variety *cantalupensis*. J Agric Food Chem 40:1385–1388

Klee HJ, Kretzmer KA, Barry GF, Kishore GM (1991) Control of ethylene synthesis by expression of a bacterial enzyme in transgenic tomato plants. Plant Cell 3:1187–1193

Ku VVV, Wills RBH, Ben-Yehoshua S (1999) 1-Methylcyclopropene can differentially affect the postharvest life of strawberries exposed to ethylene. Hortic Sci 34:119–120

Meilgard M, Civille GV, Carr BT (eds) (1999) Sensory evaluation techniques, 3rd edn. CRC Press, Boca Raton

Mistry BS, Reineccius T, Olson LK (1997) Gas-chromatography-olfactometry for the determination of key odorants in foods. In: Marsili R (ed) Techniques for analyzing food aroma. Series food science and technology. Dekker, New York, pp 265–292

Moshonas MG, Shaw PE, Baldwin EA, Yuen W (1993) Volatile and nonvolatile components in Hami melon (*Cucumis melo* L.). Lebensm Wiss Technol 26:577–589

Mullins ED, McCollum TG, McDonald RE (2000) Consequences on ethylene metabolism of inactivating the ethylene receptor sites in diseased non-climacteric fruit. Postharv Biol Technol 19:155–164

Murashige T, Skoog F (1962) A revised medium or rapid growth and bioassay with tobacco culture. Physiol Plant 15:473–497

Oeller PW, Min-Wong L, Taylor LP, Pike DA, Theologis A (1991) Reversible inhibition of tomato fruit senescence by antisense RNA. Science 254:437–439

Parliment TH (1997) Solvent extraction and distillation techniques. In: Marsili R (ed) Techniques for analyzing food aroma. Series food science and technology. Dekker, New York, pp 1–26

Porat R, Weiss B, Cohen L, Daus A, Goren R, Droby S (1999) Effects of ethylene and 1-methylcyclopropene on the postharvest qualities of 'Shamouti' oranges. Postharv Biol Technol 15:155–163

Reinecius G (1993) Biases in analytical flavor profiles introduced by isolation method. In: Ho C-T, Manley CH (eds) Flavor measurements. IFT basic symposium series. Dekker, New York, pp 61–76

Schieberle P, Ofner S, Grosch W (1990) Evaluation of potent odorants in cucumbers (*Cucumis sativus*) and muskmelons (*Cucumis melo*) by aroma extract dilution analysis. J Food Sci 55: 193–195

Schreier P (ed) (1984) Chromatographic studies of biogenesis of plant volatiles. Hüthig, Heidelberg

Serek M, Sisler EC, Reid MS (1994) Novel gaseous ethylene binding inhibitor prevents ethylene effects in potted flowering plants. J Am Soc Hortic Sci 119:1230–1233

Sides A, Robards K, Helliwell S (2000) Developments in extraction techniques and their application to analysis of volatiles in foods. Trends Anal Chem 19:322–329

Takeoka G, Buttery RG, Teranishi R, Flath RA (1991) Identification of additional pineapple volatiles. J Agric Food Chem 39:1848–1851

Takeoka RG, Buttery RG, Flath RA (1992) Volatile constituents of Asian pear (*Pyrus serotina*). J Agric Food Chem 40:1925–1929

Takeoka G, Buttery RG, Flath RA, Teranishi R, Wheeler EL, Wieczorek RL (1989) Volatile constituents of pineapple (*Ananas comosus* (L) Merr.). In: Teranishi R, Buttery RG, Shahidi F (eds) Flavor chemistry, trends and developments. ACS Symp Series 388. American Chemical Society, Washington, DC, pp 223–237

Tarun AS, Lee JS, Theologis A (1998) Random mutagenesis of 1-aminocyclopropane-1-carboxylate synthase: a key enzyme in ethylene biosynthesis. Proc Natl Acad Sci USA 95:9796–9801

Tucker GA (1993) Introduction. In: Seymour GB, Taylor JE, Tucker GA (eds) Biochemistry of fruit ripening. Chapman and Hall, London, pp 1–51

Wyllie SG, Leach DN (1992) Sulfur-containing compounds in the aroma volatiles of melons (*Cucumis melo*). J Agric Food Chem 40:253–256

Yabumoto K, Jennings WG, Yamaguchi M (1977) Volatile constituents of cantaloupe *Cucumis melo*, and their biogenesis. J Food Sci 42:32–37

9 Detection of Physiologically Active Flower Volatiles Using Gas Chromatography Coupled with Electroantennography

F.P. Schiestl[1] and F. Marion-Poll[2]

9.1 Introduction

Most insect-pollinated plants produce floral odour to attract their pollinators. Fragrances and floral morphology typically advertise the presence of food rewards, i.e. nectar, pollen, or oil (Vogel 1983). Floral volatiles, however, can also be the reward themselves: some euglossine bees and fruit flies actually collect the fragrance from certain flowers and thereby pollinate them (Dodson et al. 1969; Sazima et al. 1993; Tan and Nishida 2000). Some orchids emit floral odour, typically not detectable by humans, which mimics the sex pheromone of insects, the males of which pollinate the flowers in an attempted copulation or precopulatory routine (Kullenberg 1961; Peakall and Beattie 1996; Schiestl et al. 1999). These orchids exploit their pollinators as they do not provide any reward for them. On the other hand, herbivorous insects can utilise floral volatiles to find their host plants, particularly if they feed or oviposit on the flowers (Blight et al. 1995). Floral odour is thought to have originally evolved to deter insects that fed on reproductive parts of the plant (Pellmyr and Thien 1986). Accidental pollination by these insects may have outweighed the disadvantages of their feeding activity, thus leading to a selection for a means of increasing insect attraction to the flowers.

The chemical composition of floral fragrances has been investigated in great detail in numerous plant species (e.g. Knudsen et al. 1993; Kaiser 1995). However, to understand how plants communicate with their pollinators, or how insects exploit floral signals, we need to know more about the insects' olfactory detection of floral odour compounds and how volatiles mediate their behaviour. This knowledge increases our comprehension of the evolution of certain pollination syndromes and helps us to interpret and manipulate insect behaviour (Agelopoulos et al. 1999).

The investigation of insect chemical communication has been greatly facilitated by the use of electrophysiological methods. In insect olfaction, these techniques were pioneered by Schneider (1957) who, for the first time, recorded an "electroantennogram" (EAG), i.e. a difference in potential between the tip and the base of an excised antenna of a *Bombyx mori* male stimulated with the female sex

[1] Division of Botany and Zoology, The Australian National University, Canberra, ACT 0200, Australia
[2] INRA Station de Phytopharmacie et Médiateurs Chimiques, Route de St. Cyr, 78026 Versailles Cedex, France

pheromone. Moorhouse et al. (1969) invented the coupling of the EAG with a gas chromatograph (GC). The GC column was split, with one branch leading to the detector of the GC and the other to a micro-collector in the GC oven. The effluent was collected for 15 s and then flushed over an antenna prepared for EAG recordings. The signals of the GC detector and the EAG responses were recorded simultaneously. Arn et al. (1975) improved the method by using a capillary GC column and continuously monitoring the GC effluent for EAG active compounds using an insect antenna mounted between two silver wires. This allowed components of an odour blend to be separated and tested individually for their potential to elicit EAG responses. This setup was termed the electroantennographic detector (EAD) since it could be used selectively to detect minute amounts of pheromone components. Some authors prefer the term GC-EAG, which refers to the same technique. The GC-EAD method provided an easy and very efficient tool for identifying physiologically active components (those compounds that release receptor potentials in olfactory neurons) within a blend. It has been used widely for the identification of semiochemicals in insects (Struble and Arn 1984). A further development was achieved by coupling single sensillum recordings (SSR) with gas chromatographic analyses (GC-SSR), which allowed a more sensitive and specific detection of physiologically active odour components (Wadhams 1984).

A thorough analysis of the behaviour-mediating functions of semiochemicals, including floral odour compounds, will optimally involve the following sequential steps: scent collection, evaluation of physiologically active compounds, identification and synthesis of compounds, and behavioural tests. In this chapter, we describe the methods used for these steps, with the exception of compound identification and synthesis, providing examples particularly related to floral scents. Further, we list physiologically active compounds emitted from flowers that have been identified using GC-EAD, GC-SSR, EAG, and SSR.

9.2 Collection of Floral Scent

Floral odours can be sampled by extraction of a flower or flower parts using solvent or a polymer-coated fibre (solid-phase microextraction, SPME; see Sect. 9.2.3). Typically, flower extracts predominantly comprise compounds of low volatility. Alternatively, fragrances can be collected from living flowers over longer periods of time, providing more realistic quantitative data on volatile emissions and the ratios of airborne compounds, and avoiding artefacts attributable to extraction of plant tissue. These methods also yield greater amounts of more highly volatile compounds, since floral odour is usually not stored but emitted as it is produced (Vogel 1983). In so-called headspace collections, the compounds in the headspace above a flower can be trapped on polymer (headspace sorption), in a cold trap (headspace cold trapping), or on a SPME fibre. Methods for floral scent collection have been reviewed in great detail (e.g. Dobson 1991; Kaiser 1991; Agelopoulos and

Pickett 1998; Millar and Sims 1998; Raguso and Pellmyr 1998). In the following sections we point out some important considerations to ensure that ecologically relevant odour samples are obtained, and describe the relatively new method of SPME.

9.2.1 Location of Floral Scent Emission

Floral scents are produced in a variety of flower parts such as petals, sepals, stamens, pistil, and pollen (Vogel 1983; Dobson 1994). Additionally, some flowers bear morphologically distinct structures for scent production, the osmophores (Vogel 1962). The different flower parts can produce distinctive odours, which can trigger specific behavioural reactions in pollinator insects (Bergström et al. 1995; Dobson et al. 1996; Raguso and Pichersky 1999). Pollen odour, for example, can act as close-range attractant for pollen-eating bees or beetles (von Aufsess 1960; Dobson et al. 1999). While flowers are ordinarily the source of pollinator-attracting odours, in some instances volatiles emitted from vegetative parts have been suspected to play a role in pollinator attraction (Vogel 1983). Therefore, odour collection from different parts of a flower might be necessary to sample physiologically active compounds that trigger specific behavioural reactions. It is generally advisable to check samples/fractions by EAG and/or behavioural tests prior to GC-EAD analyses to detect which samples bear biological activity. In order to collect fragrances from specific floral organs, flowers can be dissected and the scent of specific parts sampled. This, however, disconnects the tissue from its water supply and may disrupt metabolic processes. SPME may be useful to extract from different flower parts of intact flowers, when the fibre is brought in contact with the desired floral parts.

9.2.2 Variation of Scent Emission

The production of floral volatiles can vary in both space and time, which is an important consideration. Variation can be found between individuals of one plant species and even between flowers of one plant (Dobson 1991; Ayasse et al. 2000). Scent emission can be dependent upon the age of flowers and has been shown to change after cutting and also after pollination, which can influence the pollinators' behaviour (Tollsten and Bergström 1989; Schiestl et al. 1997). Some flowers show rhythmic volatile emission, often dependent on environmental factors such as the photoperiod (Vogel 1983; Altenburger and Matile 1988, 1990). The emission often correlates with pollinator activity. In some species, rhythmic emission is only expressed when the flowers are still attached to the plant (Matile and Altenburger 1988). Choice of sampling time and the use of intact plants should therefore be considered. Particularly when dealing with quantitative analyses, repeated sampling, possibly from more than one individual plant, is recommended (Raguso and Pellmyr 1998).

9.2.3 Choice of Type and Amount of Adsorbent Material

The most frequently used method for collecting floral odours is headspace sorption with Porapak Q (Super Q being a refined version), Chromosorb, Tenax TA, or activated charcoal (e.g. Supelco, Alltech) as the adsorbent material. These adsorbents can also be used in combination (Williams and Whitten 1983). Fragrances can be desorbed using solvent or heat. Porapak Q is most commonly used and has been recommended for its superior adsorbing capacity and recovery features (Dobson 1991; Raguso and Pellmyr 1998).

Since the quantity of floral odours released differs widely within plant species, the correct amount of adsorbent material is important. In the case of large amounts of volatiles being emitted by the flower, too little adsorbent material might saturate, quickly leading to bleed-through and loss of compounds. With low scent emission, however, small amounts of adsorbent are advantageous, since the sampled compounds might be masked by contaminants emanating from the adsorbent material. Schiestl et al. (1997) collected volatiles from *Ophrys sphegodes* flowers, which emit only a total amount of only 70 $ng^{-1} h^{-1} flower^{-1}$, by using cartridges containing as little as 1.5 mg of activated charcoal (Brechbühler AG, Urdorf, Switzerland; Grob et al. 1984; Kaiser 1991). Adsorbent tubes with greater amounts of charcoal proved to be completely useless due to internal contamination. Similar effects were observed with Porapak Q and glass tubes filled with 5 mg adsorbent performed well (Schiestl, unpubl. data).

A relatively new method is SPME (Supelco; Zhang and Pawliszyn 1993). This consists of a polymer-coated fused silica fibre onto which volatiles are adsorbed from a headspace or the surface of the biological material. The compounds are desorbed from the fibre through direct exposure to a heated GC-injection port operating in the splitless mode. Different fibre coatings can be chosen according to the compounds to be sampled. Special field samplers are available where adsorbed compounds are sealed behind a replaceable septum. Technical bulletins that describe different fibre coatings, cleaning procedures, etc. can be obtained from Supelco. SPME is a solvent-free method, which has the advantage of excluding interactions of solvent with odour compounds and avoiding a solvent peak in the GC analysis, which can easily mask other peaks. However, it does not provide a liquid sample, which is necessary for adding an internal standard for quantitative analysis; repeated analyses of one sample, frequently required in GC-EAD studies, as well as cool on-column injections are also impossible with this technique. However, because of its ease and flexibility, SPME might prove to be a useful tool in GC-EAD analyses of floral scent. Few studies have yet applied the SPME technique to collection of floral volatiles; MacTavish et al. (2000) used a 100-µm polydimethylsiloxane (PDS) fibre to sample volatiles from Brown Boronia flowers. Care must be taken when comparing SPME results with traditional collection techniques. The affinity and selectivity of SPME fibres towards different classes of molecules, and their small volume, which limits their trapping capacities, still need to be carefully evaluated (Bartelt 1997). In a comparative study, Agelopoulos and Pickett (1998) found the ratios of ripe banana volatiles trapped on a 100-µm PDS fibre to differ from Porapak Q and Tenax TA collections, with no differences between the two polymers.

9.3 Gas Chromatography

For a number of reasons, including its simplicity, high resolution and reproducibility of relative retention times, capillary gas chromatography (GC) has become the method of choice for analysing complex blends of volatile organic molecules (Heath and Dueben 1998). The flame ionisation detector (FID), which is most commonly used with GC, is extremely sensitive and detects nearly all classes of compounds. Floral scents, unlike many insect pheromones, typically comprise of a large number of compounds having different carbon chain lengths, functional groups, and isomeric configuration (Knudsen et al. 1993). Furthermore, abundances of individual compounds often differ greatly. The separation of such complex blends can be difficult and requires special attention. The next sections focus on methods for preliminary fractionation and GC separation of complex samples and the coupling of the GC with the electroantennographic detector.

9.3.1 Fractionation of Samples

A method to reduce the number of components in a complex sample is to split it into different fractions (Millar and Sims 1998). This is also useful if it is necessary to remove certain compounds from a sample. For example, low-volatility lipids are often the predominant compounds in flower extracts and can mask minor peaks with possible EAG activity. A sample can be separated into fractions of different polarity using a short bed of chromatographic packing (e.g. silica gel, Waters, Milford, Massachusetts). Fractions can be eluted from the column using solvents of different polarity. Thiéry et al. (1990) used purified pentane and ether to split sunflower extracts on a silica gel column into polar and non-polar fractions. Ômura et al. (2000) separated concentrated *Osmanthus fragrans* flower extract into six fractions using silica gel and hexane, benzene, benzene-chloroform (1:1), chloroform, acetone, and methanol. Andersen and Metcalf (1986) used preparative HPLC with a radial-compression-type silica gel column and a mobile phase of 10% tetrahydrofuran in hexane to collect 13 fractions of *Cucurbita maxima* volatiles. While fractionating can aid GC-EAD analyses, too many fractions will significantly complicate the procedure, since repetitive analyses of one sample type are generally necessary to discriminate reproducible EAD responses from shifts in the baseline due to noise, which may resemble an EAD response.

9.3.2 Injector Types

Two types of GC injectors are widely employed, each offering different advantages. Split/splitless injectors involve flash vaporisation of the sample and solvent, which are then swept onto the column by the carrier gas (Heath and Dueben 1998). The injector is heated to 200–300 °C and compounds that do not volatilise at this temperature will be retained. This is advantageous when crude flower extracts are injected, as these often contain low-volatility compounds that will not elute from the column. The second injector type that has commonly been used in GC-EAD studies with floral volatiles is the cool on-column injector (Thiéry et al. 1990; Wadhams et al. 1994; Blight et al. 1995). With this technique, the sample is directly introduced onto the column at low temperatures, thus avoiding degradation of thermally sensitive compounds. A retention gap (1–20 m of uncoated, deactivated, fused silica column) inserted between the injector and the analytical column is generally recommended, since it traps non-volatile compounds before they reach the analytical column (Heath and Dueben 1998).

9.3.3 Columns

The crucial part for the successful separation of compounds is the type of column used, which should be chosen according to the type of chemicals to be analysed. Most commonly used, and a good starting point for flower volatile analyses, are non-polar columns with dimethylpolysiloxane (e.g. DB-1, J&W Scientific) or 5% phenylmethylpolysiloxane (e.g. DB-5) bonded phases or polar columns with polyethylene glycol (e.g. DB-WAX). For the separation of enantiomers, which frequently differ in their biological activity, columns with chiral phases are often essential (Jones and Oldham 1999). To detect all physiologically active compounds in a heterogeneous sample, GC-EAD recordings with different columns may be necessary (Fig. 9.1). Cork et al. (1990) describe a GC-EAD system with two injectors and two columns connected at their outlet with a Y-tube connector, followed by a short capillary tubing and another Y-tube, which splits the effluent, with one part leading to the FID and one to the antenna. In this elegant setup, one column can be used to analyse a sample, with helium at 1 ml/min as carrier gas, whereas the other column delivers make-up gas at 2 ml/min for better performance of the split (see Sect. 9.3.4.1). The flow rates can be interchanged to use the other column for analysis. For simultaneous analysis of a sample using two types of stationary phases, two columns can also be connected in series (Jones and Oldham 1999). This will, however, lead to longer retention times, which is disadvantageous for GC-EAD analyses, since it is often difficult to maintain a live antennal preparation during the whole time needed for an analysis. This should also be considered when choosing the length of a single column: although longer columns generally lead to better separation, longer retention times may be a problem for electrophysiological recordings. As a good compromise, 30-m columns are commonly used for GC-EAD

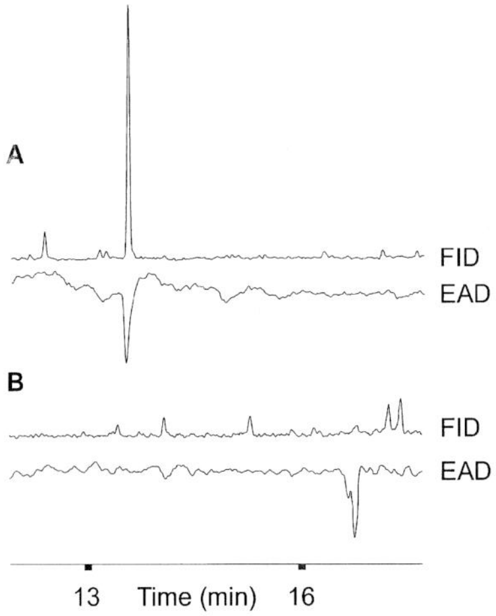

Fig. 9.1. GC-EAD analyses of the same floral odour sample on a non-polar DB-5 column (**A**) and a polar DB-WAX column (**B**). In **A**, the *big peak* seems to be the only active one. The second analysis (**B**), however, suggests that the real active peak is quite small and underlying the big one in analysis **A**. Additionally, the polar column resolves two active peaks. (Schiestl, unpubl. data)

analyses. The heating rate also influoftennces retention times and resolving power of a column. Whereas slow heating is often desirable for optimal resolution, in GC-EAD recordings, heating rates above 10 °C/min are generally recommended to shorten retention times and to ensure sharp peaks with a maximum amount of material being delivered onto the antenna at the onset of the peak.

9.3.4 Coupling the GC with the Electroantennographic Detector (EAD)

In order to record FID and EAD responses simultaneously, the end of the GC column needs to be split into two branches of equal length. One branch leads to the FID and the other to the EAG preparation through a heated transfer line inserted into the stimulus delivery tube (Fig. 9.2) Fig. 9.3.

9.3.4.1 Split

The use of a simple splitter (press-fit Y-connection, J&W Scientific) without the addition of a make-up gas generally works well (Thiéry et al. 1990; Schiestl et al. 2000). However, to reduce residence time and condensation of compounds in the transfer line, and to maximise FID responses, an addition of 30–60 ml/min nitrogen as make-up gas is advantageous (Fig. 9.2; Arn et al. 1975; Wadhams 1982). This can be achieved by using a four-way X-cross (SGE Australia Pty. Ltd.). The split ratio can be altered by varying the inner diameter of the fused silica transfer line or by using an adjustable microneedle valve effluent splitter (Scientific Glass Engineering Ltd., SGE). This feature might be useful if it is necessary to deliver more material to the antenna to obtain significant EAD responses.

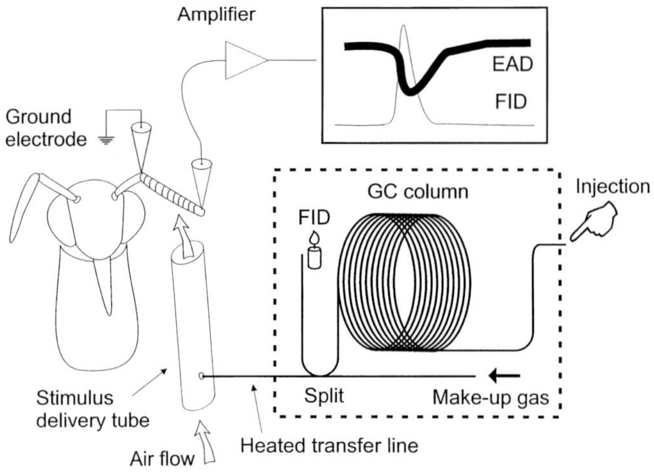

Fig. 9.2. Typical setup of a gas chromatograph with an electroantennographic detector (GC-EAD)

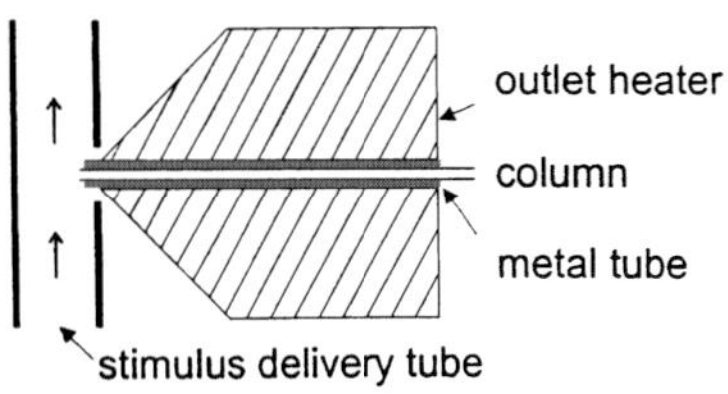

Fig. 9.3. Heating of the transfer line used by Schiestl et al. (2000) to reduce condensation of compounds on the column protruding in the cold air-flow through the stimulus delivery tube

9.3.4.2 Heating of the Transfer Line

To prevent condensation of the effluent in the tip of the split column, the transfer line into the cool air needs to be heated by an auxiliary heating system. The temperature is typically held at the maximum oven temperature or 10–15 °C above. Particularly for detecting EAD reactions to low-volatility compounds, it is crucial to properly heat the column part that leads into the cool airflow. This can be achieved by minimising the insulating air layer between the heater and the column by shielding the column in the stimulus delivery tube with a metal (optimally copper) tube, from which it protrudes only approx. 1 mm into the airflow (Fig. 9.3; Schiestl et al. 2000).

9.3.4.3 Air Flow Over the Antenna

To reduce desiccation of the antenna and avoid unwanted stimulation from contaminants in the laboratory air, the airflow (250–500 ml/min) through the stimulus delivery tube needs to be cleaned and humidified, before the effluent from the column is directed into it. This is most easily achieved by directing the air through two gas washing bottles, one filled with activated charcoal and the other with distilled water.

9.4 Electrophysiology

To identify biologically significant compounds eluting from a GC column, the coupling of a physical detector (typically FID) with an odourant-sensing organ seems an ideal approach. Two important assumptions are required, however. The first is that only chemicals that are detected by the olfactory systems are potentially relevant: if an insect is "blind" to a compound, it is unlikely that it will influence the behaviour. The second assumption is that, if a chemical has a particular biological significance, it will be detected with more sensitivity than the other components of a sample. Therefore, when presenting a series of compounds to an insect antenna, we should be able to identify the compounds most likely to influence its behaviour. In a natural flower bouquet, some compounds are emitted in large quantities whereas others are less pronounced. However, the compounds that are likely to mirror the effectiveness of the blend are not the most abundant ones, but those that are most detectable to the olfactory receptors. Compounds that are not detected by the olfactory receptor neurons (ORNs), although they might influence factors such as the emission rates of other volatiles, will only indirectly contribute to the biological activity and may be omitted from the analysis. In the following sections we briefly review the insect's olfactory system, describe techniques for recording summed receptor potentials and single sensillum activity, highlight possible solutions for sensitivity problems, and compare different methods.

9.4.1 Olfactory System

To identify which odours are detected by the olfactory system of insects, one can use electrophysiological methods and record the electrical activity from ORNs. Fortunately, most of the ORNs are grouped on the antenna, within individual sensing units called sensilla which each houses a small number of neurons. All sensilla consists of one or several ORNs, three modified epithelial cells called accessory cells, and a cuticular structure, which may differ to a great extent between insects (Kaissling 1986). The receptor cells extend their odourant-sensitive dendrites (branched or unbranched) into the sensillum lymph, while their axons project directly into the first synaptic relay (the antennal lobe) within the central nervous system. Combined anatomical and physiological observations have led to the separation of sensilla into several types. In moths, most hair-like sensilla trichodea house two or three ORNs (Steinbrecht 1987). In honeybees, pore plates (called sensilla placodea) house between 15 and 30 ORNs (Esslen and Kaissling 1976). On one antenna of a worker honeybee, there can be as many as 3,000 sensilla placodea (Lacher 1964). In all insects studied so far, the ORNs project into the antennal lobe, which is a central nervous structure, specialised to process this information (Hansson 1995). In the worker honeybee, one antenna bears about 65,000 ORNs that converge onto about 7,000 interneurons (Esslen and Kaissling 1976). The olfactory information thus encoded and processed is then sent to higher centres, mainly the mushroom bodies, where it is integrated with signals from other modalities and eventually memorised (Menzel 1993).

9.4.2 EAG

Two approaches are available to record an electrical signal from the antennae: single sensillum recording (SSR) and whole antenna electroantennography (EAG), with the latter being the most frequently used (Roelofs 1984). EAG records variations of potentials between the two ends of an insect antenna. During stimulation with an appropriate odourant molecule, this potential drops by a few millivolts. The shape of this response is similar to a receptor potential and it is claimed to be the sum of individual receptor potentials (Schneider 1962). Additional factors (air speed, temperature, variations in airflow), however, contribute to an EAG, and its true origins are still controversial (De Kramer and Hemberger 1987). The amplitude of the deflection is large when a significant number of olfactory receptors are simultaneously activated. There seems to be a good agreement between the intensity of the EAG response and the behavioural effectiveness of a given odourant. However, the converse is not true, i.e. the absence of a response does not imply that a substance is not detected. For example, a given odourant might stimulate, very efficiently, specialised neurons that are too few to induce a measurable EAG response. EAG has been used extensively in pheromone research on a number of insects, especially moths, to screen for effective pheromonal molecules (Roelofs 1984). The technique is simple and experience with different insect species has accumulated over the years. Used as an "electroantennographic detector (EAD)" it is suitable to detect ORNs reactions to compounds eluting from a GC column (Arn et al. 1975). Complete devices, including software, for EAG, SSR, GC-EAD, or GC-SSR are available from Syntech (Hilversum, The Netherlands; syntech@knoware.nl).

9.4.2.1 EAG Preparations

An adequate EAG signal can be obtained by different techniques. The simplest technique consists of maintaining a cut antenna between two silver electrodes, with the help of an electrocardiogram gel. This is adequate for large antennae and avoids the use of an expensive micromanipulator (antennal holders are available from

Peter Ockenfels, Freiburg, Germany, ockenfel@uni-freiburg.de). The gel prevents the desiccation of the antennal tissues and is convenient for a number of insects. The ends of a cut antenna can also be inserted into two glass capillaries mounted on a micromanipulator and filled with either insect's Ringer's solution or indeed any electrolyte solution, for example 0.1 M KCl. In case of problems with rapidly declining EAG responses, Bjostad (1998) recommends careful matching of ion concentration, osmolarity, and pH in the chemoreceptor cells and gives some recipes for electrolyte solutions. Arrangements with micromanipulators are quite convenient for small antennae and allow a great flexibility in the design of the recording setup.

Some insects are simply too small or too fragile to allow such approaches. Here, leaving the head connected to the antenna can be a possibility to get a good EAG signal (Cossé et al. 1995). In that case, the excised head can be mounted on the indifferent electrode, and the recording electrode placed at the end of one antenna. In some insects it is not necessary to cut the end of the antenna. A setup like this is good for recording EAG from insects as small as thrips (Pow et al. 1998) or aphids (Birkett et al. 2000). Some authors record the EAG from both antennae (Pow et al. 1998; Zhang et al. 1999). In general, the indifferent electrode can be inserted either at the base of one antenna, in the head, or even covering the other antenna. However, the best preparations are obtained by using undissected insects (Cork et al. 1990). In these cases, one has to design a system to immobilise the insect and choose an appropriate place to insert the indifferent electrode so that muscle potentials do not contaminate the EAG signal. Although the electrophysiological signal is generally better and the preparation more viable than with the simpler methods, one needs a more elaborate setup with several micromanipulators and a stereomicroscope placed close to the GC.

9.4.2.2 Recording an EAG

Recording an EAG signal is generally straightforward. The electrodes used for this purpose are of low impedance (around 1–10 MOhms), which means that EAG recordings are less prone to contamination by electrical interference than are SSR recordings. Shielding the preparation with a Faraday cage is usually not necessary, but may improve recordings in case of noise introduced by electrostatically charged air. In addition, this signal has a slow kinetic property and one can afford to sample it at a low frequency (1–10 Hz) provided that it is correctly filtered by pass-band filters (0.1–5 Hz, for example). Although many laboratories use standard electrophysiological amplifiers (with high-input impedance, typically 10^{12} Ohms) and filters, several authors have proposed custom-made amplifiers that perform reasonably well (Bjostad and Roelofs 1980; Bjostad 1998). In addition to amplifier and filters, an oscilloscope or a suitable computer with a data acquisition program is necessary to record and check the signal before and during an experiment. The EAG might be sampled along with the FID signal on a two-channel integrator. More integrated approaches (see, for example, Syntech) sample and store the electrophysiological signal along with the FID signal on a computer file (Thiéry and Marion-Poll 1998).

9.4.3 GC-SSR (GC-SCR)

GC-SSR (GC-SCR) (SSR, single sensillum recording; SCR, single cell recording) uses the same principle as GC-EAD, but recordings are made from individual sensilla (Boeckh 1962; Wadhams 1982). SSR gives a good indication of the sensitivity of individual neurons. It is particularly useful when the number of stimulated neurons is small and gives only a minute EAG. For example, one active compound from the aggregation pheromone of the bark beetle *Scolytus scolytus* was detected by GC-SSR but not by GC-EAD (Wadhams 1982, 1990).

9.4.3.1 Technique

Recordings can be made from intact ORNs by inserting finely etched tungsten electrodes into the cuticle (or at the surface of it) near the target sensilla. For some insects, it is feasible to use fine glass capillary electrodes filled with an electrolyte and to record the electrical signal just through the cuticle. If one wishes to record from sensilla trichodea, it may be simpler to cut the end of the hair and cap its tip with a capillary electrode filled with sensillum saline (Kaissling and Thorson 1980). These techniques are noticeably more difficult than EAG. They require a skilled operator and more complex electrophysiological equipment. When physical space is limited, one could make use of a portable SSR system, such as the one designed by Syntech, which can be used with the cut sensillum technique.

9.4.3.2 Signal Measurement

The electrical signal recorded with these techniques has two components: fast action potentials and slow potential changes during the stimulation. Slow potential changes are usually described as receptor potentials. A receptor potential is a depolarisation affecting the ORN's membrane when its acceptor sites are stimulated by an odour. The amplitude of this potential is directly linked to the number of sites occupied and will ultimately trigger action potentials. This term is misleading, however, since the electrodes are extracellular, and this potential possibly includes potentials from several ORNs and an active potential (called a trans-epithelial potential) generated by the accessory cells. Nonetheless, when the ORNs are stimulated by an odour, their membrane depolarises and one can record a downward-going potential deflection that has many of the characteristics of a true receptor potential. To our knowledge, this potential has never been used in the context of GC-SSR studies.

Most studies concentrate on action potentials (APs) since they carry the information that is conveyed to the central nervous system. An AP is a rapid event (2–3 ms) which usually has the shape of a small upward-trending hyperpolarisation (500 µV–3 mV) that quickly returns to the baseline. It should be stressed that APs are generated in the membranes of the individual neurons. If the electrodes were placed intracellularly, their amplitude would be much larger. An extracellular electrode picks up only a portion of the currents generated in the neurons, currents that follow different paths in the sensillum.

When a neuron is stimulated it may be excited or inhibited. With an exciting response, it discharges APs at a high rate, up to 200–300/s. When it is not stimulated, it usually remains silent or maintains a discharge rate of 1–3 APs per second. Since several ORNs may be housed within a given sensillum, one is never sure which neuron is firing. In many experimental situations, however, it is possible to discriminate APs of different neurons on the basis of their amplitude and shape. Actually, the neurons within a sensillum have slightly different sizes, leading to the differences in the signals recorded extracellularly from them.

9.4.4 Overcoming Problems of Low Sensitivity

Electroantennograms record sums of potentials from receptor cells located in the antenna. Therefore, the number of ORNs activated by a given compound is likely to influence the detectability of these reactions. The number of ORNs selective for floral odours may be lower than receptors sensitive for pheromones. In the honeybee, receptors for queen substance are the most frequent types in drones (Vareschi 1971). The worker's antenna seems to be less specialised for detecting pheromone signals (Brockman and Brückner 1998); however, Akers and Getz (1993) identified more ORNs sensitive to citral than to geraniol > linalool > limonene; citral and geraniol are constituents of the honeybee's Nasonov pheromone (Free 1987). Given that less ORNs react to a floral odour compound than to a pheromone substance, the former may be more difficult to detect in a summed recording since the signal-to-noise ratio will be poorer.

One possibility for increasing sensitivity in GC-EAD recordings is the use of more than one antenna. Antennae can be connected in series (Moore 1981) or parallel. Antennal holders for EAG recordings from more than one antenna are available from Syntech. In the honeybee, conditioning to the odour of interest prior to EAG analysis increased the EAG responses (De Jong and Pham-Delègue 1991; Wadhams et al. 1994), which might be useful in other insects as well. It should be stressed that olfactory receptors do not perform optimally when stimulated with odour compounds eluting from a GC. GC peaks are often quite long (10–20 s) and asymmetrical. Olfactory receptors respond best to fast-changing stimuli and to the logarithm of the concentration, i.e. they encode fluctuations well at low concentration but less accurately changes at higher concentrations (Marion-Poll and Thiéry 1996). In this respect, systems that use pulsed stimuli are more appropriate for EAG recordings. This has been considered by systems that collect the effluent

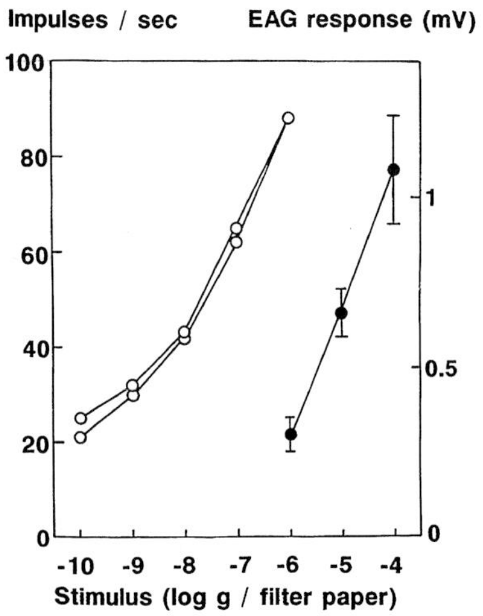

Fig. 9.4. Comparison of SCR (*open circles*) and EAG (*solid circles*) responses of honeybee antennae stimulated with 2-phenylethanol in various concentrations, demonstrating the lower threshold concentration in SSR. (Wadhams et al. 1994; C.M. Woodcock, unpubl. SSR data)

from the column and flush it over the antenna every 2–3 s (Cork et al. 1990; Gouinguené et al. 1998). These methods increase the detection threshold since the effluent is concentrated (10–100 times less molecules are needed) and reduce adaptation of antennal receptors, however, at the cost of losing resolution of the analysis.

SSRs are generally more sensitive than summed recordings, since the detection threshold for a compound is not dependent upon the number of receptors sensitive to it, and threshold concentrations for olfactory neurons have been shown to be similar, irrespective of their specificity (Fig. 9.4; Wadhams 1990). Indeed, studies employing this technique have detected ORNs sensitive for specific compounds even when no EAG reaction could be recorded (Wadhams 1984; Blight et al. 1995).

9.4.5 Comparison of EAG, GC-EAD, and GC-SSR

Although EAG has been used for the same purpose as GC-EAD by testing single synthetic compounds identified from flowers, the use of a coupled GC-EAD system has several advantages. Trace compounds with physiological activity are more likely to be detected in a coupled system, since it screens the whole sample and avoids a preselection of compounds to be tested. A GC provides a means to quantify compounds in a sample; therefore, the amounts actually being delivered onto the antenna are easier to estimate. Olfactory receptors often react very unspecifically when stimulated with large amounts of compounds (Hansson 1995). This might be one reason why many studies using EAG found all tested compounds to be physiologically active to some degree, whereas in GC-EAD investigations, typically only a subset of compounds elicited EAD reactions. When testing synthetic compounds, the GC provides a powerful device for purifying compounds, with even enantiomers being separable. Therefore, contamination, which can also contribute to an EAG signal, can usually be detected. However, in GC-EAD analyses, multiple compounds cannot be tested simultaneously, which is necessary when investigating the detection of whole blends versus single compounds; additionally, the whole setup is considerably more expensive than an EAG device only.

The GC-SSR technique generally provides lower detection thresholds for active compounds. Additionally, information on the specificity and abundance of receptor neurons can be obtained. Recording from single sensilla is considerably more difficult, however, and hence requires more training of the experimenter. Moderately expensive electrophysiological and microscopic equipment is necessary. The method is also more time consuming, since the recorded electrophysiological signal usually needs additional processing before being used (for example to count and discriminate spikes into classes). The approach requires the sampling of many sensilla before one gets a good image of the potentialities of the olfactory system.

9.5 Behavioural Tests

In order to evaluate the behavioural activity of specific odour compounds, they need to be tested in bioassays, since electroantennographic reactions are not necessarily correlated with behavioural responses. A problem in behavioural tests with floral volatiles is that reactions to these compounds may not be hard wired, i.e. they may not elicit innate reactions in the insects, as with sex pheromone compounds. Instead, associative learning as well as other signals emitted from flowers (e.g. colours or heat) are often involved and mediate behavioural responses, thus complicating the interpretation of a test result. To complete this chapter, we briefly overview behavioural tests that have been used in connection with EAG and GC-EAD studies.

9.5.1 Attraction Tests

In some systems, behavioural responses can be elicited by providing an odour stimulus only. Synthetic, physiologically active compounds were used to trap chrysomelid beetles, *Diabrotica virgifera virgifera* (Andersen and Metcalf 1986), and the cabbage seed weevil *Ceutorhynchus assimilis* (Blight et al. 1992) in the field. Gabel et al. (1992) trapped European gravevine moths, *Lobesia botrana*, in field cages. In other insects it may be important to also present appropriate visual cues, taking into account that UV reflection can be significant. In attraction tests with the butterfly *Pieris rapae*, Honda et al. (1998) used an Erlenmeyer flask covered with green paper and a doughnut-shaped, yellow filter-paper disk dyed with 0.1% Tartrazine on top of it. Stimulus chemicals were dispersed in water in the flask or applied on a white paper towel inserted within the flask. Schiestl et al. (2000) used dried, odourless, female bees (*Andrena nigroaenea*) onto which a blend of synthetic, pheromone-mimicking compounds identified in the early spider orchid (*Ophrys sphegodes*) was applied. These dummies were offered to male bees in the field.

9.5.2 Proboscis Extension

A commonly used bioassay involves recording the extension of the mouthparts in response to an odour stimulus, which can be interpreted as a feeding response. Such tests can be done either with unconditioned insects (proboscis extension reflex, PER) or with insects that had previously been conditioned to an odour stimulus (conditioned proboscis extension, CPE). In combined PER and attraction tests in *Pieris rapae* butterflies, EAG-active compounds with high PER activity stimulated visiting of artificial flowers more than compounds with low PER activity (Honda et al. 1998; Ômura et al. 1999b). PER was also shown to predict the deterrency potential of certain compounds in *P. rapae* better than EAG reactions (Ômura et al. 2000). In honeybees, however, individuals that were not previously conditioned to a stimulus showed no proboscis extension reaction to physiologically active odour compounds (Wadhams et al. 1994). Using the conditioned proboscis extension bioassay coupled with EAG and GC, these authors showed that most honeybees extended their proboscis only in response to three compounds out of a blend of six EAG active compounds, to which they had been conditioned (Fig. 9.5).

9.6 Compilation of Results

Tables 9.1–9.3 summarise studies on floral volatiles using GC-EAD or GC-SSR (Table 9.1), and EAG or SSR (Table 9.2). In Table 9.3, physiologically active floral odour compounds that have been identified in these studies are listed.

9.7 Concluding Remarks

Gas chromatography with electroantennographic detection provides an extremely useful tool for identification of physiologically active volatiles from complex blends of compounds. Although it has been used extensively in studies dealing with pheromones and general plant volatiles, surprisingly few investigations applied this technique to study the detection of floral odour by insects. The great diversity of plant–pollinator interactions, from generalised to extremely specialised systems,

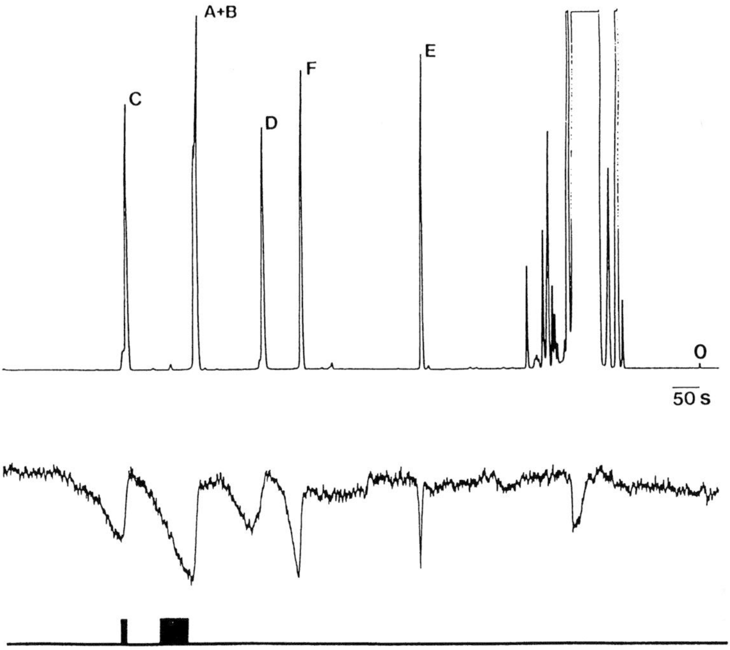

Fig. 9.5. Coupled GC-EAD-CPE recording with a six-component mixture using a conditioned honeybee. *Upper trace* represents the FID, *middle trace* EAD, and *lower trace* CPE responses. Tested compounds are *A* linalool, *B* 2-phenylethanol, *C* methyl salicylate, *D* benzyl alcohol, *E* (E)-2-hexenal, and *F* 1-octen-3-ol. (Wadhams et al. 1994; with kind permission from Kluwer Academic Publisher)

including mutual benefit and exploitation, is likely to be based on a variety of chemical communication systems. Knowledge about these mechanisms may therefore contribute to our general understanding of odour detection, insect behaviour, and the evolution of plant–insect interactions. We hope this review will stimulate GC-EAD research on floral volatiles, combined with chemical and behavioural studies.

Acknowledgements. We thank all authors that sent us their publications, especially Dainius Plepys (Department of Ecology, University of Lund), and Christine Woodcock (IACR-Rothamsted, Harpenden) who provided unpublished material. Further, we are grateful for valuable comments on the manuscript provided by Manfred Ayasse (Department of Evolutionary Biology, University of Vienna), Fernando Ibarra (Department of Organic Chemistry, University of Hamburg), Rod Peakall (Division of Botany and Zoology, The Australian National University), Jan van der Pers (Syntech, Hilversum), Dainius Plepys, Robert Raguso (Department of Biology, University of South Carolina), Denis Thiéry (Institute National de la Recherche Agronomique, Villenave d'Ornon cedex), and Christine Woodcock.

Table 9.1. GC-EAD and GC-SSR studies on floral odour compounds: list of investigated plant and insect species

No.	Plant sp. (Family)	Insect sp. (Order)	Motivation for flower visit	Odour collection[a]	Electrophysiology[b] Behavioural assay	Reference
1	*Brassica napus* (Brassicaceae)	*Ceutorhynchus assimilis* (Coleoptera)	Herbivory	HS	GC-EAD+GC-SCR, field trapping	Blight et al. (1992, 1995)
2	*Chrysanthemum* (*Dendranthema morifolium*, Asteraceae)	*Frankliniella occidentalis* (Thysanoptera)	Herbivory	HS	Olfactometer, glasshouse trapping	Pow et al. (1998)
3	*Cirsium arvense* (Asteraceae)	*Autographa gamma* (Lepidoptera)	Food reward	HS	Wind tunnel	Plepys (unpubl.); Plepys et al. (2000)
4	*Centaurea scabiosa* (Asteraceae)	*A. gamma*	Food reward	HS	Wind tunnel	Plepys (unpubl.)
5	*Helianthus annuus* (Asteraceae)	*Apis mellifera* (Hymenoptera)	Food reward	FE	–	Thiéry et al. (1990)
6	*Malus* x *domestica* (Rosaceae)	*Anthonomus pomorum* (Coleoptera)	Herbivory	HS	–	Kalinová et al. (2000)
7	*Medicago sativa* (Fabaceae)	*A. mellifera*	Food reward	Synthetics	–	Henning and Teuber (1992)
8	*Nepeta faasseni* (Lamiaceae)	*A. gamma*	Food reward	HS	Wind tunnel	Plepys (unpubl.)-
9	*Ophrys sphegodes* (Orchidaceae)	*Andrena nigroaenea* (Hymenoptera)	Mating behaviour	FE	Field tests	Schiestl et al. (2000)
10	*Platanthera bifolia* (Orchidaceae)	*A. gamma*	Food reward	HS	Wind tunnel	Plepys (unpubl.)-; Plepys et al. (2000)
11	*Saponaria officinalis* (Caryophyllaceae)	*A. gamma*	Food reward	HS	Wind tunnel	Plepys (unpubl.)-; Plepys et al. (2000)
12	*Tanacetum vulgare* (Asteraceae)	*Lobesia botrana* (Lepidoptera)	Unknown	FE	Attraction in field cage	Gabel et al. (1992)
13	*Trifolium pratense* (Fabaceae)	*A. gamma*	Food reward	HS	Wind tunnel	Plepys (unpubl.); Plepys et al. (2000)
14	General floral odour	*A. mellifera*	–	–	CPE	Wadhams et al. (1994); Pham-Delègue et al. (1997)

[a] HS: Headspace collection, FE: flower extracts
[b] GC-EAD if not otherwise mentioned
CPE: Conditioned proboscis extension

Table 9.2. EAG and SSR studies on floral odour compounds: list of investigated plant and insect species. *HS* Headspace collection; *FE* flower extracts

No.	Plant sp. (Family)	Insect sp. (Order)	Motivation for flower visit	Odour collection	Electrophysiology[a] Behavioural assay	Reference
1	*Brassica napus* (Brassicaceae)	*Apis mellifera* (Hymenoptera)	Food reward	HS	CPE	Pham-Delègue et al. (1992)
1.1		*Ceutorhynchus assimilis* (Coleoptera)	Herbivory	HS	–	Evans and Allen-Williams (1992)
2	*B. rapa* (Brassicaceae)	*Pieris rapae* (Lepidoptera)	Food reward	HS, FE	PER, attraction test	Ômura et al. (1999b)
3	*Clarika breweri* (Onagraceae)	*Sphinx elegans* (Lepidoptera)	Food reward	HS[b]	–	Raguso and Light (1998)
3.1		*Hyles lineata* (Lepidoptera)				Raguso et al. (1996)
4	*Cucurbita maxima* (Cucurbitaceae)	*Diabrotica undecimpunctata howardi* (Coleoptera)	Food reward	HS	Field trapping	Andersen and Metcalf (1986)
5	*Ligustrum japonicum* (Oleaceae)	*P. rapae*	Food reward	FE	PER, attraction tests	Honda et al. (1998)
6	*Ophrys insectifera* (Orchidaceae)	*Argogorytes mystaceus*	Mating behaviour	HS, FE	–	Ågren and Borg-Karlson (1984)
6.1	O. insectifera	*A. fargei*				
6.2	O. insectifera	*Gorytes quinquecinctus*				
6.3	O. insectifera	*G. quadrifasciatus* (all Hymenoptera)				
7	*Ophrys* spp.	*Eucera tuberculata* (Hymenoptera)	Mating behaviour	FE	–	Priesner (1973)
8	*Osmanthus fragrans* (Oleaceae)	*P. rapae*	Repelled	FE	PER, attraction tests	Ômura et al. (2000)
9	*Prunus yedoensis* (Rosaceae)	*Luehdorfia japonica* (Lepidoptera)	Food reward	FE	PER	Ômura et al. (1999a)
10	General floral odour	*Bombus hypnorum*, *B. terrestris* (Hymenoptera)	–	–	–	Fonta and Masson (1984)
11	General floral odour	*Spodoptera littoralis* (Lepidoptera)	–	–	SCR	Anderson et al. (1995)
12	General floral odour	*S. elegans*	–	–		Raguso and Light (1998)
13	General floral odour	*H. lineata*	–	–		Raguso et al. (1996)

[a]EAG if not otherwise mentioned
[b]According to Raguso and Pichersky (1995)
CPE: Conditioned proboscis extension; PER: Proboscis extension reflex; SCR: Single cell recording

Table 9.3. Physiologically active floral odour compounds, which have been identified by GC-EAD, GC-SSR, EAG, and SSR. *Superscript numbers* refer to numbers of references listed in Tables 9.1 and 9.2

GC-EAD and GC-SSR	EAG and SSR
Fatty acid derivatives	
Alkanes	**Alkanes**
Octadecane[13]	Undecane[6, 6.1]
Heneicosane[9]	Tridecane[6, 6.1]
Docosane[9]	Pentadecane[6, 6.1]
Tricosane[9]	Nonadecane[6]
Tetracosane[9]	
Pentacosane[9]	**Alkenes**
Hexacosane[9]	1-Heptadecene[6, 6.1]
Heptacosane[9]	
	Aldehydes
Alkenes	(*E*)-2-Hexenal[5]
(*Z*)-9-Pentacosene[9]	
(*Z*)-12+(*Z*)-11-Heptacosene[9]	**Ketones**
(*Z*)-9-Heptacosene[9]	(*Z*)-Jasmone[12, 13]
(*Z*)-12+(*Z*)-11-Nonacosene[9]	
(*Z*)-9-Nonacosene[9]	**Alcohols**
	Butanol[10]
Acids	(*E*)-2-Hexen-1-ol[5]
Hexadecanoic acid[5]	(*Z*)-3-Hexen-1-ol[1.1, 5]
	Octan-1-ol[5, 6, 6.1, 6.2, 6.3]
Aldehydes	1-Octen-3-ol[5]
Hexanal[1]	
(*E*)-2-Hexenal[1, 5, 14]	**Esters**
Octadecanal[9]	3-Methylbutyl acetate (Isoamyl acetate)[10]
Eicosanal[9]	(*Z*)-3-Hexen-1-yl acetate[1.1]
	Methyl jasmonate[13]
Ketones	γ-Decalactone[8]
3-Octanone[7]	
Alcohols	
1-Penten-3-ol[5]	
1-Pentanol[1]	
2-Methylbutanol[7]	
3-Methyl-1-butanol[5]	
1-Hexanol[1, 5]	
(*Z*)-3-Hexen-1-ol[1, 4]	
1-Octen-3-ol[1, 14]	
Tetradecanol[5]	
Esters	
Decyl acetate[7]	
Dodecyl acetate[7]	
(*Z*)-3-Hexenyl acetate[1, 2, 7]	
Methyl (D)-lactate[4]	
Methyl hexadecanoate[13]	
2-Nonyl tetradecanoate[9]	
2-Nonyl hexadecanoate[9]	
Dodecyl dodecanoate[9]	
Dodecyl tetradecanoate[9]	

Table 9.3. *Continued*

Aromatic compounds

GC-EAD and GC-SSR

Aldehydes
Benzaldehyde[1,3,10,13]
4-Methoxybenzaldehyde (p-Anisaldehyde)[1]
Phenylacetaldehyde[1,11]
Vanillin (4-Hydroxy-3-methoxybenzaldehyde)[5]
4-Hydroxy-2-methoxy cinnamaldehyde[5]

Alcohols
Benzyl alcohol[1,10,11,14]
2-Methoxyphenol (guaiacol)[1]
2-Phenylethanol[1,14]
Cinnamic alcohol[10]

Esters
Benzyl acetate[10]
Benzyl benzoate[3,10,11,13]
Benzyl salicylate[10]
Methyl benzoate[10,11]
Methyl 2-methoxybenzoate[3]
Methyl salicylate[1,2,3,7,10,14]

EAG and SSR

Acids
Benzoic acid[9]

Aldehydes
4-Methoxybenzaldehyde[9]
Benzaldehyde[2,5,9,12,13]
Phenylacetaldehyde[2,5,9,11,12,13]
Vanillin (4-Hydroxy-3-methoxy benzaldehyde)[3,3.1,10]
Veratraldehyde[3,3.1]
(*E*)-Cinnamic aldehyde[12,13]

Alcohols
Benzyl alcohol[1,2,5,9,12,13]
4-Methoxybenzylalcohol[9]
2-Phenylethanol[1,2,5,8,9,12,13]
Eugenol[3,3.1,5,10]
Methyl isoeugenol[3,3.1]

Esters
Amyl salicylate[12,13]
Benzyl acetate[3,3.1]
Benzyl benzoate[3,3.1]
Benzyl salicylate[12,13]
Ethyl phenylacetate[5,9]
Methoxy-2-methyl benzoate[12,13]
Methyl benzoate[5,12,13]
Methyl phenylacetate[5]
Methyl salicylate[1,3,3.1,5]
Methyl 4-methoxybenzoate[9]
Methyl cinnamate[12,13]
2-Phenylethyl acetate[5]

Isoprenoids

GC-EAD and GC-SSR

Monoterpenes
Bornyl acetate[5]
3-Carene[6]
Chrysanthenone[2]
1,8-Cineole[1,5]
p-Cymene[2,12]
d-Limonene[12]; Limonene[7]
Linalool oxide[3,13]
(±) Linalool[1,3]; Linalool[7,14]
Myrcene[7]
Myrtenal[5]
Myrtenol[5]
Ocimene[7,8]
Perillene[6]
Piperitone[12]
Sabinene[1]

EAG and SSR

Monoterpenes
1,8-Cineole[12,13]
Citronellol[6,6.1,6.2]
Geranial[10]
Geraniol[5,10,11,12,13]
Geranyl + neryl acetone[6,6.1]
Linalool[1,1.1,5,6.1,8,9,12,13]
Linalool oxides (*E*/*Z*-Furanoid)[3,3.1,8]
Linalool oxide (*Z*-Pyranoid)[3.1,8]
Linalool oxide (*E*-Pyranoid)[8]
d-Limonene[5]; Limonene[1.1,6,6.1,10,12,13]
β-Myrcene[5]; Myrcene[1.1,6,6.2,12,13]
Nerol[10]
(*E*)-β-Ocimene[12,13]
allo-Ocimene[13]
α-Pinene[1.1]

Table 9.3. *Continued*

GC-EAD and GC-SSR	EAG and SSR
(*E*)-Sabinene hydrate[5]	β-Pinene[1.1]
Terpinene-4-ol[12]	Sabinene[1.1]
(*Z*)-Thujenol[5]	Terpinene-4-ol[6, 6.1]
α-Thujene[12]	
α-Thujone[12]	**Sesquiterpenes**
β-Thujone + Thujyl alcohol[12]	γ-Cadinene[7]
(*Z*)-Verbenol[12]; verbenone, verbenol[1]	δ-Cadinene[7]
	Caryophyllene[1.1]
Sesquiterpenes	α-Cedrene[1.1]
(±)-Caryophyllene[4,6,13], β-Caryphyllene[2]	α-Copaene[7]
Caryophyllen oxide[5]	Cyclosativene + cyclocopacamphene[6, 6.1]
β-Elemene[5]	α-Farnesene[1.1]
(*E,E*)-α-Farnesene[1, 2, 6]; α-Farnesene[3, 11, 13]	Farnesol[6, 12, 13]
(*E*)-β-Farnesene[2]	Germacrene D[12]
Farnesol[9]	γ-Muurolene[7]
Farnesyl acetone[13]	Nerolidol[3, 6]
Farnesyl hexanoate[9]	(*E*)-Nerolidol[5]
Germacrene D[5]	α-Ylangene[7]
	Irregular terpenoids
	α-Ionone[8]
	β-Ionone[8]
	6-Methyl-5-hepten-2-one[5]
Nitrogen-containing compounds	
Allyl isothiocyanate[1]	2,5-Dimethyl-3-isoamylpyrazine[6, 6.1, 6.2, 6.3]
Benzyl cyanide (Phenylacetonitrile)[1]	2,6-Dimethyl-3-isoamylpyrazine[6, 6.2]
3-Butenyl isothiocyanate[1]	5,6-Dimethyl-2-isoamylpyrazine[6, 6.1, 6.2, 6.3]
Indole[1, 13]	Indole[2, 4, 12, 13]
Goitrin[1]	2-Methyl-butyraldoxime[12, 13]
4-Pentenyl isothiocyanate[1]	Phenylacetonitrile[2]
2-Phenylethyl isothiocyanate[1]	Methyl anthranylate[12, 13]
Others	
2,3,3-Trimethyl-epoxy cyclopentyl acetaldehyde[5]	
Naphtalene[13]	
Propiovallinone[5]	

References

Agelopoulos NG, Pickett JA (1998) Headspace analysis in chemical ecology: effects of different sampling methods on ratios of volatile compounds present in headspace samples. J Chem Ecol 24(7):1161–1172

Agelopoulos N, Birkett MA, Hick AJ, Hooper AM, Pickett JA, Pow EM, Smart LE, Smiley DWM, Wadhams LJ, Woodcock CM (1999) Exploiting semiochemicals in insect control. Pestic Sci 55:225–235

Ågren L, Borg-Karlson A-K (1984) Responses of *Argogorytes* (Hymenoptera: Sphecidae) males to odour signals from *Ophrys insectifera* (Orchidaceae). Preliminary EAG and chemical investigation. Nova Acta Reg Soc Sci Ups Ser 5(C3):111–117

Akers RP, Getz WM (1993) Response of olfactory receptor neurons in honeybees to odorants and their binary mixtures. J Comp Physiol A 173:169–185

Altenburger R, Matile P (1988) Circadian rhythmicity of fragrance emission in flowers of *Hoya carnosa* R. Br Planta 174:248–252

Altenburger R, Matile P (1990) Further observations on rhythmic emissions of fragrance in flowers. Planta 180:194–197

Andersen JF, Metcalf RL (1986) Identification of a volatile attractant for *Diabrotica* and *Acalymma* spp. from blossoms of *Cucurbita maxima* Duchnesne. J Chem Ecol 12(3):687–699

Anderson P, Hansson BS, Löfquist J (1995) Plant-odour-specific receptor neurons on the antennae of female and male *Spodoptera littoralis.* Physiol Entomol 20:189–198

Arn H, Städler E, Rauscher S (1975) The electroantennographic detector – a selective and sensitive tool in the gas chromatographic analysis of insect pheromones. Z Naturforsch 30c: 722–725

Ayasse M, Schiestl FP, Paulus HF, Löfstedt C, Hansson BS, Ibarra F, Francke W (2000) Evolution of reproductive strategies in the sexually deceptive orchid *Ophrys sphegodes*: how does flower-specific variation of odor signals influence reproductive success? Evolution 54(6):1995–2006

Bartelt RJ (1997) Calibration of a commercial solid-phase microextraction device for measuring headspace concentrations of organic volatiles. Anal Chem 69(3):364–372

Bergström G, Dobson HEM, Groth I (1995) Spatial fragrance pattern within the flowers of *Ranunculus acris* (Ranunculaceae). Plant Syst Evol 195:221–242

Birkett MA, Campbell CAM, Chamberlain K, Guerrieri E, Hick AJ, Martin JL, Matthes M, Napier JA, Petterson J, Pickett JA, Poppy GM, Pow EM, Pye BJ, Smart LE, Wadhams GH, Wadhams LJ, Woodcock C (2000) New roles for *cis*-jasmone as an insect semiochemical and in plant defense. Proc Natl Acad Sci USA 97(16):9329–9334

Bjostad LB (1998) Electrophysiological methods. In: Millar JG, Haynes KF (eds) Methods in chemical ecology, vol 1. Chemical methods. Chapman and Hall, Norwell, pp 339–375

Bjostad LB, Roelofs WL (1980) An inexpensive electronic device for measuring electroantennogram responses to sex pheromone components with a voltmeter. Physiol Ent 5:309–314

Blight MM, Hick AJ, Pickett JA, Smart LE, Wadhams LJ, Woodcock CM (1992) Volatile plant metabolites involved in host-plant recognition by the cabbage seed weevil, *Ceutrhynchus assimilis.* In: Proc 8th Int Symp Insect–Plant Relationships. Kluwer, Dordrecht, pp 105–106

Blight MM, Pickett JA, Wadhams LJ, Woodcock CM (1995) Antennal perception of oilseed rape, *Brassica napus* (Brassicaceae), volatiles by the cabbage seed weevil *Ceutorhynchus assimilis* (Coleoptera, Curculionidae). J Chem Ecol 21(11):1649–1664

Boeckh J (1962) Elektrophysiologische Untersuchungen an einzelnen Geruchsrezeptoren auf den Antennen des Totengräbers (*Necrophorus.* Coleoptera). Z Vergl Physiol 46:212–248

Brockmann A, Brückner D (1998) The EAG response spectra of workers and drones to queen honeybee mandibular gland components: the evolution of a social signal. Naturwissenschaften 85:283–285

Cork A, Beevor PS, Gough AJE, Hall DR (1990) Gas chromatography linked to electroantennography: a versatile technique for identifying insect semiochemicals. In: Proc 1st Int Symp on Chromatography and Isolation of Insect Hormones, Pheromones and Related Compounds. Plenum Press, New York, pp 271–279

Cossé AA, Todd JL, Millar JG, Martinez LA, Baker TC (1995) Electroantennographic and coupled gas chromatographic electroantennographic responses of the Mediterranean fruit fly, *Ceratitis capitata*, to male-produced volatiles and mango odor. J Chem Ecol 21(11):1823–1836

De Jong R, Pham-Delègue M-H (1991) Electroantennogram responses related to olfactory conditioning in the honey bee (*Apis mellifera ligustica*). J Insect Physiol 374:319–324

De Kramer JJ, Hemberger J (1987) The neurobiology of pheromone reception. In: Prestwich GD, Blomquist GJ (eds) Pheromone biochemistry. Academic Press, Orlando, pp 433–472

Dobson HEM (1991) Analysis of flower and pollen volatiles. In: Linskens HF, Jackson JF (eds) Modern methods of plant analysis, new series, vol 12. Springer, Berlin Heidelberg New York, pp 231–251

Dobson HEM (1994) Floral volatiles in insect biology. In: Bernays EA (ed) Insect–plant interactions, vol V. CRC Press, Boca Raton, pp 47–81
Dobson HEM, Groth I, Bergström G (1996) Pollen advertisement: chemical contrasts between whole-flower and pollen odors. Am J Bot 83(7):877–885
Dobson HEM, Danielson EM, Van Wesep ID (1999) Pollen odor chemicals as modulators of bumble bee foraging on *Rosa rugosa* Thunb (Rosaceae). Pl Spec Biol 14:153–166
Dodson CH, Dressler RL, Hills HG, Adams RM, Williams NH (1969) Biologically active compounds in orchid fragrances. Science 164:1243–1249
Esslen J, Kaissling K-E (1976) Zahl und Verteilung antennaler Sensillen bei der Honigbiene (*Apis mellifera* L.). Zoomorphologie 83:227–251
Evans KA, Allen-Williams LJ (1992) Electroantennogram responses of the cabbage seed weevil, *Ceutorhynchus assimilis*, to oilseed rape, *Brassica napus* ssp. *oleifera*, volatiles. J Chem Ecol 18(9):1641–1659
Fonta C, Masson C (1984) Comparative study by electrophysiology of olfactory responses in bumblebees (*Bombus hypnorum* and *Bombus terrestris*). J Chem Ecol 10(8):1157–1169
Free JB (1987) Pheromones of social bees. Chapman and Hall, London
Gabel B, Thiéry D, Suchy V, Marion-Poll F, Hradsky P, Farkas P (1992) Floral volatiles of *Tanacetum vulgare* L. attractive to *Lobesia botrana* Den. et Schiff. females. J Chem Ecol 18(5):693–701
Gouinguené S, de Cruz I, van der Pers J, Wadhams L, Marion-Poll F (1998) A new method to improve olfactory responses to GC effluents. Chem Sens 23:647–652
Grob K, Grob G, Habich A (1984) Overcoming background contamination in closed loop stripping analysis (CLSA). J High Resolut Chromatogr Commun 7:340–342
Hansson BS (1995) Olfaction in Lepidoptera. Experientia 51:1003–1027
Heath RR, Dueben BD (1998) Analytical and preparative gas chromatography. In: Millar JG, Haynes KF (eds) Methods in chemical ecology, vol 1. Chemical methods. Chapman and Hall, Norwell, pp 185–125
Henning AJ, Teuber RL (1992) Combined gas chromatography–electroantennogram characterization of alfalfa floral volatiles recognized by honey bees (Hymenoptera: Apidae). J Econ Entomol 85(1):226–232
Honda K, Ômura H, Hayashi N (1998) Identification of floral volatiles from *Ligustrum japonicum* that stimulate flower-visiting by cabbage butterfly, *Pieris rapae*. J Chem Ecol 24(12):2167–2180
Jones GJ, Oldham NJ (1999) Pheromone analysis using capillary gas chromatographic techniques. J Chromatogr A 843:199–236
Kaiser R (1991) Trapping, investigation and reconstitution of flower scents. In: Müller PM, Lamparsky (eds) Perfumes – art, science and technology. Elsevier, London, pp 213–250
Kaiser R (1995) New or uncommon volatile compounds in floral scents. In: Proc 13th Int Congr of Flavours, Fragrances and Essential Oils, Istanbul, pp 1–34
Kaissling K-E (1986) Chemo-electrical transduction in insect olfactory receptors. Annu Rev Neurosci 9:121–145
Kaissling K-E, Thorson J (1980) Insect olfactory sensilla: structural, chemical and electrical aspects of the functional organization. In: Sattelle DB, Hall LM, Hildebrand JG (eds) Receptors for neurotransmitters, hormones and pheromones in insects. Elsevier/North-Holland, Amsterdam, pp 261–282
Kalinová B, Stránsky K, Harmantha J, Čtvrtečka R, Žd' árek J (2000) Can chemical cues from blossom buds influence cultivar preference in the apple blossom weevil (*Anthonomus pomorum*)? Entomol Exp Appl 95:47–52
Knudsen JT, Tollsten L, Bergström G (1993) Floral scents – a checklist of volatile compounds isolated by headspace techniques. Phytochemistry 332:253–280
Kullenberg B (1961) Studies in *Ophrys* pollination. Almquist and Wiksells Boktrykeri, Uppsala
Lacher V (1964) Elektrophysiologische Untersuchungen an einzelnen Rezeptoren für Geruch, Kohlendioxyd, Luftfeuchtigkeit und Temperatur auf den Antennen der Arbeitsbiene und der Drone (*Apis mellifica* L.). Z Vergl Physiol 48:587–623
MacTavish HS, Davies NW, Menary RC (2000) Emission of volatiles from brown Boronia flowers: some comparative observations. Ann Bot 86:347–354
Marion-Poll F, Thiéry D (1996) Dynamics of EAG responses to host plant volatiles delivered by a gas chromatograph. Entomol Exp Appl 80(1):120–123
Matile P, Altenburger R (1988) Rhythms of fragrance emission in flowers. Planta 174:242–247
Menzel R (1993) Associative learning in honey bees. Apidologie 24(3):157–168

Millar JG, Sims JJ (1998) Preparation, cleanup, and preliminary fractionation of extracts. In: Millar JG, Haynes KF (eds) Methods in chemical ecology, vol 1. Chemical methods. Chapman and Hall, Norwell, pp 1–37

Moore I (1981) Biological amplification for increasing electroantennogram discrimination between two female sex pheromones of *Spodoptera littoralis* (Lepidoptera: Noctuidae). J Chem Ecol 7(5):791–798

Moorhouse JE, Yeadon R, Beevor PS, Nesbitt BF (1969) Method for use in studies of insect chemical communication. Nature 223:1174–1175

Ômura H, Honda K, Nakagawa A, Hayashi N (1999a) The role of floral scent of the cherry tree, *Prunus yedoensis*, in the foraging behavior of *Luehdorfia japonica* (Lepidoptera: Papilionidae). Appl Entomol Zool 34(3):309–313

Ômura H, Honda K, Hayashi N (1999b) Chemical and chromatic bases for preferential visiting by the cabbage butterfly, *Pieris rapae*, to rape flowers. J Chem Ecol 25(8):1895–1906

Ômura H, Honda K, Hayashi N (2000) Floral scent of *Osmanthus fragrans* discourages foraging behavior of cabbage butterfly, *Pieris rapae*. J Chem Ecol 26(3):655–666

Peakall R, Beattie AJ (1996) Ecological and genetic consequences of pollination by sexual deception in the orchid *Caladenia tentactulata*. Evolution 50(6):2207–2220

Pellmyr O, Thien LB (1986) Insect reproduction and floral fragrances: keys to the evolution of the angiosperms? Taxon 351:76–85

Pham-Delègue MH, Blight MM, Le Métayer M, Marion-Poll F, Picard AL, Pickett JA, Wadhams LJ, Woodcock CM (1992) Plant chemicals involved in the honeybee–rapeseed relationships: behavioural, electrophysiological and chemical studies. In: Proc 8th Int Symp Insect–Plant Relationships. Kluwer, Dordrecht, pp 129–130

Pham-Delègue MH, Blight MM, Kerguelen V, Le Métayer M, Marion-Poll F, Sandoz JC, Wadhams L (1997) Discrimination of oilseed rape volatiles by the honeybee: combined chemical and biological approaches. Entomol Exp Appl 83:87–92

Plepys D, Ibarra F, Löfstedt C, Hansson B, Francke W (2000) Behavioural and electrophysiological responses of silver Y moth, *Autographa gamma* (Lepidoptera: Noctuidae), to floral volatiles. In: Proc 21st Int Congr on Entomology, Foz do Iguassu, p 180

Pow EM, Bennison JA, Birkett MA, Luszniak MJ, Manjunatha M, Pickett JA, Segers IS, Wadhams LJ, Wardlow LR, Woodcock CM (1998) Behavioural responses of western flower thrips (*Frankliniella occidentalis*) to host plant volatiles. In: Proc 6th Int Symp on Thysanoptera, Antalya, pp 121–128

Priesner E (1973) Reaktionen von Riechrezeptoren männlicher Solitärbienen (Hymenoptera, Apoidea) auf Inhaltsstoffe von *Ophrys*-Blüten. Zoon Suppl 1:43–54

Raguso AR, Light MD (1998) Electroantennogram responses of male *Sphinx perelegans* hawkmoths to floral and "green-leaf volatiles". Entomol Exp Appl 86:287–293

Raguso AR, Pellmyr O (1998) Dynamic headspace analysis of floral volatiles: a comparison of methods. Oikos 81:238–254

Raguso AR, Pichersky E (1995) Floral volatiles from *Clarika breweri* and *C. concinna* (Onagraceae): recent evolution of floral scent and moth pollination. Plant Syst Evol 194:55–67

Raguso AR, Pichersky E (1999) A day in the life of a linalool molecule: chemical communication in a plant–pollinator system, part 1: linalool biosynthesis in flowering plants. Plant Spec Biol 14:95–120

Raguso AR, Light MD, Pickersky E (1996) Electroantennogram responses of *Hyles lineata* (Sphingidae: Lepidoptera) to volatile compounds from *Clarika breweri* (Onagraceae) and other moth-pollinated flowers. J Chem Ecol 22(10):1735–1767

Roelofs WL (1984) Electroantennogram assays: rapid and convenient screening procedures for pheromones. In: Hummel HE, Miller TA (eds) Techniques in pheromone research. Springer, Berlin Heidelberg New York, pp 131–159

Sazima M, Vogel S, Cocucci A, Hausner G (1993) The perfume flowers of *Cyphomandra* (Solanaceae)-pollination by euglossine bees, bellows mechanism, osmophores, and volatiles. Pl Syst Evol 187(1–4):51–88

Schiestl FP, Ayasse M, Paulus HF, Erdmann D, Francke W (1997) Variation of floral scent emission and postpollination changes in individual flowers of *Ophrys sphegodes* subsp. *sphegodes*. J Chem Ecol 23 (12):2881–2895

Schiestl FP, Ayasse M, Paulus HF, Löfstedt C, Hansson BS, Ibarra F, Francke W (1999) Orchid pollination by sexual swindle. Nature 399:421–422

Schiestl FP, Ayasse M, Paulus HF, Löfstedt C, Hansson BS, Ibarra F, Francke W (2000) Sex pheromone mimicry in the early spider orchid (*Ophrys sphegodes*): patterns of hydrocarbons as the key mechanism for pollination by sexual deception. J Comp Physiol A 186:567–574
Schneider D (1957) Elektrophysiologische Untersuchungen von Chemo- und Mechanorezeptoren der Antenne des Seidenspinners (*Bombyx mori* L.). Z Vergl Physiol 40:8–41
Schneider D (1962) Electrophysiological investigation on the olfactory specificity of sexual attracting substances in different species of moths. J Insect Physiol 8:15–30
Steinbrecht RA (1987) Functional morphology of pheromone-sensitive sensilla. In: Prestwich GD, Blomquist JG (eds) Pheromone biochemistry. Academic Press, New York, pp 353–383
Struble DL, Arn H (1984) Combined gas chromatography and electroantennogram recording of insect olfactory responses. In: Hummel HE, Miller TA (eds) Techniques in pheromone research. Springer, Berlin Heidelberg New York, pp 161–178
Tan K-H, Nishida R (2000) Mutual reproductive benefits between a wild orchid, *Bulbophyllum patens*, and *Bactrocera* fruit flies via a floral synomone. J Chem Ecol 262:533–546
Thiéry D, Marion-Poll F (1998) Electroantennogram responses of Douglas-fir seed chalcids to plant volatiles. J Insect Physiol 44:483–490
Thiéry D, Bluet JM, Pham-Delègue MH, Etiévant P, Masson C (1990) Sunflower aroma detection by the honeybee; study by coupling gas chromatography and electroantennography. J Chem Ecol 163:701–711
Tollsten L, Bergström J (1989) Variation and post-pollination changes in floral odours released by *Platanthera bifolia*. Nord J Bot 9(4):359–362
Vareschi E (1971) Duftunterscheidung bei der Honigbiene – Einzelzell-Ableitungen und Verhaltensreaktionen. Z Vergl Physiol 75:143–173
Vogel S (1962) Duftdrüsen im Dienste der Bestäubung; über Bau und Funktion der Osmophoren. Akad Wiss Lit (Mainz) Abh Math-Nat Kl (10):1–165
Vogel S (1983) Ecophysiology of zoophilic pollination. In: Lange OL, Nobel PS, Osmond CB, Ziegler H (eds) Encyclopedia of plant physiology, new series vol 12C. Springer, Berlin Heidelberg New York, pp 560–611
von Aufsess A (1960) Geruchliche Nahorientierung der Biene bei entomophilen und ornithophilen Blüten. Z Vergl Physiol 43:129–137
Wadhams LJ (1982) Coupled gas chromatography – single cell recording: a new technique for use in the analysis of insect pheromones. Z Naturforsch 37c:947–952
Wadhams LJ (1984) The coupled gas chromatography–single cell recording technique. In: Hummel HE, Miller TA (eds) Techniques in pheromone research. Springer, Berlin Heidelberg New York, pp 179–189
Wadhams LJ (1990) The use of coupled gas chromatography-electrophysiological techniques in the identification of insect pheromones. In: McCaffery AR, Wilson ID (eds) Chromatography and isolation of insect hormones and pheromones. Plenum Press, New York, pp 289–298
Wadhams LJ, Blight MM, Kerguelen V, Le Métayer M, Marion-Poll F, Masson C, Pham-Delègue MH, Woodcook CM (1994) Discrimination of oilseed rape volatiles by honey bee: novel combined gas chromatographic–electrophysiological behavioral assay. J Chem Ecol 20(12): 3221–3231
Williams NH, Whitten WM (1983) Orchid floral fragrances and male euglossine bees: methods and advances in the last sesquidecade. Biol Bull 164:355–395
Zhang ZY, Pawliszyn J (1993) Analysis of organic compounds in environmental samples by headspace solid phase microextraction. HRC J High Resolut Chromatogr 16(12):689–692
Zhang AJ, Linn C, Wright S, Prokopy R, Reissig W, Roelofs W (1999) Identification of a new blend of apple volatiles attractive to the apple maggot, *Rhagoletis pomonella*. J Chem Ecol 25(6): 1221–1232

10 Analysis of Rhythmic Emission of Volatile Compounds of Rose Flowers

J.P.F.G. Helsper, J.A. Davies, and F.W.A. Verstappen

10.1 Introduction: Rhythmicity in Emission of Volatile Compounds, How and Why

Many phenomena in plants and animals show rhythmic patterns in activity and/or appearance. This rhythmicity may be controlled directly by periodic changes in environmental conditions such as temperature or light, e.g. generation of electrons by photosynthetic pigments in relation to daily changes in exposure to light. Many other processes, which are dependent on this activity, will show rhythmicity with the same interval but with a delay from which the length of time is dependent on intermediary reactions. When there is no further regulatory mechanism other than the abundance or intensity of the external regulatory signal, the rhythmicity of the phenomena will disappear when the signal loses its rhythmicity, e.g. when the plant is kept in constant dark or light.

Some phenomena, which follow a rhythmically changing environment, continue to show rhythmicity even when the external signal becomes constant. Such endogenous or "circadian" rhythms were first discovered for plants by De Mairan (1729) and later more extensively described by Bünning and Stern (1930) who observed that movements of bean leaves continued under constant light or dark. Jones and Mansfield (1975) formulated two additional characteristics which are accepted as typical for circadian rhythmicity. These additional requirements are: (1) a cycle time upon exposure to constant environmental conditions, the so-called free-running period, which is slightly different – mostly longer – than 24 h, and (2) the possibility to induce a phase shift in rhythmicity by changing the periodicity of the signal that controls the cycle, e.g. by introduction of a single 24-h period of constant darkness into a 12-h light /12-h dark regime.

One of the phenomena that is strongly regulated by light–dark cycles is the emission of volatile compounds by flowers. Rhythmicity in emission is related to the activity pattern of pollinating insects (Gimenes et al. 1996). Therefore, plant species that are pollinated by night-flying insects such as moths show nocturnal rhythmicity, with maximum emission during the dark period, while others that are pollinated by bumble bees or honey bees exhibit a diurnal pattern with maximum

Plant Research International, Business Unit Cell Cybernetics, Bornsesteeg 65, 6708 PD Wageningen, The Netherlands emission during day light. Nocturnal emission of volatiles is often observed to be circadian in nature, while diurnal emission by flowers is thought to be controlled directly by light levels (Jakobsen and Olsen 1994; Jakobsen et al. 1994). However, Helsper et al. (1998) observed that emission of virtually all the volatiles by rose flowers shows a diurnal, circadian rhythmicity. The rationale for a controlled emission is the high metabolic cost associated with the biosynthesis of most volatile compounds (Gershenzon 1994). The majority of these belong to three major chemical classes: terpenoid, aromatic and aliphatic compounds (Knudsen et al. 1993). Some sulphur-containing volatile compounds are also observed (Flament et al. 1992). Recently, emission of volatile compounds has also been found to have a function in plant resistance to pest insects. In these tripartite interactions, the herbivory insect itself induces the emission by the attacked host plant of volatiles which attract carnivorous mites that are predatory to the pest insect (Takabayashi et al. 1994; Loughrin et al. 1995, 1997; Bouwmeester et al. 1999).

Accurate quantitative analysis of volatile compounds has developed gradually from semiquantitative to fully quantitative observations when GCMS analysis became accessible as a common laboratory facility. Overland (1960) described a sensorically detectable rhythmicity in fragrance strength as emitted by flowers of *Cestrum nocturnum*. More accurate but still semiquantitative analyses, giving quantities as detector responses on GCMS analyses, have been reported since Harada and Mihara (1984) showed the predominant emission of linalool by Freesia flowers. Another early semiquantitative report on rhythmicity was presented by Mookherjee et al. (1990) who described differences between day and night in composition of emitted volatiles, expressed as proportion of total emitted compounds, for several other plant species. Early quantitative data on volatile emission by flowers, given as weight unit, mostly in nanograms or picomoles of emitted compound per flower per day or hour, came from Matile and Altenburger (1988) and from Patt et al. (1988). The latter way of expressing meets the standards for complete quantification of volatile emission. Jakobsen and Olsen (1994) and Jakobsen et al. (1994) were among the first authors to quantitatively describe circadian rhythmicity in volatile emission by *Brassica* and *Trifolium* species. Whether this rhythmicity was due to petal movement was not described. Also losses, due to incomplete recovery of volatiles between the moment of emission by the flower and injection into the GC or GCMS, have not been accounted for.

In 1991 Dobson gave a systematic survey of analytical aspects, such as containment of plant samples, adsorption and desorption of volatile compounds and GCMS analysis using whole plants, flowers and distinctive flower parts (stamen, pollen, petals) in *Modern Methods of Plant Analysis*, the predecessor of this series. The present paper will update this survey, with special emphasis on quantitation of volatiles emitted from flowers of *Rosa hybrida*, which is the most important ornamental species for various market sections, such as florists and the perfume industry. Rose flowers do not show a daily rhythmicity in petal movement. Therefore, the circadian nature of the rhythmic emission patterns cannot be attributed to this physical rhythmic phenomenon, but is more likely a reflection of the chemical composition of the emitting tissue. Therefore, we investigated the concentrations of glycosylated and non-glycosylated precursors of these volatile compounds in petals.

10.2 Rhythmicity in Emitted Volatiles

10.2.1 Methods

10.2.1.1 Plant Containment

Dependent on the size of the plant material and the desired level of environmental control, several experimental designs for collection of emitted volatiles have been described. The size of plant material may vary from very small inflorescences to whole plants. Under laboratory conditions small experimental objects, e.g. flowers on a flower stem, can be contained in a three-necked round-bottom flask sealed with a Teflon stopper. The flower stem is guided through a trough in the stopper placed in the middle neck and sealed with paraffin. The two other necks are used for the in- and outgoing airflows. Larger plant samples can be contained in glass jars with a smooth and flat lower rim which is placed on a Teflon or glass platform to give an airtight seal (for examples see Flament 1991/1992; Jakobsen and Olsen 1994). The use of paraffin in seals should be minimized because of the possible release of contaminating volatile alkanes. The ingoing air flow can be purified by the same adsorbing material used to collect the volatiles emitted by the flower. Leakages in the containment can be detected by measuring the air flows at the inlet and outlet side.

10.2.1.2 Environmental Conditions

Control of environmental conditions, in particular temperature, is very important to obtain reliable quantitative results. Petal temperature can be affected considerably by accumulation of radiation energy in the flower containment. This will affect the amount of volatiles released by the plant and will confuse effects attributed to light only. We observed a temperature increase of 2 °C at the petal surface of jar-enclosed rose flowers after the light was switched on in a climatic chamber held at a constant temperature of 20 °C. The petal temperature can be adjusted by surrounding the lower side of the round-bottom flask with tubing with running water cooled to 14 to 16 °C provided by a thermostatic waterbath. This waterbath is switched on and off together with the light source. The daily rise in petal temperature upon light exposure, which due to unfolding increases during flower development, was measured using a thermocouple in direct contact with the petal surface. Petal temperature is adjusted daily via the thermostatic waterbath to the temperature observed during the dark period. This approach enabled us to maintain temperature fluctuations at less than 0.5 °C throughout flower development.

Relative humidity can also be controlled but is thought to have little or no influence on volatile emission (Jakobsen and Olsen 1994). We kept relative humidity constant by placing the entire experimental setup in a climatic chamber.

10.2.1.3 Volatile Adsorption

Headspace measurements can be carried out in the dynamic or in the direct mode. In the dynamic mode volatiles are accumulated via a continuous air flow onto an adsorbent during a given time period before analysis. In the direct mode air samples are taken from the containing jar and injected directly into the GCMS. Mostly dynamic measurements are applied since the accumulation of volatiles over a longer time period allows measurement of smaller emission activities. Several adsorbents have been used in the past such as Porapak Q, charcoal and Tenax TA or Tenax GR. Besides accumulation on solid adsorbents, also cold trapping by conducting volatiles through glass spirals or tubes cooled in solid CO_2 or liquid N_2 or ice has been applied, but to the author's knowledge no reports have mentioned this approach since Dobson's survey in 1991. Of the solid adsorbents, Tenax TA has been used most commonly. In our study with rose flowers (Helsper et al. 1998) we used glass cartridges (140 mm long, 4 mm diameter) containing a weighed amount of 150 mg Tenax TA. The adsorption capacity for aromatic, aliphatic and terpenoid volatiles is very high; we observed no breakthrough to a second tube, connected in series to a first one, during 1 week of continuous accumulation of rose flower volatiles. However, if the volatiles are either very apolar (like pinene) or small molecules (less than six carbons) breakthrough has been observed (Simon et al. 1995). Flow rates of the air stream (mostly natural air is used to mimic the physiological plant environment) are dependent on the size of plant containment. In accordance with earlier studies (Dobson 1991) we used a 2.5-l, three-necked flask to which an air flow of 100 ml/min was applied regulated by a calibrated flow meter placed in series after the adsorbent cartridges.

A recent development for volatile collection is the use of solid phase microextraction (SPME) where the adsorbing material is placed in the headspace in the form of an exposed needle tip which, after adsorption is completed, can be introduced directly in the injection port of a GCMS. As described by Elmore et al. (1997), this technique is reproducible. For quantitative measurements of volatiles this technique appears less appropriate since diffusion is the only driving force that accomplishes the required contact between the passive needle and the volatile. It is, however, very time-efficient and therefore certainly feasible for screening and semi-quantitative studies, e.g. in breeding programs.

10.2.1.4 Volatile Desorption

The volatiles adsorbed to Tenax, SPME or otherwise should be desorbed before analysis on GCMS. The simplest, fastest and most sensitive method is thermal desorption which is achieved by heating the adsorbing material, followed by condensation of the volatiles in a cold trap and their subsequent injection after heating for analysis. Using SPME the condensation is omitted and volatiles are injected directly after thermal desorption. Volatiles adsorbed to Tenax or other adsorbents can also be subjected to solvent desorption. Internal standards to correct for errors occurring during GCMS analysis can be added to this solvent. Internal standards for plant samples can be chosen after confirmation of their absence in the headspace using the SCAN mode of the GCMS. When hexane and/or pentane-diethylether (4:1, v/v) are used, more than 99% of all components are desorbed from 150 mg Tenax with 5 ml of each of both solvents to which 1 ng/μl limonene was added as an internal standard. The use of hexane as the first solvent is advisable since its low volatility will result in better reproducible injections in GCMS analyses.

10.2.1.5 GC and GCMS Analysis

Gas liquid chromatography for emitted volatiles should result in separation of all components in the headspace. In our experiments with flowers of *Rosa hybrida* cv. Honesty we observed 13 detectable compounds in the headspace (see Table 10.1 for further information): three monoterpenols geraniol, nerol and citronellol, three oxidized monoterpenols *E*-citral, *Z*-citral and methyl geranylate, two sesquiterpenes *trans*-caryophyllene and β-cubebene, the carotenoid metabolite dihydro-β-ionone, three aromatic compounds 3,5-dimethoxytoluene, 2-phenylethanol and 2-phenylethylacetate and the fatty acid derivative hexylacetate. Sometimes traces of rose oxide were also detected. With the exception of methyl geranylate these are all observed earlier in the headspace of rose flowers (Knudsen et al. 1993). In other rose cultivars a larger variety of volatiles is observed like hydrocarbon monoterpenes and aldehydes (Flament et al. 1992). Since these components sometimes vary widely in polarity and boiling point the use of a slightly polar packing material and a temperature gradient in GCMS analysis may be required. In our system we used an HP 5890 (series II) gas chromatograph, connected to a HP 5972 A mass selective detector, equipped with an HP-Innowax capillary column (30 m long, 0.25 mm i.d., 0.25 μm film thickness). HP5 columns did not allow separation of nerol and citronellol. Columns with characteristics of HP-Innowax, associated with its polyethylene glycol phase, e.g. AT-Wax, Carbowax, Supelcowax, ZB-Wax, are provided by several manufacturers. The column oven was programmed at an initial temperature of 45 °C for 1 min with a ramp of 7.5 °C/min to 220 °C and a final time of 1 min. Helium was used as a carrier gas at 0.7 ml/min. The injection (splitless mode), interface and MS source temperatures were 250, 290 and 178 °C, respectively. In GCMS analysis it is important to first identify all volatiles emitted by the flower. An aliquot obtained from longer continuous headspace sampling in order to obtain large quantities can be analysed in the SCAN mode where entire mass spectra are obtained for all compounds. The spectra are used for identification using a library, e.g. Wiley 138.1 commonly provided with the GCMS software. Identification can be

Table 10.1. Composition (% of total) and amount of volatile compounds (μg/24 h) emitted by flowers of *Rosa hybrida* cv. Honesty attached to the intact plant and as cut flowers in vase water containing sucrose at 0, 15 or 30 g/l. Results are obtained during the day of maximal emission and expressed as mean±SD of three independent experiments. *Rt* Retention time (min) in GCMS analysis; m/z values are used for selective ion mode detection in mass spectrometry. (Helsper et al. 1998)

Component	Rt – m/z values	Intact plant	Cut flower		
			Sucrose concentration		
			0	15	30
		Composition (%)			
Monoterpenols					
Geraniol	17.39–41; 69; 53	10.1 ± 4.1	11.2 ± 1.6	9.1 ± 7.2	9.9 ± 0.3
Nerol	16.74–55; 69; 93	25.1 ± 8.3	29.5 ± 0.8	34.0 ± 4.7	29.4 ± 2.6
Citronellol	16.21–55; 69; 81	21.3 ± 1.3	15.0 ± 1.7	17.8 ± 2.4	18.1 ± 3.6
Oxidized monoterpenols					
E-Citral	15.80–84; 94; 109	2.2 ± 1.1	3.4 ± 0.9	4.5 ± 0.3	3.2 ± 0.5
Z-Citral	15.05–84; 94; 109	1.8 ± 0.2	2.5 ± 0.7	3.1 ± 0.3	2.4 ± 0.7
Methylgeranylate	15.25–41; 83; 114	7.0 ± 5.7	3.8 ± 2.5	3.0 ± 1.7	2.6 ± 1.1
Sesquiterpenes and dihydro-β-ionone					
trans-Caryophyllene	13.79–79 ; 93; 133	0.4 ± 0.7	0.7 ± 0.3	0.5 ± 0.3	0.5 ± 0.1
β-Cubebene	15.51–105; 119; 161	1.4 ± 1.7	2.4 ± 0.4	2.4 ± 0.9	2.5 ± 0.2
Dihydro-β-ionone	17.28–121; 136; 161	0.7 ± 2.2	1.2 ± 0.1	1.5 ± 0.3	2.1 ± 0.9
Aliphatic and aromatic compounds					
Hexylacetate	8.32–56; 61; 84	10.2 ± 3.5	9.7 ± 1.6	11.2 ± 3.5	14.4 ± 2.5
2-Phenylethanol	18.40–65; 91; 122	6.8 ± 2.3	4.2 ± 1.8	2.2 ± 0.5	3.7 ± 2.3
2-Phenylethylacetate	17.05–78 ; 91; 104	2.5 ± 0.8	2.4 ± 0.8	2.0 ± 0.5	1.8 ± 1.5
3,5-Dimethoxytoluene	17.53–77; 123; 152	10.6 ± 3.8	14.1 ± 2.8	8.7 ± 3.8	9.4 ± 5.6
Total amount (μg)		604 ± 199	804 ± 324	720 ± 294	617 ± 91

substantiated on the basis of retention time and mass spectra by subjecting authentic reference samples to GCMS analysis under the same conditions.

10.2.1.6 Calibration Curves

After identifying the volatiles in the SCAN mode, calibration curves should be obtained for each individual component to calculate response factors, mostly expressed as area units on the MS detector per weight or molar unit of component. For quantitative analyses measurement in the SIM (selective ion monitoring) mode is preferred since this is more sensitive and because the calibration shows linearity for a larger range. The ions for quantification in the SIM mode (Table 10.1) can be selected from the mass spectra obtained in the SCAN mode and further defined by measuring ion abundance at selected m/z values with smaller intervals around the SCAN-derived values.

10.2.1.7 Quantification of Compounds for Which No Authentic Standard Is Available

If components are present from which authentic samples are not available it is possible to estimate, but not accurately measure, the response factor in SIM. This method is based on the assumption that molecules with similar chemical structure have similar response factors when measured in the SCAN mode. Thus, the amount of β-cubebene in the headspace of flowers of *Rosa hybrida* cv. Honesty could be estimated in the SCAN mode using the SCAN-derived response factor of commercially available α-cubebene (Fluka, ICN). Hereafter, the area units for the same amount of β-cubebene were measured in the SIM mode. The ratio of the amount of β-cubebene calculated with the SCAN mode and the area units obtained in the SIM mode gives the estimated response factor for β-cubebene in SIM. This value can be statistically evaluated by repeating the approach with several plant samples.

10.2.1.8 Recovery of Volatiles in the Experimental Setup from Plant to GCMS

Losses in emitted volatiles, occurring from the point of emission by the flower up to injection for GCMS analysis, e.g. by leakage, adsorption to surfaces, incomplete desorption from Tenax, etc., may lead to underestimation of these components in quantitative headspace analysis. Therefore, for each volatile component this recovery should be quantified and corrected for. This can be achieved by placing a vial containing a small amount of volatile solvent, e.g. hexane, in which all components are included at the position of the flower in the experimental system. Since recovery can vary with the quantity and type of component, such recovery measurements should be performed using amounts in the same range as emitted during an experiment. In our system we observed recoveries of about 55% for monoterpenols (citronellol, nerol), 82% for oxidized monoterpenols (like rose oxide), 90% for aromatic volatiles (2-phenylethanol, 3,5-dimethoxytoluene) and 74% for aliphatic compounds (hexylacetate) in the experimental design after accumulation during 3 h using Tenax TA as an adsorbent and 5 ml *n*-hexane as desorbing solvent.

10.2.2 Circadian Rhythmicity in Emission of Volatile Compounds by Rose Flowers: Experimental Results and Discussion

A nocturnal rhythmicity in emission of volatiles by flowers, probably controlled by petal movement, was first described for *Cestrum nocturnum* (Overland 1960). Although rhythmicity was shown to be maintained under conditions of constant darkness, not all requirements for circadian rhythmicity, as given by Jones and Mansfield (1975), were investigated. Therefore, a claim as "endogenous" cannot be extended to "circadian" for this study. Since then, studies on nocturnal, circadian rhythmicity in volatile emission have been described which fully meet the requirements for circadian control, i.e. for *Hoya carnosa* and *Stephanotis floribunda* (Altenburger and Matile 1988, 1990; Matile and Altenburger 1988) and for *Nicotiana sylvestris* (Loughrin et al. 1991). Diurnal rhythmicity in volatile emission has hitherto been thought to be non-circadian in nature and directly regulated by the level of radiation, e.g. in *Nicotiana suaveolens* (Loughrin et al. 1991), *Trifolium repens* (Jakobsen and Olsen 1994), *Brassica napus* (Jakobsen et al. 1994) and *Odontoglossum constrictum* (Altenburger and Matile 1990).

Light, in particular red light, is observed to be a driving force (oscillator) for petal movement (Engelsmann et al. 1997), where phytochrome may function as an intermediate in the signal transduction cascade. Phytochrome is also known as a possible zeitgeber for circadian rhythms in the production of fragrant terpenoids like thymol and carvacrol in thyme (Tanaka et al. 1989; Yamaura et al. 1989, 1991). Circadian rhythmicity in diurnally emitting flowers was not reported until 1998 when our laboratory described such an endogenously controlled emission for flowers of *Rosa hybrida* cv. Honesty, which emits a strong and sweet fragrance consisting of the components listed in Table 10.1. In contrast to results obtained by Mookherjee et al. (1990), we observed no difference in fragrance composition of cut flowers and flowers on the intact plant (Table 10.1). Moreover, the generally assumed increase in emission of volatiles upon addition of sucrose to the vase water was not found for flowers of *Rosa hybrida* cv. Honesty. The period of ornamental value lasts about 7 days starting at the point where the pink petals start to curl outward until the discoloration. During this period, the composition of the headspace also changes (Fig. 10.1); dihydro-β-ionone is only emitted until day 3 (Fig. 10.1C). At day 3 monoterpenols (Fig. 10.1A) and oxidized monoterpenols (Fig. 10.1B) show maximal emission, while sesquiterpenes show maximal emission at day 5. Rhythmicity in emission by these flowers was particularly pronounced for the monoterpenols (Fig. 10.1A) and their oxidized derivatives (Fig. 10.1B), 2-phenylethanol and its acetate derivative and hexylacetate (Fig. 10.1D), but less so for the sesquiterpenes (Fig. 10.1C). Rhythmicity is almost absent for 3,5-dimethoxytoluene (Fig. 10.1D). For 2-phenylethanol maximum emission was first observed during the light period but shifted slowly to an emission in the dark at the end of the experimental period (Fig. 10.1D).

When rose flowers were exposed to continuous darkness during flower bud opening no rhythmicity was observed (Fig. 10.2) and the level of emission was much lower or completely absent, e.g. for the oxidized monoterpenols (Fig. 10.2B), caryophyllene and dihydro-β-ionone (Fig. 10.2C). When flowers are subsequently exposed to a 12-h photoperiod, rhythmicity is installed very quickly, although some compounds like β-cubebene show aberrant periodicity with maximum emission in the dark. Obviously, light enhances emission of volatiles. When flowers, exposed for 4 days to a 12-h photoperiod, were held in constant light (Fig. 10.3) or darkness (Fig. 10.4) rhythmicity continues for at least two cycles, although the rhythmic pattern gradually becomes less pronounced. When the periods between maximum emission are studied by taking measurements at 1-h intervals instead of 3-h intervals it becomes obvious that the separation between peaks is not exactly

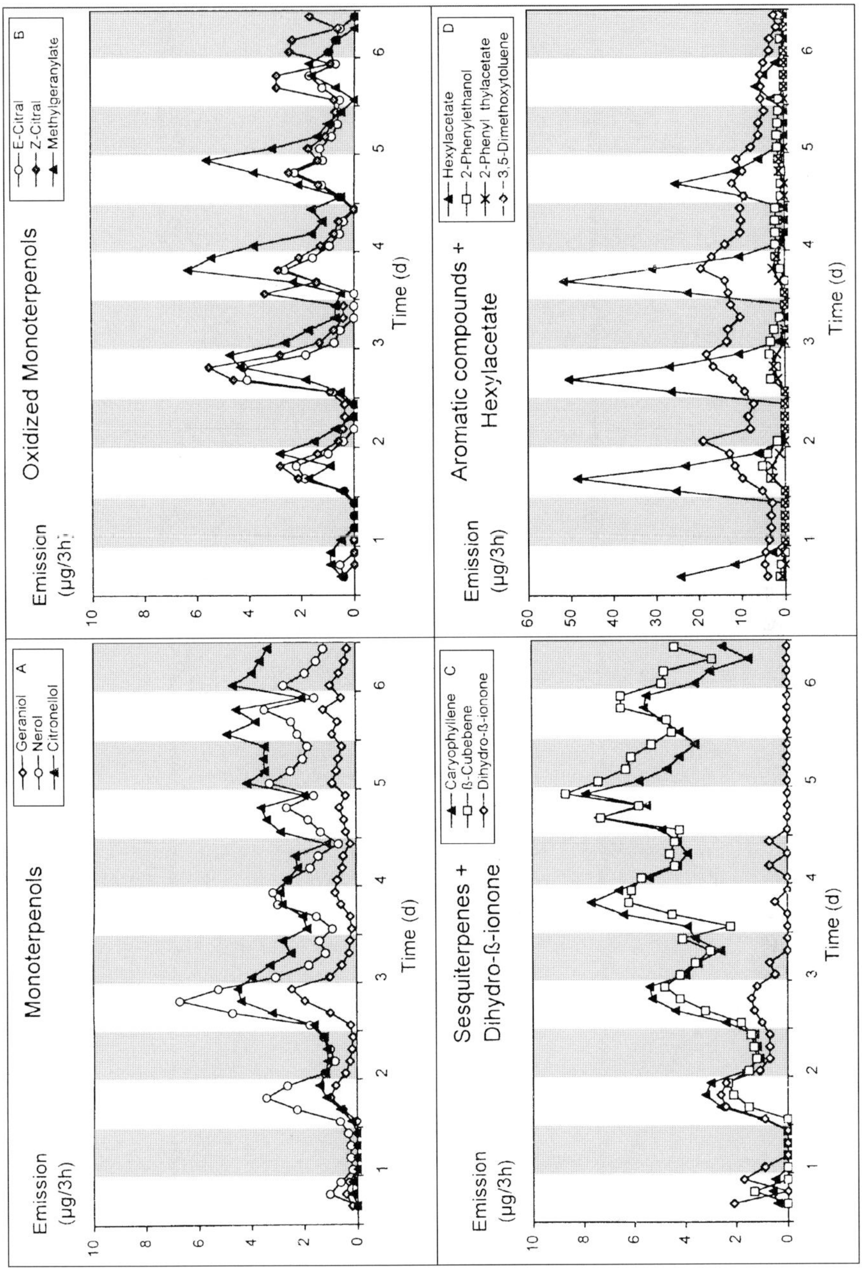

Fig. 10.1. Emission of volatile compounds by flowers of *Rosa hybrida* cv. Honesty upon exposure of the intact plant to a 12-h photoperiod at constant temperature (20 °C) and relative humidity (75%). *Shaded* and *non-shaded areas* correspond to exposure to darkness and light, respectively. (Helsper et al. 1998)

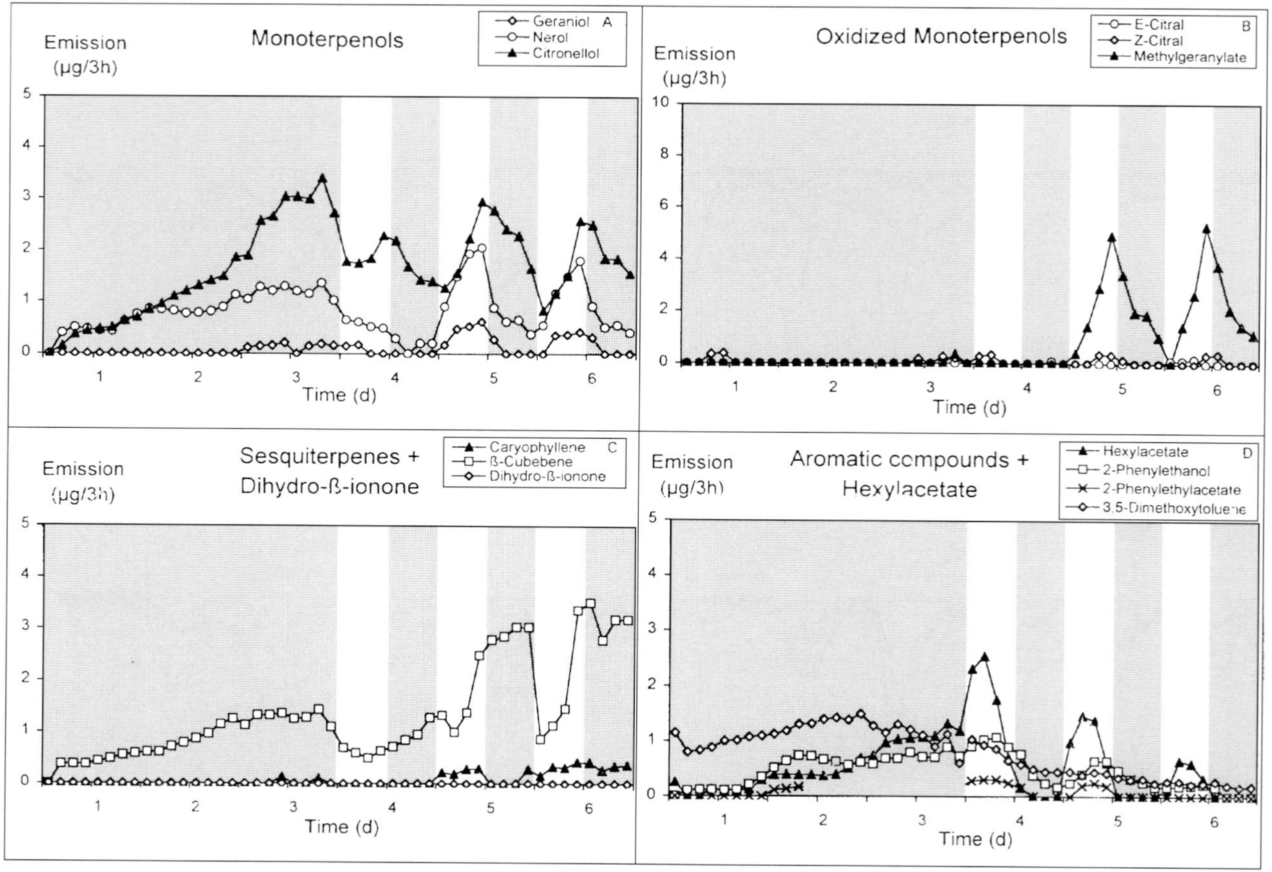

Fig. 10.2. Emission of volatile compounds by flowers of *Rosa hybrida* cv. Honesty upon exposure of the intact plant to continuous darkness from 5 days before to 3 days after flower bud opening, which was followed by a 12-h photoperiod. For further details see Fig. 1. (Helsper et al. 1998)

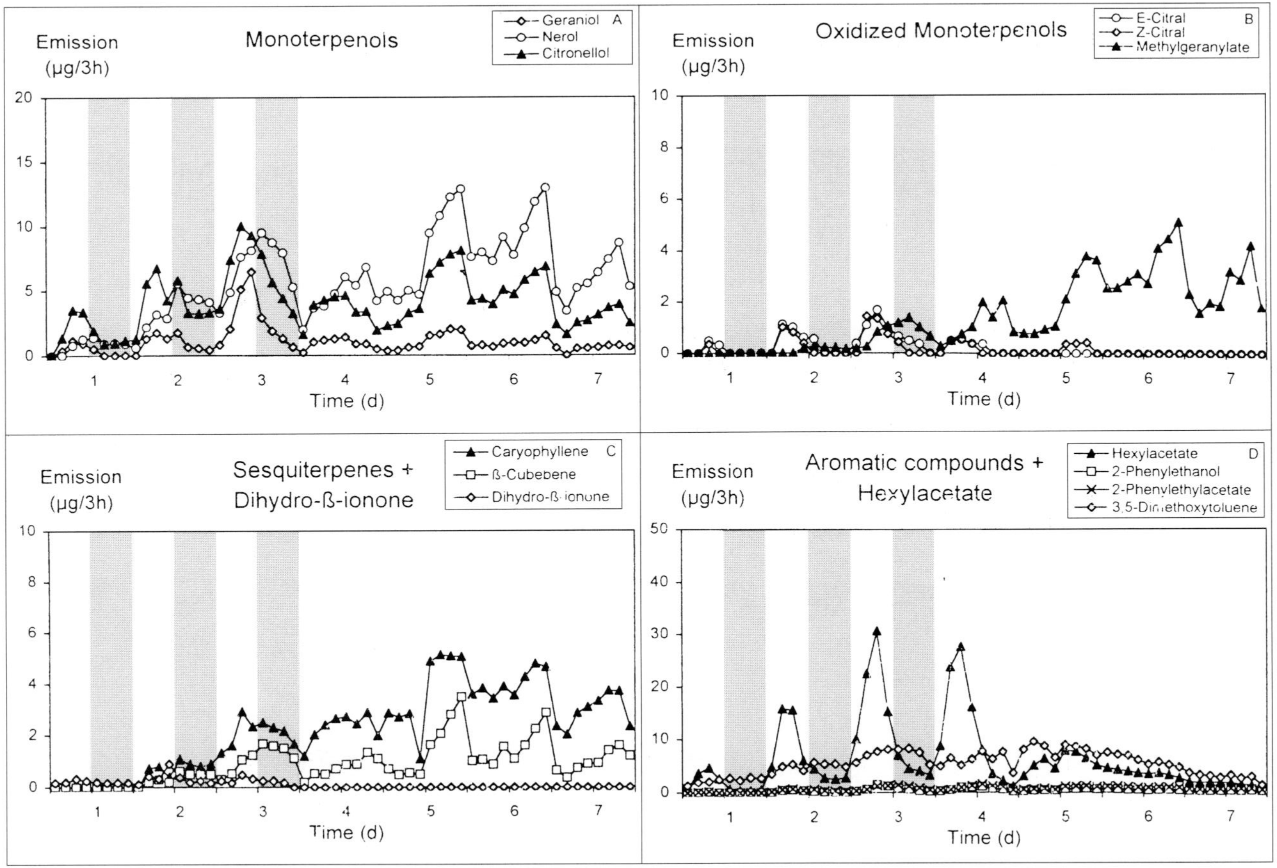

Fig. 10.3. Emission of volatile compounds by flowers of *Rosa hybrida* cv. Honesty upon exposure of the intact plant to a 12-h photoperiod, followed by exposure to continuous light. For further details see Fig. 1. (Helsper et al. 1998)

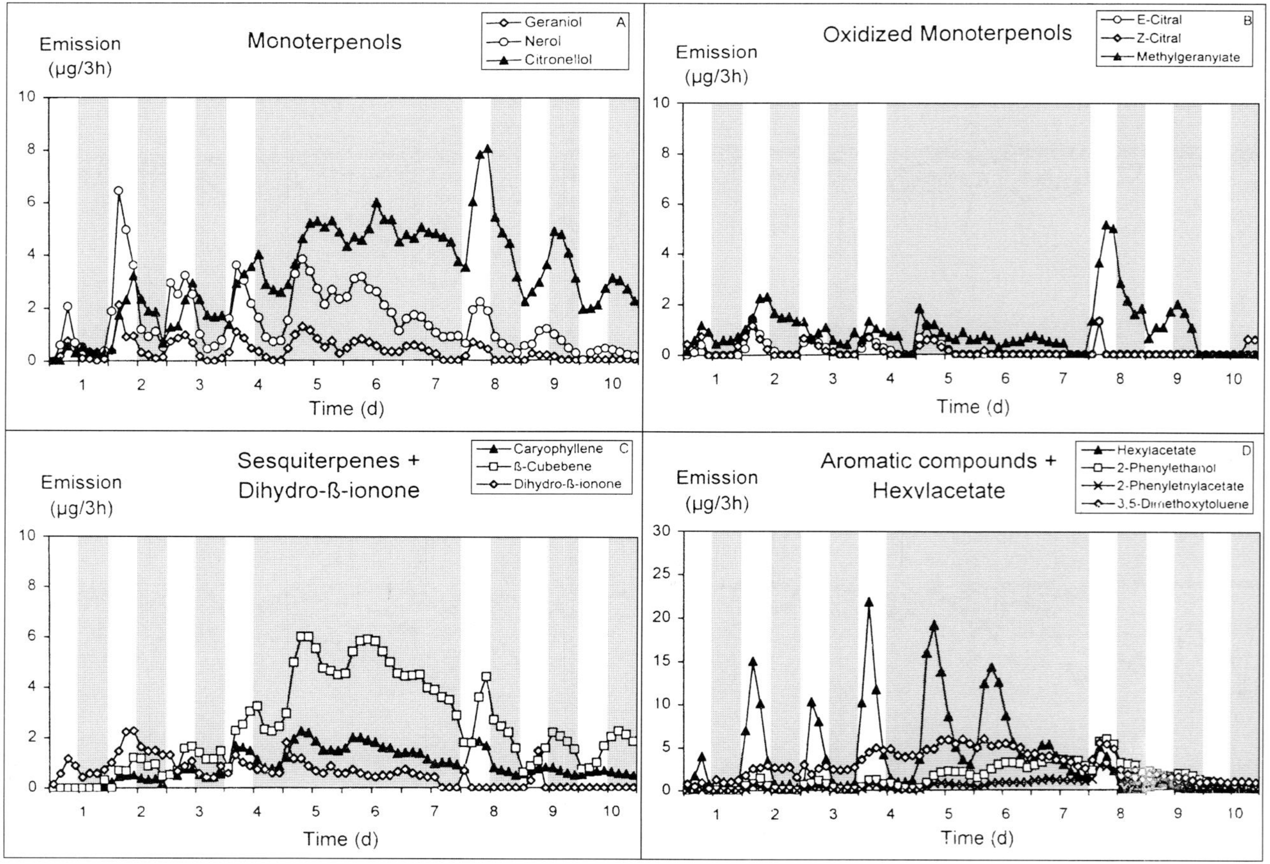

Fig. 10.4. Emission of volatile compounds by flowers of *Rosa hybrida* cv. Honesty upon exposure of the intact plant to a 12-h photoperiod, followed by exposure to continuous darkness for 84 h and subsequently again to a 12-h photoperiod. For further details see Fig. 1. (Helsper et al. 1998)

24 h but slightly longer: 25–28 h (not shown). This has been described as one of the principal characteristics of circadian rhythmicity (Jones and Mansfield 1975) in addition to a continuation of rhythmicity upon exposure to constant light or darkness. Also the third and last criterion for circadian regulation of rhythmicity, the possibility to create a phase shift by changing the pattern of the external (light/dark) signalling, is observed (Fig. 10.5); when a 12-h photoperiod is interrupted by a 24-h continuous dark treatment, rhythmicity in volatile emission still shows a maximum in the second phase of the second 12 h of this 24-h dark period, as expected for a circadian rhythmicity, but then shifts almost immediately upon re-exposure to light to the newly installed photoperiod with a 12-h shift in peak emission.

These data show several interesting aspects of rhythmicity in emission of volatiles by roses. It is the first report of a diurnal rhythmicity, after several reports on nocturnal rhythmicity, that meets all characteristics of circadian regulation as formulated by Jones and Mansfield (1975). Moreover, since roses do not show petal movement but a continuous unfolding of the petals, the circadian rhythmicity in volatile emission cannot be associated with the physical barrier provided by the petal tissue. Therefore, biochemical processes in the petal tissue itself should control the rhythmicity (see below). The zeitgeber for this biochemical process, the external signal regulating the phenomenon, is obviously the alternating incidence of light and dark. Besides its role as zeitgeber for rhythmicity, the mere presence of light also enhances the emission of volatiles, as illustrated by the absence or reduction of volatile emission when flowers open in the dark (Fig. 10.2).

10.3 Rhythmicity in Precursors of Emitted Volatiles in Rose Petal Tissue

10.3.1 Introduction

Since rose petals do not show petal movement, the circadian rhythmicity in volatile emission described in the former paragraph cannot be attributed to rhythmic petal movement. As temperature and other climatic conditions were kept constant, these conditions also cannot have caused the rhythmicity in emission during constant light or darkness. In a study by Altenburger and Matile (1990) the height of the boiling points of emitted components did not explain their concentration in the headspace. A very probable explanation for rhythmic emission is a rhythmic pattern in the concentration in petal tissue of the compounds to be emitted or their immediate precursors.

Probable precursors may be the non-acetylated forms of hexylacetate and 2-phenylethylacetate and the glycosyl derivatives of the alcohols, like the monoterpenols and 2-phenylethanol, and aldehydes, like *E*- and *Z*-citral. Glycosylated forms have been described for a wide variety of flavour and fragrance compounds in

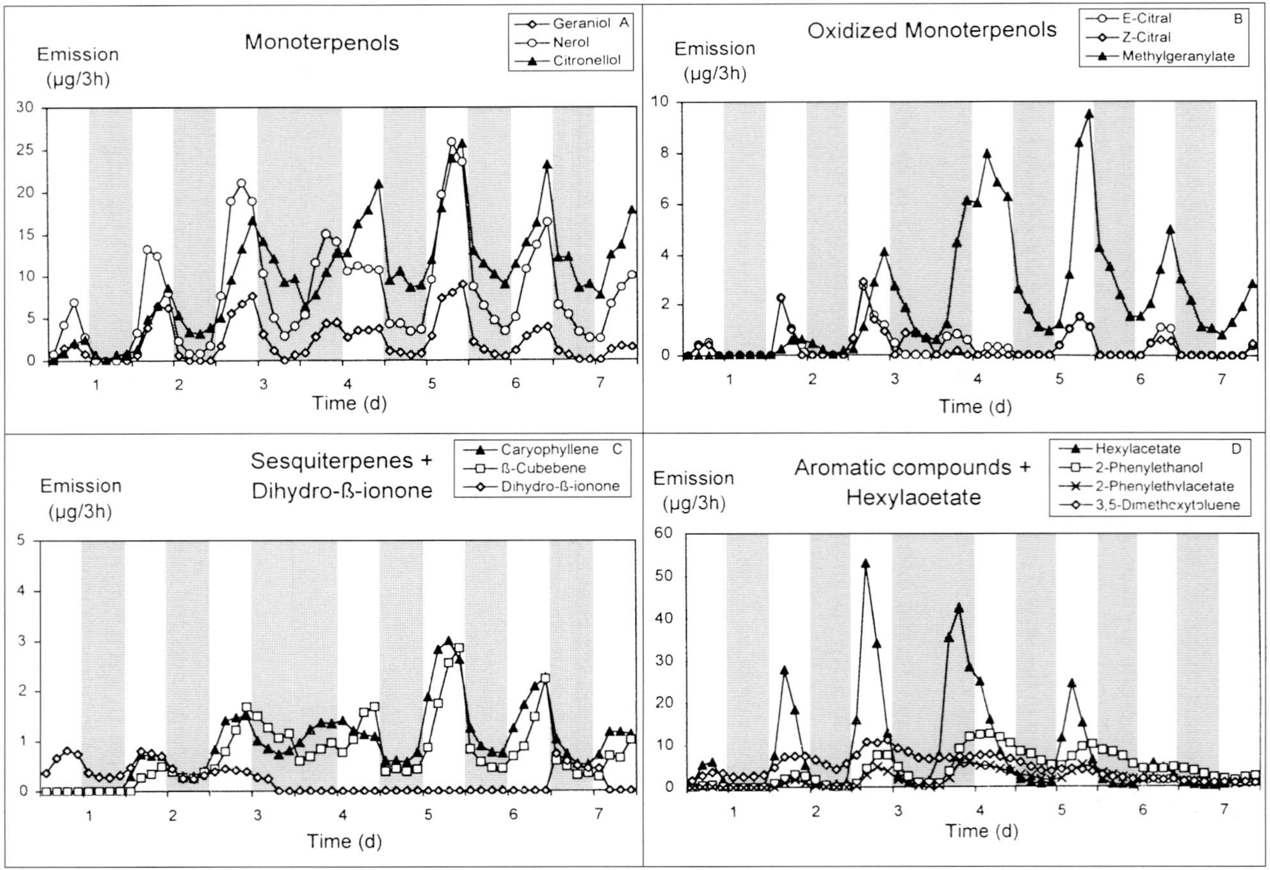

Fig. 10.5. Emission of volatile compounds by flowers of *Rosa hybrida* cv. Honesty upon exposure of the intact plant to a 12-h photoperiod, followed by exposure to 24 h of darkness and subsequently again to a 12-h photoperiod (inverted light/dark regime). For further details see Fig. 1. (Helsper et al. 1998)

flowers and fruits of various species like rose (Francis and Allcock 1969; Ackermann et al. 1989), *Nicotiana* species (Loughrin et al. 1992), raspberry (Pabst et al. 1992), apple (Schwab et al. 1989) and grape (Voirin et al. 1990). Loughrin et al. (1992) investigated, but did not observe, a correlation between concentrations of glycosylated and emitted aglycons. Altenburger and Matile (1990) found no correlation between rhythmic emission and concentrations of free forms of fragrance compounds in the tissue and the rhythmic emission pattern in *Citrus medica* flowers.

To gain insight into the regulation of emission pattern by petal concentrations we quantitatively investigated the fluctuations in free and glycosylated forms first at 6-h intervals and eventually, when rhythmic pattern becomes evident, at 3-h intervals.

10.3.2 Methods

10.3.2.1 Plant Material

Plants of *Rosa hybrida* cv. Honesty were grown as described for headspace measurements. Since assay for tissue concentrations of free and glycosylated fragrance compounds and their precursors involved destructive measurement, flowers were harvested at 3- or 6-h intervals during only day 4 after flower bud opening, the day of maximal emission for most volatiles. On this day the petals were removed from the bud, weighed, frozen in liquid nitrogen and stored at −20 °C. On the day of extraction of tissue-derived fragrance compounds and their precursors the petals were ground to a fine powder under liquid nitrogen using a mortar and pestle. From this powder a fixed amount was weighed out for analysis of free and glycosylated fragrance compounds.

10.3.2.2 Assay of Non-glucosylated Fragrance Compounds in Petal Tissue

Aliquots of 0.5 g of powdered petal tissue were further homogenized in 150 ml water, containing 18.75 g sodium chloride, using an all-glass potter. The suspension was transferred to a round-bottom flask, which was incorporated in a distillation setup according to Stahl (Fig. 10.6) and after addition of 5 ml ethylacetate and glass beads the mixture was distilled for about 1 h. The ethylacetate fraction, distilled to the calibrated right arm of the setup, contains volatile lipophilic components and is free of non-volatile phospholipids and triglycerides. Its volume was adjusted to 5 ml and subjected to GCMS analysis as described for emitted volatiles.

Recovery during distillation for authentic standards was more than 85% for all components, with the exception of only 2-phenylethanol for which the recovery also showed a large variability (46 ± 27%). We did not observe matrix effects, due to the presence of homogenised petal tissue, on recovery of volatile compounds during steam distillation. Such matrix effects can be calculated by mea-

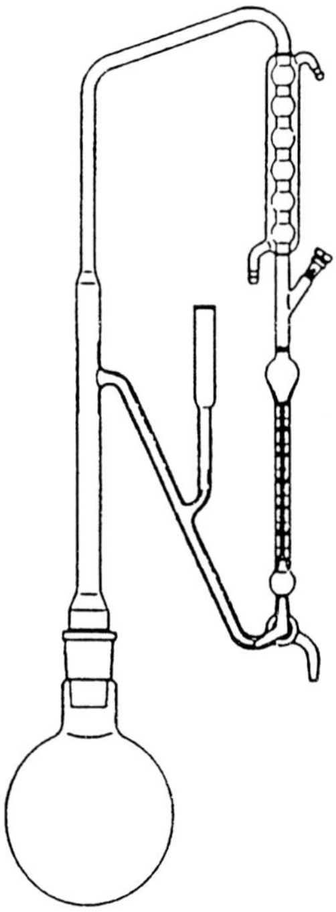

Fig. 10.6. Distillation setup according to Stahl

suring the area units (AU) from GCMS analysis of ethylacetate fractions, obtained by distillation of (1) the authentic standard ($AU_{standard}$), (2) the same compound from 0.5 g petal tissue alone (AU_{petals}) and (3) mixtures ($AU_{standard \pm petals}$) using the formula:

$$\text{Matrix effect (in \%)} = \frac{(AU_{standard+petals}) - (AU_{petals})}{(AU_{standard})} \times 100\%$$

10.3.2.3 Assay of Glucosylated Fragrance Compounds in Petal Tissue

A procedure modified from Loughrin et al. (1992) was used, which started with separation of glycosides from aglycones, followed by enzymolysis and solvent partitioning of the glycoside-derived aglycon. Shortly, 0.5 g powdered petal tissue was weighed and suspended in 50 ml 0.2 M sodium phosphate buffer (pH 7.4), containing 1.8 g of *D*-gluconolactone to inhibit endogenous glucosidases and 0.5 mg of phenylglucoside as an internal standard. The slurry was homogenised using a Sorval omnimixer (1 min), centrifuged at 2,800 × *g* for 30 min and filtered over a Büchner funnel. The filtrate was applied to an Amberlite XAD-2 column (20 cm length, 0.5 cm diameter) equilibrated with 0.2 M sodium phosphate buffer (pH 7.4) which was eluted with subsequently 30 ml of water, 45 ml of hexane and 65 ml of methanol. The methanol fraction was washed by solvent partitioning with 20 ml of hexane to remove aglycons and evaporated to dryness under reduced pressure at 35 °C. The residue was redissolved and incubated in 3 ml of 0.2 M sodium acetate (pH 5.0), containing 37 mg of almond glucosidase (Sigma, 200 units) for 24 h at 37 °C. Longer incubation did not result in higher yields. After cooling to 4 °C the mixture was extracted twice with 1 ml ethylacetate. The two extracts were combined and the total volume adjusted to 3 ml. This sample was subjected to GCMS analysis as described for emitted volatiles.

10.3.3 Rhythmicity in Petal Concentrations of Precursors of Volatile Compounds: Experimental Results and Discussion

When exposed to a 12-h photoperiod, tissue concentrations of non-glycosylated fragrance components in flower petals of *Rosa hybrida* cv. Honesty show a rhythmic pattern (Fig. 10.7) as was observed for the headspace (Fig. 10.1). The most pronounced rhythmicity is observed for the oxidized monoterpenols methylgeranylate, *E*- and *Z*-citral (Fig. 10.7B). Rhythmicity is less pronounced for the monoterpenols geraniol, nerol and citronellol (Fig. 10.7A) and for 2-phenylethylethanol (Fig. 10.7D). Rhythmicity is shallow for β-cubebene (Fig. 10.7C) and 2-phenylethylacetate (Fig. 10.7D), and is absent for *trans*-caryophyllene, dihydro-β-ionone (Fig. 10.7C) and 3,5-dimethoxytoluene (Fig. 10.7D). Hexylacetate, one of the major headspace components which also showed the most pronounced rhythmicity, was not observed in petal tissue (detection limit 0.01 μg/g FW) and neither was its putative non-acetylated precursor n-hexanol. Two monoterpenols, linalool and α-terpineol, were only observed in the petal tissue, without rhythmicity in concentration, and not in the headspace (data not shown).

When flowers exposed to a 12-h photoperiod for the first 3 days of flower opening were kept in the dark for 24 h on the fourth day the rhythmicity in petal concentration of headspace components was maintained (Fig. 10.8). Maximum emission occurred slightly later than during a normal 12-h photoperiod. These two phenomena, continuation of rhythmicity and a slight delay in maximum concentration, are characteristics typical for circadian rhythmicity, as formulated by Jones and Mansfield (1975). Investigations required to meet all the requirements formulated in their report, i.e. the possibility of a phase shift in maximum concentration by exposure to an inverted photoperiod, are very laborious since flowers have to be sacrificed for each of the quadruplet measurements at every time point.

Due to the requirement of an alcohol or aldehyde group, seven of the components observed in the headspace of flowers of *Rosa hybrida* cv. Honesty may also occur in glycosylated form. All the potentially possible glucosides – geraniol, nerol, citronellol, *E*-citral and *Z*-citral – were observed, most of them at a concentration similar to that of the free form (Table 10.2). The glucoside of the aromatic compound 2-phenylethanol was observed at a five-fold higher concentration as compared to the aglycon.

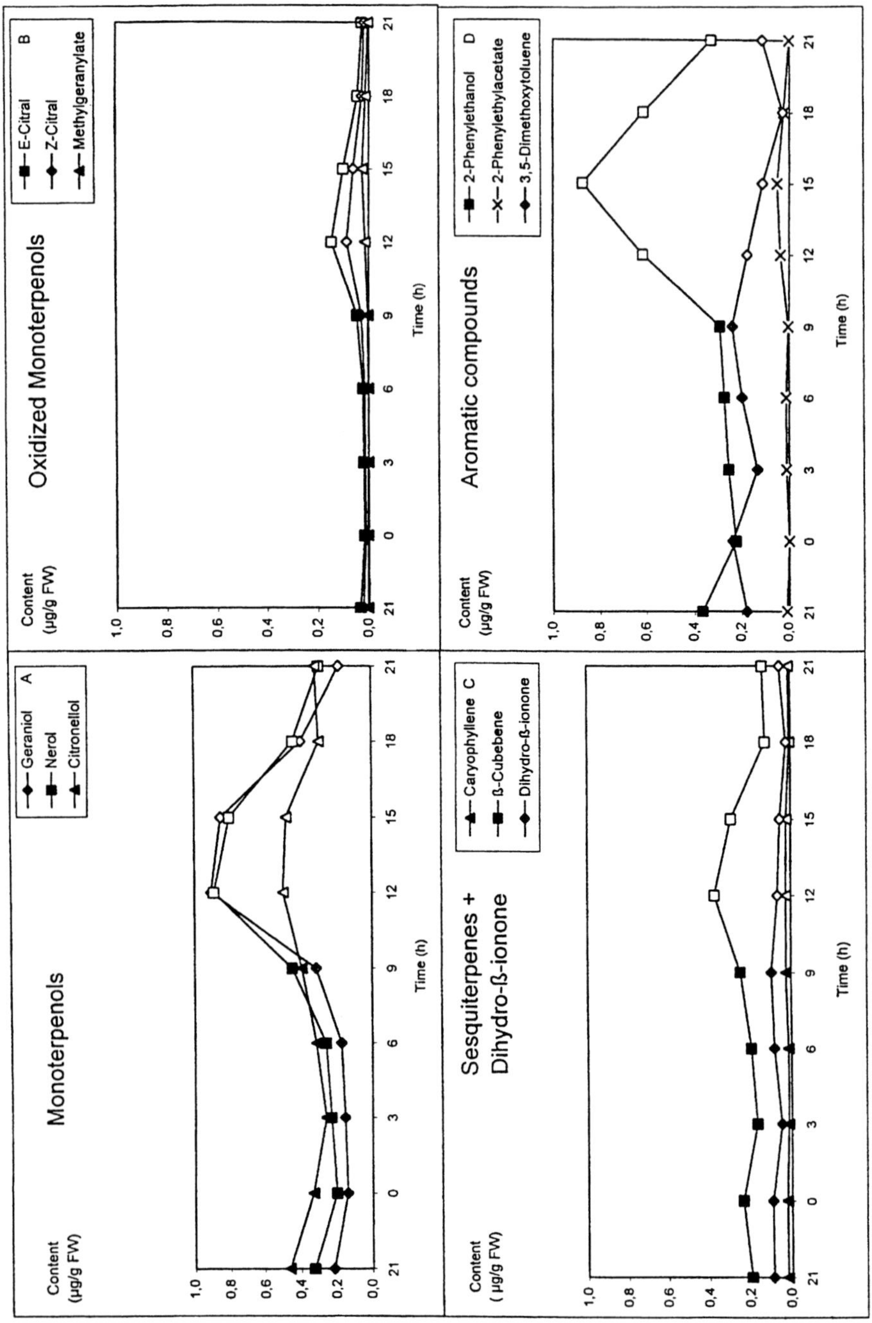

Fig. 10.7. Concentration patterns of volatile compounds (aglycons) in petal tissue of *Rosa hybrida* cv. Honesty at day 4 after flower bud opening. Intact plants were exposed to a 12-h photoperiod. *Closed* and *open symbols* represent flowers exposed to darkness and light, respectively

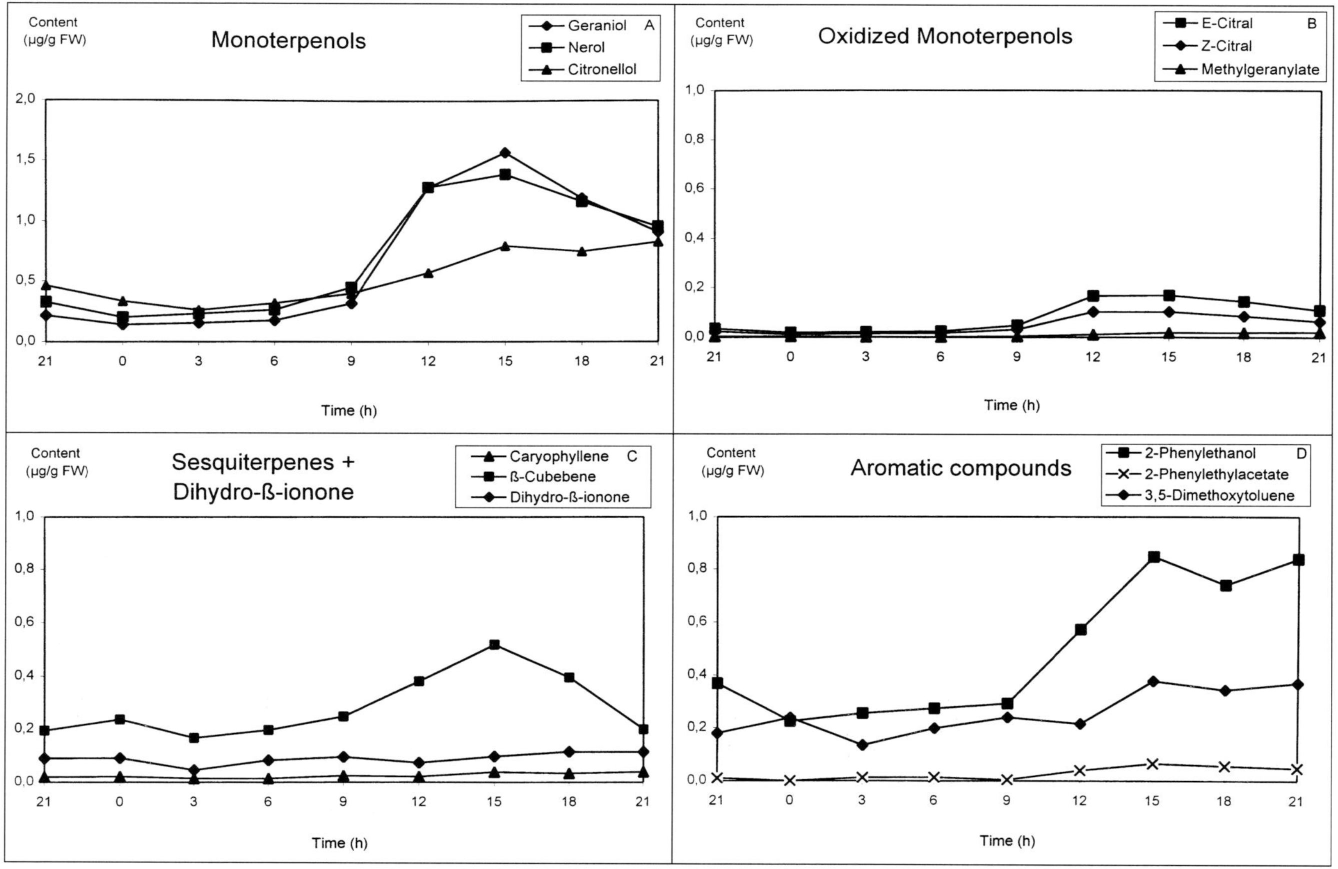

Fig. 10.8. Concentration patterns of volatile compounds (aglycons) in petal tissue of *Rosa hybrida* cv. Honesty at day 4 after flower bud opening. Intact plants were exposed to a 12-h photoperiod until day 3 and to constant darkness at day 4

Table 10.2. Tissue concentrations of free and glucosylated forms of volatile compounds observed in the headspace of flowers of *Rosa hybrida* cv. Honesty. Values expressed are means±SD in µg/g FW from three individual flowers. *n.d.* Not detectable (detection limit is 0.1 µg/g FW)

Compound	Tissue concentration	
	Free form	Glucosylated form
Monoterpenols		
Geraniol	5.2 ± 0.6	5.4 ± 0.1
Nerol	18.1 ± 1.1	16.9 ± 0.6
Citronellol	7.7 ± 0.3	1.1 ± 0.1
Oxidized monoterpenols		
E-Citral	8.1 ± 0.5	8.3 ± 0.9
Z-Citral	3.3 ± 0.1	4.3 ± 0.5
Aliphatic and aromatic compounds		
Hexanol (precursor for hexylacetate)	n.d.	5.0 ± 0.5
2-Phenylethanol	17.5 ± 2.5	92.2 ± 4.4

Flowers, harvested at 6-h intervals, did not show any rhythmicity in concentrations of glucosides (Fig. 10.9). Therefore, we did not pursue the rhythmicity in glucoside concentrations at 3-h intervals as a possible regulatory mechanism for circadian rhythmic emission of volatile compounds by rose flowers.

10.4 General Conclusion

Quantification of rhythmicity in emission of volatile compounds by plants or parts hereof involves an experimental design which minimises losses of volatiles from the point of release by the plant (part) until quantitative GC or GCMS analyses. For each of the volatiles investigated the losses occurring should be quantified at the concentration observed in the headspace or in the tissue. The criteria, given by Jones and Mansfield (1975), can be taken as a basis to establish whether this rhythmicity is under circadian control.

Using this approach we observed that flowers of *Rosa hybrida* cv. Honesty emit volatile compounds following a circadian, diurnal rhythmicity (Figs. 10.1 to 10.5). Given the results in Figs. 10.7 and 10.8 the conclusion that petal concentrations of the free forms of lateron-emitted volatiles show a rhythmic pattern which is circadian in nature seems justified. The glucosylated forms do not show rhythmicity (Fig. 10.9). This is the first report of such a rhythmicity in petal concentrations of components which upon emission contribute to the fragrance of rose flowers. It

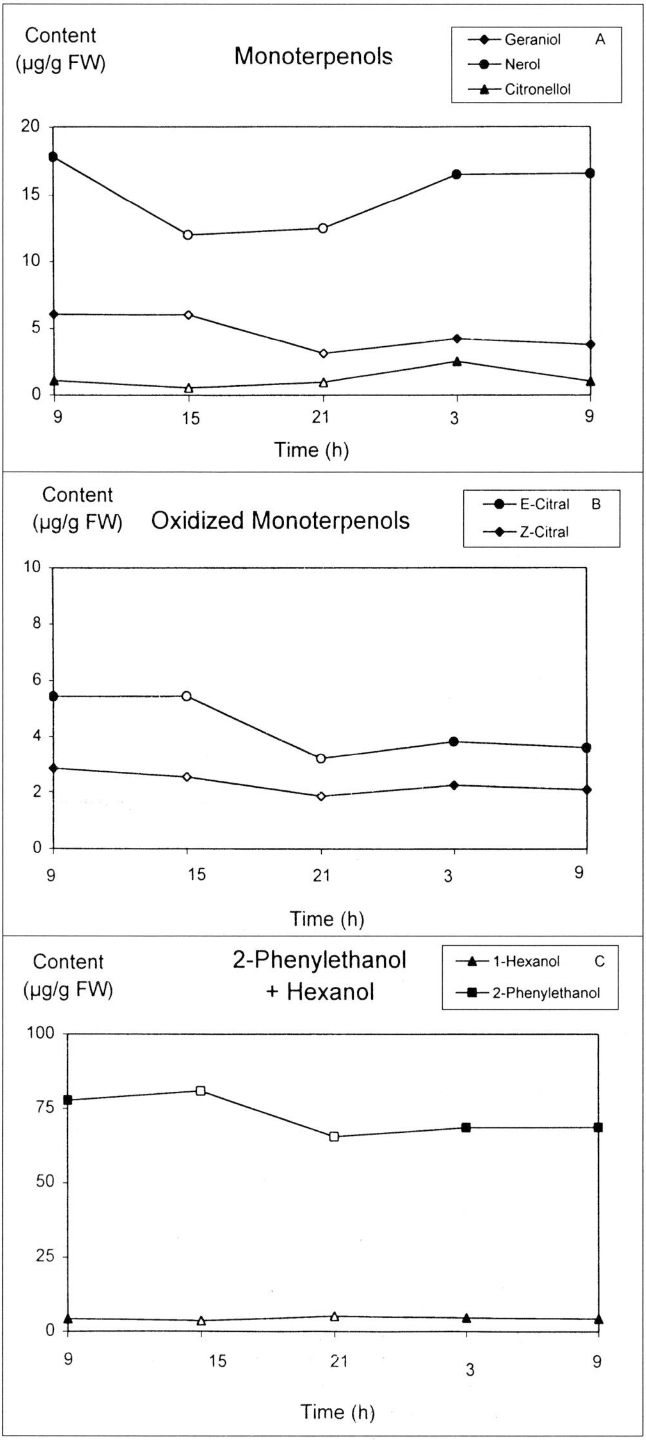

Fig. 10.9. Concentration patterns of glucosylated volatile compounds in petal tissue of *Rosa hybrida* cv. Honesty at day 4 after flower bud opening. Intact plants were exposed to a 12-h photoperiod. *Closed* and *open symbols* represent flowers exposed to darkness and light, respectively

implies that there is a biochemical basis for this circadian rhythmicity in emission. In other plant species petal movement may also form (one of) the regulatory mechanism(s), as is observed for nocturnal, circadian rhythmicity of fragrance emission, e.g. in *Cestrum nocturnum* (Overland 1960).

References

Ackermann IE, Bahnthorpe DV, Fordham WD, Kinder JP, Poots I (1989) β-Glucosides of aroma components from petals of *Rosa* species: assay, occurrence and biosynthetic implications. J Plant Physiol 134:567–571

Altenburger R, Matile P (1988) Circadian rhythmicity of fragrance emission in flowers of *Hoya carnosa*. Planta 174:248–252

Altenburger R, Matile P (1990) Further observations on rhythmic emission of fragrance in flowers. Planta 180:194–197

Bouwmeester HJ, Verstappen FWA, Posthumus MA, Dicke M (1999) Spider mite-induced (3S)-(E)-Nerolidol synthase activity in cucumber and lima bean. The first dedicated step in acyclic homoterpene biosynthesis. Plant Physiol 121:173–180

Bünning E, Stern K (1930) Ber Dtsch Bot Ges 48:227

De Mairan (1729) Histoire de l'Academie Royale des Sciences, Paris, p 35

Dobson HEM (1991) Analysis of flower and pollen volatiles. In: Linskens HF, Jackson JF (eds) Modern methods of plant analysis, vol 12. Essential oils and waxes. Springer, Berlin Heidelberg New York, pp 231–251

Elmore JS, Esbahadin MA, Mottram DS (1997) Comparison of dynamic headspace concentration on Tenax with solid phase microextraction for the analysis of aroma volatiles. J Agric Food Chem 45:2638–2641

Engelsmann W, Sommerkamp A, Veit S, Hans J (1997) Methyl-jasmonate affects the circadian petal movement of Kalonchoe flowers. Biol Rhythm Res 28:377–390

Flament I (1991/1992) Chromatography – a 3D-analysis of the volatile constituents of living flowers. Cosmet Toilet, pp 114–122

Flament I, Sauvegrain P, Sauberli U (1992) De la science a l'art structure et odeur des composants volatiles de la rose. Quadrinestale Inst Tetrahedron 3:1–13

Francis MJO, Allcock C (1969) Geraniol β-D-glucoside occurrence and synthesis in rose flowers. Phytochemistry 8:1339–1347

Gershenzon J (1994) Metabolic costs of terpenoid accumulation in higher plants. J Chem Ecol 20:1281–1328

Gimenes M, Benedicto-Silva AA, Marques MD (1996) Circadian rhythms of pollen and nectar collection by bees on the flowers of *Ludwigia elegans* (Onagraceae). Biol Rhythm Res 27:281–290

Harada K, Mihara S (1984) The volatile constituents of Freesia flower (*Freesia hybrida* Hort). Agric Biol Chem 48:2843–2845

Helsper JPFG, Davies JA, Bouwmeester HJ, Krol AF, van Kampen MH (1998) Circadian rhythmicity of volatile compounds by flowers of *Rosa hybrida* L. cv. Honesty. Planta 207:88–95

Jakobsen HB, Olsen CE (1994) Influence of climatic factors on emission of flower volatiles in situ. Planta 192:365–371

Jakobsen HB, Friis P, Nielsen JK, Olsen CE (1994) Emission of volatiles from flowers and leaves of *Brassica napus* in situ. Phytochemistry 37:695–699

Jones MB, Mansfield TA (1975) Circadian rhythms in plants. Sci Prog Oxf 62:103–125

Knudsen JT, Tolsten L, Bergstrom LG (1993) Floral scents – a checklist of volatile compounds isolated by headspace techniques. Phytochemistry 33:253–280

Loughrin JH, Hamilton-Kemp TR, Andersen RA, Hildebrand DF (1991) Circadian rhythm of volatile emission from flowers of *Nicotiana sylvestris* and *N. suaveolens*. Physiol Plant 83: 492–496

Loughrin JH, Hamilton-Kemp TR, Burton HR, Andersen RA, Hildebrand DF (1992) Glycosidically bound volatile compounds of *Nicotiana sylvestris* and *N. suaveolens* flowers. Phytochemistry 31:1537–1540

Loughrin JH, Potter DA, Hamilton-Kemp TR (1995) Volatile compounds induced by herbivory act as aggregation kairomones for the Japanese beetle (*Papilla japonica* Newman). J Chem Ecol 21:1457–1467

Loughrin JH, Potter DA, Hamilton-Kemp TR (1997) Response of Japanese beetles (Coleoptera, scarabiacidae) to leaf volatiles of susceptible and resistant maple species. Environ Entomol 26:334–342

Matile P, Altenburger R (1988) Rhythms of fragrance emission in flowers. Planta 174:242–247

Mookherjee BD, Trenkle RW, Wilson RA (1990) The chemistry of flowers, fruits and spices: live vs dead, a new dimension in fragrance research. Pure Appl Chem 62:1357–1364

Overland L (1960) Endogenous rhythm in opening and odor of flowers of *Cestrum nocturnum.* Am J Bot 47:378–382

Pabst A, Barron D, Semon E, Schreier P (1992) A 4-hydroxy-β-ionone disaccharide glycoside from raspberry fruits. Phytochemistry 31:3105–3107

Patt JM, Rhoades DF, Corkill JA (1988) Analysis of floral fragrance of *Platanthera stricta.* Phytochemistry 27:91–95

Schwab W, Mahr C, Schreier P (1989) Studies on the enzymic hydrolysis of bound aroma components from *Carica papya* fruit. J Agric Food Chem 37:1009–1012

Simon V, Riba M-L, Waldhart A, Torres L (1995) Breakthrough volume of monoterpenes on Tenax TA: influence of temperature and concentration for α-pinene. J Chromatogr 704:465–471

Takabayashi J, Dicke M, Posthumus MA (1994) Volatile herbivore-induced terpenoids in plant-mite interactions: variation caused by biotic and abiotic factors. J Chem Ecol 20:1329–1354

Tanaka S, Yamaura T, Shigemoto R, Tabata M (1989) Phytochrome-mediated production of monoterpenes in thyme seedlings. Phytochemistry 28:2955–2957

Voirin S, Baumes R, Bitteur S, Gunata Z, Bayonove C (1990) Novel monoterpene disaccharide glycosides of *Vitis vinifera* grapes. J Agric Food Chem 38:1373–1378

Yamaura T, Tanaka S, Tabata M (1989) Light-dependent formation of glandular trichomes and monoterpenes in thyme seedlings. Phytochemistry 28:741–744

Yamaura T, Tanaka S, Tabata M (1991) Participation of phytochrome in the photoregulation of terpenoid synthesis in thyme seedlings. Plant Cell Physiol 32:603–607

11 Odour Intensity Evaluation in GC-Olfactometry by Finger Span Method

P.X. Etiévant

11.1 Introduction

In the last 10 years, gas-chromatography coupled with olfactometry (GC-o) has been increasingly used by flavour chemists, as reviewed recently by Grosch (1993) and Mistry et al. (1997). The success of the method is due essentially to Acree et al. (1984) and Ullrich and Grosch (1987) who both improved the protocol of sniffing by rationalizing it. This protocol allows the experimenter to obtain reliable results because of the simplicity of the task asked of the panellists, and because of a validation of the final result obtained from the multiple detection of the same odour in different dilutions of the same extract. Moreover, the aromagrams obtained by sniffing successive dilutions of the same extract give valuable information on the number of odour units of each flavour constituent eluted from the column (Acree et al. 1984).

Nevertheless, the major drawbacks of the dilution approach are, first, the difficulty of using more than one panellist, as is advisable in sensory analysis, because the method is very time-consuming. A typical aromagram obtained by one subject for one sample without repetition corresponds to ten olfactory evaluations of GC runs lasting about 30 min each. A second drawback is that the results thus obtained are based on detection thresholds and not on real intensities. The third problem is that it is almost impossible to validate differences observed between aromagrams corresponding to different products by conventional statistical tests.

In order to overcome these serious limitations, two different solutions were proposed by different authors. The first solution, developed respectively by van Ruth and Roozen (1994), van Ruth et al. (1995) and Pollien et al. (1997a,b, 1999), overcomes the problem of the restrictive number of panellists, thus both obtaining an aromagram validated on a number of people and allowing us to calculate statistical differences between aromagrams obtained from the evaluation of different samples. The protocol used has the big advantage of requiring no particular training of the panellists, and overcomes the problems of the inherent different individual sensitivities with larger panels. The authors of the method (Pollien et al. 1997a) postulated that the frequency of detection obtained by this method is

I.N.R.A., Laboratoire de Recherches sur les Arômes, 17, rue Sully, BP 86510, 21065 Dijon Cedex, France related to the actual intensity perceived, and conversely to the dilution factors. This postulate was demonstrated by van Ruth et al. (1996a–c) and Pollien et al. (1999) and then corroborated by significant correlation between intensity scores and frequency of detection (van Ruth and Roozen 1994; van Ruth et al. 1996c; Blank 1997). The reason for this relation is found in the Gaussian repartition of the detection thresholds observed in a given population. One limitation of this postulate was, however, pointed out. When the intensity is high enough and still increasing, the frequency of detection can no longer follow this increase because all panellists detect the odour at this level. However, their olfactory system can still perceive the increase because it is far from saturation (Bernet 2000).

The second solution, proposed by McDaniel (da Silva et al. 1994), should overcome all problems, since it allows a direct estimation of the intensity of the odours using a reasonable number of panellists. The method, named OSME by its authors, is based on a magnitude estimation of the odour intensity. The panellist makes this evaluation using a variable resistor with a pointer moving along a 150-mm-long category scale with 16 points. A simultaneous computerized graphic feedback of the settled position of the cursor helps the panellist to adjust this position to the perceived intensity. Using a synthetic solution, the authors clearly demonstrated in the paper cited above that the estimation of the intensity of odours detected in a chromatographic effluent can be well correlated for each panellist with the concentration of the corresponding constituents, thus proving the feasibility of a GC-o intensity measurement.

The only published application of the method on food systems concerns Pinot noir wines of different vintages and maturity (Miranda-Lopez et al. 1992). Surprisingly, the authors interpreted the differences between wine OSME aromagrams using the frequency of detection of the odours as proposed by Pollien et al. (1997a), but not using the estimation of their actual intensities as originally planned. This limited utilization of the data obtained by the OSME technique is questionable. It is perhaps due to the large discrepancy observed in the number and the quality of the substances detected between panellists, as observed by the same authors in previous papers (McDaniel et al. 1989; da Silva et al. 1994), associated with a reduced number of panellists evaluating the wine extracts. It can also be interpreted as if the OSME analysis of a complex extract was a too difficult task to be performed by the panellists.

11.2 Description of the Finger Span Cross-Modality Matching Principle

It was observed in our laboratory from 1,176 odour detections with GC-o that in the chromatographic conditions chosen, the duration of 52% of these detections was less than 4 s. As a consequence, it appeared questionable to use in GC-o a method of evaluation of the odour intensity requiring a systematic feedback of the estimated value, since the panellist probably does not have enough time to adjust or to correct this value before the stimulus ends. Another possibility to evaluate the intensity of a stimulus is to adjust its perception with the intensity of a different stimulus, as, for example, the loudness of a sound with the intensity of an electric shock (Stevens 1975). This technique, called cross-modality matching evaluation, has the advantage of requiring no feedback information from the panellist. Furthermore, it does not require special thinking, training or learning, as demonstrated by the performance of 5-year-old children asked to evaluate the intensity of a light by cross-modality matching with the intensity of a sound (Stevens 1975). It seems therefore more adapted to the short time allowed for the GC-o evaluation.

Among the numerous sensory functions that were used successfully in cross-modality matching to evaluate the intensity of olfactory stimulus (Stevens 1975), the finger span appears attractive (Stevens and Stone 1959; Ekman et al. 1967; Guichard et al. 1995) because it does not disturb the environment. Several people can therefore work in the same room without reciprocally perturbing their mental concentration.

Practically, the finger span, or more precisely the distance between the thumb and the major finger of the panellists, was registered using a prototype adapted from Ekman et al. (1967) and schematized in Fig. 11.1. The output signal was sampled using a PC acquisition board developed in the laboratory (Almanza and Mielle 1990). The panellists were instructed to match the intensity perceived with the finger span, and eventually to describe the quality of the odour detected using a tape recorder with an automatic voice setoff. After the analysis, the data were visualized and processed using software developed in the laboratory (Almanza et al. 1989). The evaluations were realized in isolated conditions in a quiet room with a natural light and temperature regulated between 20 and 22 °C.

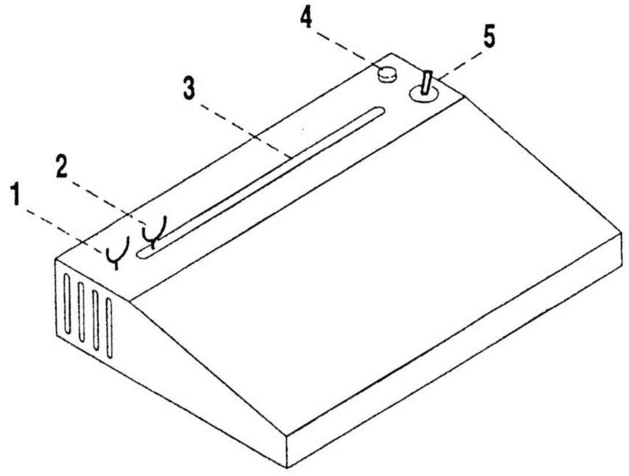

Fig. 11.1. Finger span prototype used to measure finger span (Etiévant et al. 1999; copyright 1999 American Chemical Society). *1* Fixed ring for the thumb; *2* mobile ring for the major or index finger connected to a 195-mm-long rheostat; *3* cursor track: *4* signal lamp; *5* on/off switch

11.3 Selection and Training

After two familiarization sessions, the selection of the panellists is made from the results of three evaluation sessions. During these sessions, each panellist is asked to match with the finger span different intensities of ethyl butyrate perceived at the sniffing port of a gas-chromatograph (high intensity corresponding to a large finger span and weak intensity to a small finger span). One session corresponds to four series of stimulations, and each series consists of six ethyl butyrate stimulations at different intensities. During familiarization, the panellist is informed of the relative intensity of the stimulus he or she will have to evaluate, which is naturally not the case during the evaluation (Etiévant et al. 1999).

The criteria for the selection performed are as follows. First, the panellists must feel comfortable with the prototype at the end of the familiarization. Second, the finger span estimate versus stimulus concentration plots must show the expected logarithmic increase. Third, the F value of the ANOVA comparing the mean estimations of the six different concentrations must be higher that the critical F. Fourth, the number of consecutive concentration pairs (CCP) significantly discriminated must be higher than 3.

Figure 11.2 shows the raw data obtained by two different subjects with extreme opposite performances. Panellist SD is able to estimate the different perceived intensities of the stimulus with high reproducibility as estimated by the high F value of the ANOVA and also by the number of CCPs significantly discriminated (Table 11.1). Panellist SH showed conversely poor results, and was not selected to participate in a panel because of a CCP score lower that 3 (Table 11.1).

In order to test the hypothesis made by Stevens (1975) that cross-modality matching does not require any particular training, the effect of training on judges' performance was evaluated from the evolution of the F value calculated on the concentration factor after each session for five panellists. Figure 11.3 shows that the performance of the judges varies from one session to another, but that this variation seems stable for most judges, as for judge CA, or decreases, as for judge MD. It can also be noted that some panellists, such as OC and LB, always show better performances than other panellists, such as MH and MD. The same observations are made when the number of CCPs is used as an index of the judges' performance (not shown). Thus, it can be concluded that there is no obvious training effect on the performance of the subjects, as stated by Stevens.

It is therefore practically wise to perform an initial selection of the subjects, which can be made with ethyl butyrate solutions as described above, and then to directly start the measures without any particular training. This selection takes 5 h per panellist. Our 3 years' experience shows that the average percentage of panellists rejected is lower than 30%. The selection test is rather simple since it only requires the preparation of six ethyl butyrate solutions, and their injection in the gas-chromatograph. For details, the reader may refer to Etiévant et al. (1999).

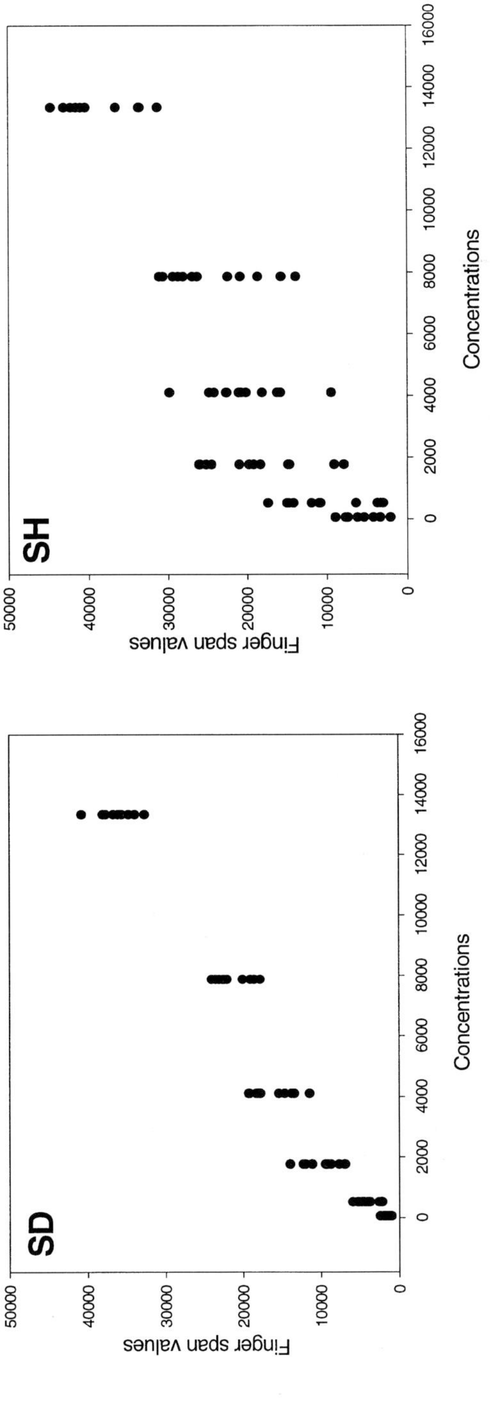

Fig. 11.2. Example of relation between ethyl butyrate concentrations and finger span raw evaluations for two subjects

Table 11.1. Scores obtained by different panellists after the training sessions

Panellist code	F value[a]	CCP[b]
SH	48	2
MB	121	5
FG	170	5
IG	170	5
LP	182	5
PP	205	4
LB	228	5
LG	327	5
SD	520	5

[a] Critical F value at $p = 0.05$: 2.35
[b] Average number of consecutive concentration pairs discriminated at $p = 0.05$: maximum CCP score is 5

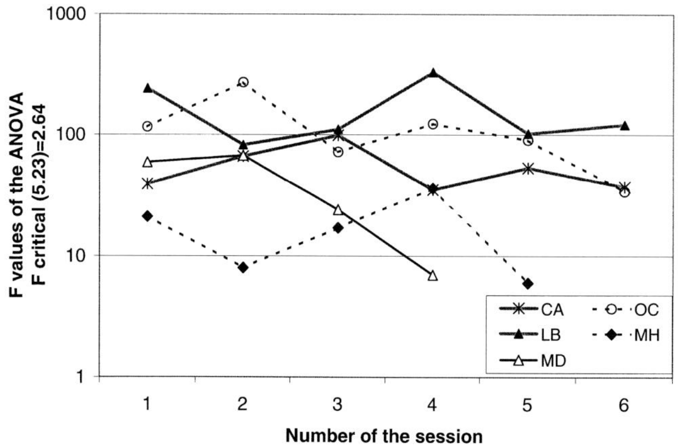

Fig. 11.3. Evaluation of training on judges' performance. (Adapted from Etiévant et al. 1999)

This large difference in judge performances can be attributed to different factors, one of which was recently tested by Bernet (2000). Working on Gewürztraminer wines, this author showed that the coefficient of variation measured for six judges on their finger span estimations of 15 wine flavour active constituents was inversely correlated with their mean finger span, as shown in Fig. 11.4.

This relation means that the precision of the judges is better when they use the largest part of the scale, and lower when they use only the small part of it, that means when they do not dare open their fingers to evaluate high intensities. This observation suggests that one should use this parameter, which can be easily calculated during the selection test with ethyl butyrate, to select subjects.

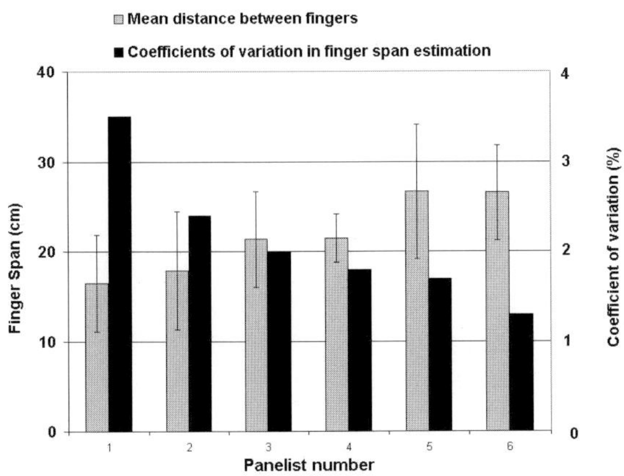

Fig. 11.4. Relation between mean finger span distance and judges' performance

11.4 Performance of the Method

Our target was to evaluate the performance of a selected panel estimating the actual intensities of constituents of a mixture. Five panellists were therefore asked to estimate in triplicate the perceived intensity of 11 constituents during the GC-o analyses of four solutions containing them at different concentrations. Typical flame ionization detector (FID) and finger span evaluation traces of one of these solutions is given in Fig. 11.5. These substances were chosen mainly for their published Stevens' exponents, their different volatility and chemical functionality, and also because they were detected at the sniffing port by all panellists at a concentration lower than 15 g/l, the GC saturation concentration.

The ability of the panel to evaluate the difference in intensity produced by the individual constituents in the synthetic solutions was first estimated from a two-way ANOVA with interaction (panellist ◊ solution), considering panellist as a random factor. Besides the expected subject effect, a clear solution (concentration) effect is demonstrated for the group (Table 11.2), showing that the mean finger span values given for each compound in the different solutions are found to be significantly different ($p < 0.01$). Moreover, the absence of interaction between the two factors tested shows that the panellists agree in their judgement of the individual odour intensities detected in the four solutions.

In order to visualize this conclusion, a PCA of the finger span data was realized. The biplot representation of the location of the solutions evaluated by the four panellists and of the 11 variables (finger span odour estimation of the 11 solution constituents) is given along the first two principal components (PCs) which explain 56% of the variance (Fig. 11.6). This figure shows clearly that our panellists are able to differentiate between the four solutions using a finger span

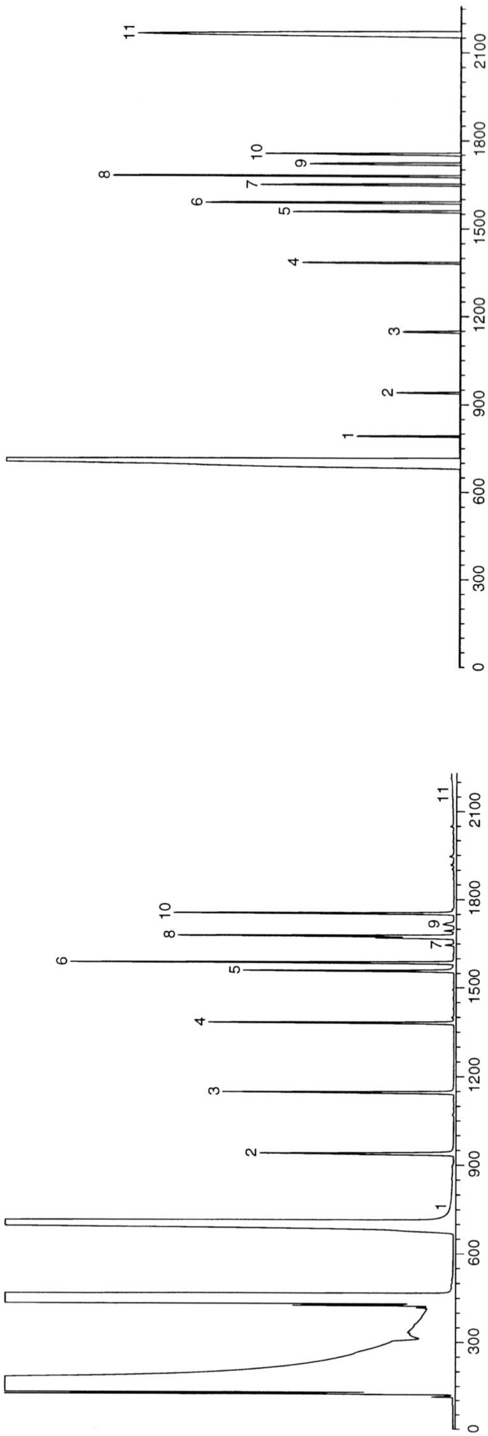

Fig. 11.5. FID and finger span profiles of a synthetic solution

Table 11.2. Two-way ANOVA of finger span values obtained from GC-o of synthetic solutions (Etiévant et al. 1999; copyright 1999 American Chemical Society)

Variable	Panellist factor	F values Solution factor	Interaction
3-Methylbutanethiol (MBT)	8***	19.9***	0.8
Hexan-2-one	3.3*	10***	0.9
Furfural	1.7	14.9***	0.7
Benzaldehyde	0.5	15.7***	1.2
Octan-1-ol	4.9**	5**	1.6
Nonanal	7.8***	9.6***	1.4
2-Methoxyphenol (guaiacol)	0.9	70***	1.8
Citronellal	17.4***	12.3***	1.3
2-Phenyl-1-ethanol	2.9*	18.3***	0.4
Decanal	5.3**	6.7***	0.8
Vanillin	4.4*	67.4***	1.7

$^{***}p<0.001$; $^{**}p<0.01$; $^{*}p<0.1$

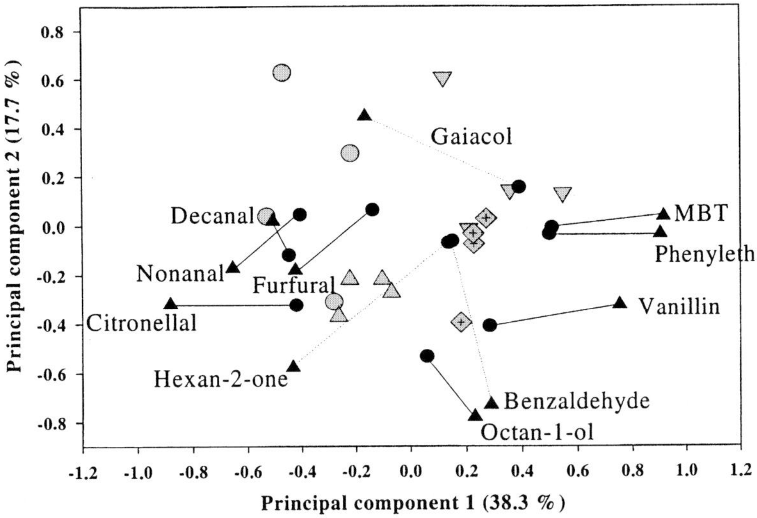

Fig. 11.6. PCA biplot of a finger span estimation of the intensity of four synthetic solution constituents. *Grey symbols* Solutions 1 to 4 (*identical grey symbols* correspond to different panellist evaluations); *black triangles* measured intensity; *black circles* theoretical intensity (illustrative variables). (Adapted from Etiévant et al. 1999)

estimation of the intensities perceived: solutions 1 (grey triangles) and 4 (grey circles) are well separated from solutions 2 (grey inverted triangles) and 3 (grey lozenges) along the first PC, and solutions 1 and 2 are separated from solutions 3 and 4 along the second PC. It also shows the absence of interaction between the solution and panellist factors since the four points corresponding to the evaluation of the same solution by the four different panellists are reasonably grouped in the same area of the graph. The figure clearly shows that the panel was able to find most of the characteristics of the different solutions, including (1) a highest intensity for MBT, 2-phenyl-1-ethanol and vanillin in solutions 2 and 3, (2) a highest intensity for nonanal and citronellal in solutions 1 and 4, (3) a highest intensity for octanol in solutions 3 and 4, and finally (4) a highest intensity for decanal and furfural in solution 1, and of hexan-2-one and benzaldehyde in solution 4. However, some characteristics of the solutions were hardly found, such as the high content of hexan-2-one in solution 1, furfural in solution 3, benzaldehyde in solution 2 and guaiacol in solution 3.

When looking at the relative position of the black triangles and circles corresponding for each variable to its estimated and theoretical intensity (determined from known concentration in solutions and from Stevens exponent found in the literature), it can be concluded that the estimated values are close to the predicted ones for MBT, phenyl-2-ethanol, vanillin, octan-1-ol, citronellal, nonanal, decanal and even furfural.

The finger span estimations of the intensity of 3 of the 11 compounds tested, hexan-2-one, benzaldehyde and guaiacol, appear different from the predicted values. We postulated that this difference may be due to literature Stevens exponents being different from the actual exponents of our panellists (Etiévant et al. 1999; Salles et al. 1999). If this hypothesis is true, the apparent poor estimation for these three particular compounds should be overcome by using a larger panel, or using coefficients obtained by the subjects who made the estimations.

11.5 Applications

11.5.1 Sample Discrimination Based on Odour Intensity of Constituents

The GC-o finger span estimation of the intensity of the odour of individual constituents of food extracts allows us to identify the flavour constituents that impact on the whole aroma, which differs according to the different processes, variety, storage condition, etc. Since the method allows us to quantify directly the flavour intensity of these compounds, it is therefore no longer necessary to quantify them.

In a recent work, four beers were studied for their difference. In order to gain information that could be more easily related to the perception of the beer flavour, the extracts submitted to GC-o with finger span estimation were obtained by concentration and injection of nose space samples taken directly from the nose of ten people drinking the different beers. In order to eliminate the noise coming from odorant molecules released in the breath of these people as a consequence of their basal metabolism, blank samples were also analysed, as described by Etiévant et al. (1999).

The GC-o analyses were realized in triplicate by eight selected panellists. Figure 11.7 shows the FID and finger span traces of a particular sample and sniffer (see also Fig. 11.8). Figure 11.7 illustrates the fact that the sensitivity and selectiv-

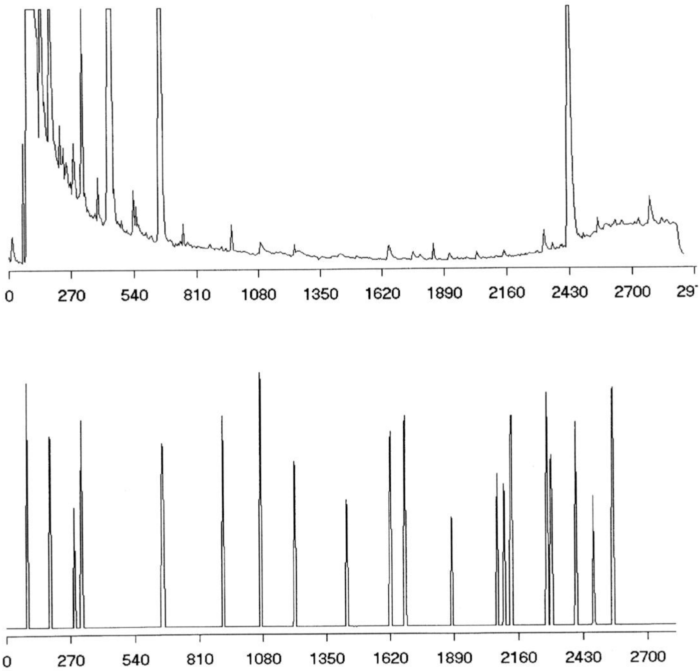

Fig. 11.7. FID (*above*) and finger span (*below*) traces of a beer nose space

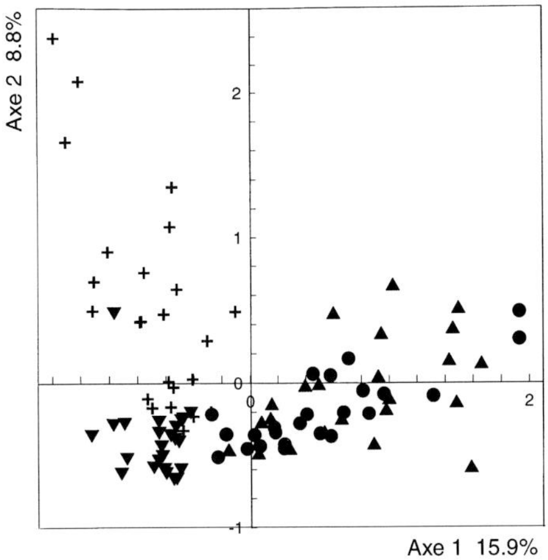

Fig. 11.8. Beer sample space of a PCA made with 49 selected finger span variables

ity of the nose is such that it is possible to get fine results on an extract that is not suitable for GC analysis with conventional detectors.

The raw finger span data table was obtained after adjustments of the retention times of the different profiles [eight panellists × (four beers + four blanks) ×

three sniffings = 192 finger span profiles] by simple subtraction of the blank samples from the beer samples. The different panellists detected between 81 and 169 different odours in the extracts they analysed. In total, 202 different odours were detected in the four beers, amongst which 49 were found to be discriminant between beers at $p < 0.05$ from a two-way ANOVA, using sniffer as a random factor. Using these discriminant variables, a PCA was realized with the 24 estimations (three replicates and eight panellists) made for each beer. The sample space along the two first axes shows a good agreement between panellists and between replications in the estimation of the odour intensities. The dispersion of the estimations made by the panellists for each beer varies very much with the beer type. For example, it is large for the beer symbolized with black circles and much narrower for the beer symbolised with inverted black triangles. To summarize, the PCA shows a clear difference along the first axis between beers + and ▼ on the negative side and beers ● and ▲ on the positive side. It also shows a clear separation of beers + and ▼ along the second axis. The two beers ▲ and ● are described with similar characteristics on this plan. A stepwise discriminant analysis allows the total discrimination of the 4 beers using 40 variables, 19 common to the former ANOVA selection and 21 different (not shown).

11.5.2 Determination of Stevens' Coefficients

Another possible application of the technique is to determine the psychophysical functions of odour potency as demonstrated by da Silva et al. (1994). A first advantage to using the coupling with GC in such evaluations is that we overcome a major problem in psychophysics which is the olfactory purity of the substances tested. In solutions, the odour of impurities often modifies the threshold or the Stevens' coefficients measured. In GC-o, the risk of coelution of the odour-active impurities with the compounds under study is considerably decreased when they are injected pure in the gas chromatograph, and the risk of coelutions of such impurities with other compounds in synthetic solutions can be easily evaluated. A second advantage of this technique is that it allows a faster estimation of a large number of substances compared to an evaluation with solutions. The third advantage is that it allows the estimation of compounds that are not available pure, as soon as they can be found in an aromatic extract. It can even be possible to determine the Stevens' coefficients for unknown substances, and in particular for substances with a very high odour potency, but present at trace level in food extracts.

An experiment was conducted with six trained panellists selected out of seven using the ethyl butyrate test developed above. The selected panellists were asked to score five times the intensity of seven compounds present at different concentrations in five different solutions. The duration of each GC-o analysis was 25 min. The results of the regressions adapted from the Stevens' equation (log finger span $= n \log (C\text{-}C°) + k$) are reported in Table 11.3. They confirm that most panellists are able to use the finger span to score the perceived intensities and to produce

Table 11.3. Log-log regression of finger span on concentration

Compound	Panellist code					
	CEL	LAU	SEB	MAT	JER	ARN
Butan-1-ol	0.65**	0.63**	0.23*	ns	0.26**	ns
Hexanal	0.33**	0.28**	0.17**	0.24**	0.15**	ns
Methanethiol butyrate	0.59**	0.71**	0.32**	0.15*	0.36**	ns
Furfural	0.63*	0.63**	0.3**	ns	0.31**	0.44*
Cyclohexanol	0.5**	0.45**	0.4**	ns	0.17**	0.4**
Dimethyltrisulfide	0.13**	0.12**	0.11**	0.1**	0.06**	0.04*
Trimethyl-2,3,5-pyrazine	0.21**	0.4**	0.14**	0.11**	0.08**	ns

ns, Not significant; *p<0.05; **p<0.01

Table 11.4. ANOVA results and slopes values obtained by the panel. Slopes with different letters are significantly different ($p < 0.05$, t-test)

Compound	Slope values (Confidence interval at 95%)	Two-way ANOVA		
		Concentration	Subject (random factor)	Interaction
Butan-1-ol	0.35^{a} (0.21)	<0.001	<0.01	ns
Hexanal	0.22^{bc} (0.1)	<0.001	<0.001	ns
Methanethiol butyrate	0.38^{a} (0.17)	<0.001	<0.05	<0.001
Furfural	0.41^{a} (0.32)	<0.001	ns	ns
Cyclohexanol	0.48^{a} (0.31)	<0.01	<0.05	<0.001
Dimethyltrisulfide (DMT)	0.1^{bc} (0.03)	<0.001	<0.01	ns
Trimethyl-2,3,5-pyrazine (DMP)	0.17^{c} (0.1)	<0.001	<0.001	<0.001

ns, Not significant: $p > 0.05$

data suitable to determine significant regressions. Two of them, ARN and MAT, could not estimate all the power functions significantly. This lower performance was related to an observed lack of interest in their task after a few sessions. If one looks at the slope values obtained by the different panellists, large differences are observed which are generally attributed to differences between individual sensitivities (Stevens 1991; da Silva et al. 1994). Despite this expected variation, a two-way ANOVA with interaction shows that the panel discriminates the different concentrations for each stimulus (Table 11.4).

A comparison of these slopes shows that three groups of compounds are found to be significantly different (Table 11.4). This preliminary result indicates that a finger span cross-matching of the intensity perceived in GC-o allows the determination of psychophysical function parameters, and the variability between panellists suggests using a large number of subjects in order to get a stabilised value of Stevens' coefficients, or a limited number of people when precise relations need to be established between theoretical and perceived intensities.

11.6 Conclusion

The cross-modality estimation of odour intensity with the finger span in GC-o allows the experimenter to work with untrained panellists since the method requires no particular thinking, training or learning. The other advantage of the method is that it is adapted to fast phenomena such as GC-olfactometry (GC-o). Reliable results can be obtained after a double selection of the subjects used in the evaluation. The first part of the selection involves subject variability in the finger span estimation of series of ethyl butyrate stimulations at the sniffing port of a GC. The second part deals with subjects' capabilities to use most of the scale of the rheostat with their fingers. Eventually, the selection could be completed on the capability of the subjects to concentrate on a particular task.

A comparison of estimated and theoretical intensities of the flavour constituents of different synthetic solutions analysed by five selected panellists shows very close results for most constituents. Moreover, experiments conducted on more complex beer nose space extracts show a good discrimination of the samples studies, with a high number of discriminant odours. The technique is also successful in determining individual Stevens' coefficients of compounds in a mixture. This application offers the advantage of a fast determination compared to a presentation of pure substances in solutions, and the advantage of a high purity of the compounds tested compared to a determination from solutions as used in a conventional olfactometer.

Acknowledgements. The results described were obtained with the financial support of Tepral (research branch of Kronembourg), Moët and Chandon Company (Epernay, France), and the "Conseil Régional de Bourgogne". Sincere thanks are expressed to D. Langlois, G. Callement and B. Atanasova for their collaboration.

References

Acree TE, Barnard J, Cunningham DG (1984) A procedure for the sensory analysis of gas chromatographic effluents. Food Chem 14:273–286

Almanza R, Mielle P (1990) Système d'acquisition de données chromatographiques multicanal "coconut". In: INRA (ed) Deuxième journée de la mesure INRA. INRA, Port-Leucate

Almanza R, Couton Y, Mielle P, Nicolardot B (1989) Traitement informatique du signal électrique provenant de calorimètres à flux continu. Cah Tech INRA 20:49–56

Bernet C (2000) Contribution à la connaissance des composés d'arôme clés des vins du cépage Gewurztraminer cultivé en Alsace. PhD Thesis, Université de Bourgogne

Blank I (1997) Gas chromatography–olfactometry in food aroma analysis. In: Marsili R (ed) Techniques for analyzing food aroma. Dekker, New York, pp 293–329

da Silva MAAP, Lundahl DS, McDaniel MR (1994) The capability and psychophysics of OSME: a new GC-olfactometry technique. In: Maarse H, Van der Heij DG (eds) Trends in flavour research. Elsevier, Amsterdam, pp 191–209

Ekman G, Berglund B, Berglund U, Lindvall T (1967) Perceived intensity of odor as a function of time of adaptation. Scand J Psychol 8:177–186

Etiévant P, Callement G, Petka J, Villanueva S, Langlois D (1999) Odor intensity evaluation of mouth space constituents by GC-olfactometry with finger span cross-matching. In: Schieberle P, Engel K-H (eds) Frontiers of flavour science. Deutsche Forschungsanstalt für Lebensmittelchemie, Garching, pp 29–33

Grosch W (1993) Detection of potent odorants in foods by aroma extract dilution analysis. Trends Food Sci Technol 4:68–73

Guichard H, Guichard E, Langlois D, Issanchou S, Abbott N (1995) GC sniffing analysis: olfactive intensity measurement by two methods. Z Lebensm Unters Forsch 201:344–350

McDaniel MR, Miranda-Lopez R, Watson BT, Micheals NJ, Libbey LM (1989) Pinot noir aroma: a sensory/gas chromatographic approach. In: Charalambous G (ed) Flavors and off-flavors. Elsevier, Rethymmon, pp 23–36

Miranda-Lopez R, Libbey LM, Watson BT, McDaniel MR (1992) Odor analysis of Pinot noir wines from grapes of different maturities by a gas chromatography-olfactometry technique (OSME). J Food Sci 57:985–993, 1019

Mistry BS, Reineccius T, Olson L (1997) Gas-chromatography-olfactometry for the determination of key odorants in foods. In: Marsili R (ed) Techniques for analysing food aroma. Dekker, New York, pp 265–292

Pollien P, Ott A, Montigon F, Baumgartner M, Munoz Box R, Chaintreau A (1997a) Hyphenated headspace-gas chromatography-sniffing technique: screening of impact odorants and quantitative aromagram comparisons. J Agric Food Chem 45:2630–2637

Pollien P, Ott A, Montigon F, Baumgartner M, Munoz-Box R, Chaintreau A (1997b) Hyphenated headspace-GC-sniffing technique: the "sniff" method. In: Rivista Italiana EPPOS (ed) Proc Int Symp on Flavours and Sensory Aspects. Nestec, Lausanne, pp 126–134

Pollien P, Fay LB, Baumgartner M, Chaintreau A (1999) First attempt of odorant quantitation using gas chromatography-olfactometry. Anal Chem 71:5391–5397

Salles C, Etiévant P, Bouchet M, Langlois D (1999) Odor intensity measurements in GC/O using modality matching: evaluation of training effects. In: Leland JV, Schieberle P, Buettner A, Acree T (eds) Gas chromatography-olfactometry. The state of the art. Oxford University Press, Cary

Stevens DA (1991) Individual differences in taste and smell. In: Lawless HT, Klein BP (eds) Sensory science theory and applications in foods. Dekker, New York, pp 295–316

Stevens SS (1975) Cross-modality matching. In: Stevens SS (ed) Psychophysics: introduction to its perceptual, neural, and social prospects. Wiley, New York, pp 99–133

Stevens SS, Stone G (1959) Finger span: ratio scale, category scale, and JND scale. J Exp Psychol 57:91–95

Ullrich F, Grosch W (1987) Identification of the most intense volatile flavour compounds formed during autoxidation of linoleic acid. Z Lebensm Unters Forsch 184:277–282

van Ruth SM, Roozen JP (1994) Gas chromatography/sniffing port analysis and sensory evaluation of commercially dried bell peppers (*Capsicum annuum*) after rehydration. Food Chem 51:165–170

van Ruth SM, Roozen JP, Cozijnsen JL, Posthumus MA (1995) Volatile compounds of rehydrated French beans, bell peppers and leeks. 2. Gas chromatography sniffing port analysis and sensory evaluation. Food Chem 54:1–7

van Ruth SM, Roozen JP, Cozijnsen JL (1996a) Gas chromatography/sniffing port analysis evaluated for aroma release from rehydrated French beans (*Phaseolus vulgaris*). Food Chem 56: 343–346

van Ruth SM, Roozen JP, Nahon DF, Cozijnsen JL, Posthumus MA (1996b) Flavour release from dehydrated French beans (*Phaseolus vulgaris*) influenced by composition and volume of artificial saliva. Z Lebensm Unters Forsch 203:1–6

Van Ruth SM, Roozen JP, Hollman ME, Posthumus MA (1996c) Instrumental and sensory analysis of the flavour of French beans (*Phaseolus vulgaris*) after different rehydration conditions. Z Lebensm Unters Forsch 203:7–13

12 Solid Phase Microextraction Application in GC/Olfactometry Dilution Analysis

K.D. Deibler, E.H. Lavin, and T.E. Acree

12.1 Introduction

12.1.1 Aroma Chemistry

Aroma chemicals, perceived on the olfactory epithelium, impart an odor character to food. Among the hundreds of volatile compounds in natural foods, only a small fraction contribute to the perceived aroma because aroma compounds must be present above their detection threshold to impart a sensation. Additionally, aromas are volatiles, but not all volatiles cause aroma sensation. Over 500 compounds that cause an aroma sensation have been identified as described on the Flavornet (http://www.nysaes.cornell.edu/fst/faculty/acree/flavornet/index.html) and the characteristic flavor of a particular food comes from the pattern of aroma compounds volatilizing from food.

Olfactometry or the measure of odor response to chemicals is often combined with gas chromatography. Called gas chromatography-olfactometry (GC/O), the technique has been in use for many years and its applicability and theory have been recently reviewed (Acree 1997). Human olfaction is selective for odor-active compounds and is orders of magnitude more sensitive than flame ionization detection and mass spectrometry for some compounds. GC/O analysis of unknown odorants provides measurements of odor quality and chromatographic retention times (RT) which, taken together, can be used to identify the odorant. Matching of retention index and odor character on two different gas chromatograph (GC) column materials of an unknown with a standard is a convincing identification of an odorant.

Since investigation of aroma involves differentiation between odor-active and non-odor-active compounds, GC/O has become essential to the study of flavor chemistry. Dilution methods such as CharmAnalysis™ and aroma extraction dilution analysis (AEDA) produce threshold information through GC/O measurements of a dilution series (Acree et al. 1984; Grosch 1993). These dilution to threshold-based methods produce estimates of potency or titer (i.e. charm values or dilution factors). Potency is a measure of the power or ability of the odorant to affect the

---

NYSAES-Cornell University, Department of Food Science and Technology, Geneva, New York 14456, USA olfactory sensors and is a function of concentration. A list of the most potent aroma compounds contributing to aroma is a result of a dilution analysis.

12.1.2 Mouth Simulators

Aromas reach the receptors on the olfactory epithelium via two routes: orthonasal and retronasal. In the orthonasal route, aroma compounds enter through the nostrils prior to consumption. For example, sniffing a glass of wine is an orthonasal experience. When the food is placed in the mouth, aroma compounds travel to the olfactory epithelium via the throat and lungs: the retronasal route. The difference in perception between ortho- and retronasal aroma release is due to different ratios of aromas being released in the mouth, influenced by chewing, saliva, body heat, and air-flow. Mouth simulators are the collection of devices that simulate parameters of the mouth that affect volatilization of compounds from food during eating (Lee 1986; DeRoos and Wolswinkel 1994; Roberts et al. 1994; van Ruth et al. 1994; Roberts and Acree 1995; Elmore and Langley 1996; Springett et al. 1999). Mouth simulators account for air flow, mixing, heat, and dilution. Most models have much less variation than is exhibited by human subjects and eliminate concern for palatability and safety. Mouth simulators including the Roberts (RAS) and the van Ruth (Model Mouth) systems work with a broad range of foods including all liquids and most solid foods. The retronasal aroma simulator (RAS), developed by Roberts and Acree in 1994, is unique in the use of blades for shearing, stainless steel to reduce adsorption, large size to increase sensitivity, and gas flow rates similar to human breathing.

Using breath-by-breath analysis, a correlation greater than 99% has been observed between the volatiles produced from the RAS and volatiles released in humans while eating cheese (Deibler et al. 2000). The RAS (Fig. 12.1) is composed of a 1-l stainless-steel blender container and assembly, a voltage controller and high torque variable speed motor to give precise control of blender speed to simulate chewing and crushing of food, a controlled gas supply to sweep over the food, the addition of artificial saliva, and an external copper coil with a water flow to control the temperature. Table 12.1 shows the parameters used for simulating mouth conditions. Mouth simulators are not intended to simulate all of the conditions and dimensions of the human mouth. They simulate only those conditions most likely to affect volatile release from food. Likewise, the time course is not intended to simulate the time during eating, but meant to show the dynamics of flavor volatilization. Volatiles in the RAS effluent can be collected by various methods including porous polymer, solid phase microextraction (SPME), and static syringe. High precision (CV<5%) and sensitivity (μg/l) have been observed from the RAS using a static syringe (Roberts and Acree 1996a,b).

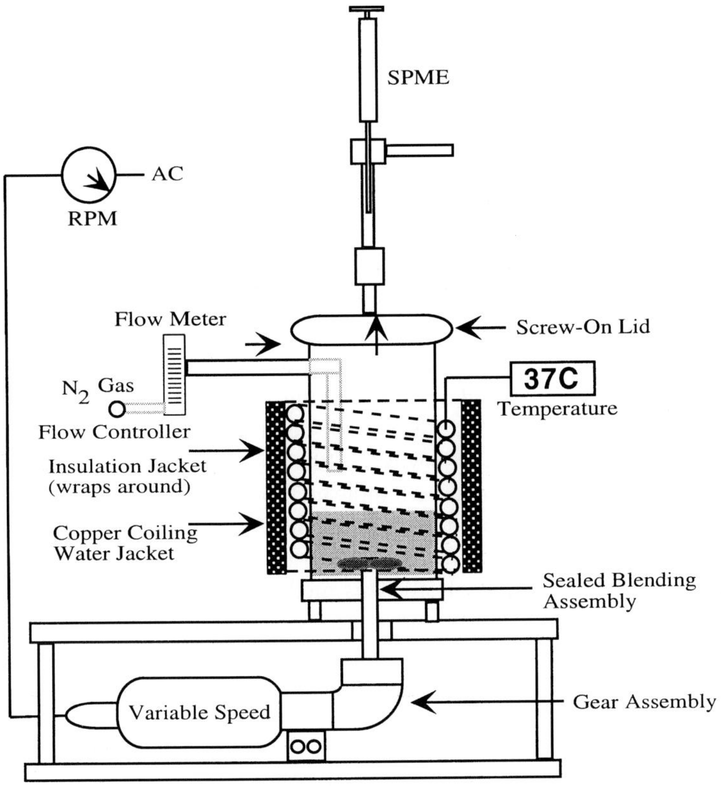

Fig. 12.1. Retronasal aroma simulator. *AC* Alternating current; *RPM* revolutions per minute

Table 12.1. Mouth simulation conditions of the RAS

Condition	Setting	
Temperature	370 °C	Conductive heat to maintain constant temperature
Air or N_2 flow	20 ml/s	Less air flow than during inhaling (100 ml/s)
Blending rate	2.5 rps	Within the shear range of various foods in mouth
Saliva to food ratio	1/5 v/w	Average stimulated saliva flow rate (2 ml/min) for 30 s of eating time

12.1.3 Solid Phase Microextraction

Since aromas are volatiles, it is most useful to study the volatile composition above the food of interest (the headspace), ideally in a simulated mouth system. Determining headspace aroma composition presents both sampling and detection challenges. Odorants are typically found in the headspace at very low concentration (parts per billion) and require concentration before analysis. Solvent extraction gives a representation of what is present in the food; however, this is usually quite different from what volatilizes and actually comes into contact with the olfactory epithelium to induce a sensation.

One simple method for concentrating and sampling the headspace is solid phase microextraction (SPME). SPME is based on absorption, which is useful for extraction and concentration of analytes either by submersion in a liquid phase or by exposure to a gaseous phase (Aurthur et al. 1992). Following exposure of the fiber to the sample, absorbed analytes can be thermally desorbed in a conventional GC injection port. The fiber behaves as a liquid solvent selectively extracting analytes with more polar fibers having a greater affinity for polar analytes. Extraction is based on the diffusion and partitioning of the analytes (Zhang and Pawliszyn 1993, 1995). SPME has been commercially available since 1993 and is available with various adsorbent materials and various coating thickness. The developer of SPME, Januz Pawlizyn, wrote an excellent book describing preliminary steps necessary for using SPME for analysis (Pawliszyn 1997).

Combined with GC/O, SPME provides a simple method to estimate odor headspace composition. Combined with a linear dilution analysis using SPME fibers of different thickness and exposure lengths, headspace samples can be diluted by almost two orders of magnitude (1 to 1/50). This method is especially useful for the analysis of fatty foods, solid foods, and aqueous foods where non-volatiles create artifacts during solvent extraction. The greatest benefit of this technique to gas phase dilution analysis is the elimination of solvent extraction and concentration. Additionally, SPME allows for headspace measurements with a greater sensitivity than other conventional methods of analysis, such as purge and trap (Chaintreau et al. 1995). The quantitative aspect of flavor dilution analysis such as CharmAnalysis makes this an important application of SPME.

12.2 Description of Methods

12.2.1 SPME Initialization

Conditions for using SPME must be defined for each type of experiment. Variables that must be defined include sample parameters (i.e., temperature, size), fiber coating, absorption time, desorption time, and desorption depth in the GC injector. The SPME fiber must be conditioned prior to its initial use and heated for a short time (i.e., 2 min) to clean prior to each use. The coating degradation over time can be monitored by sampling a standard solution periodically and watching for a reduction in the amount being extracted. Saturation of the fiber may occur when using concentrated samples and can be tested by simply seeing if the response increases with a sample of greater concentration. Saturation must be avoided for dilution analysis and for comparison of samples. When sampling from a closed system, SPME comes to equilibrium with its surroundings resulting in

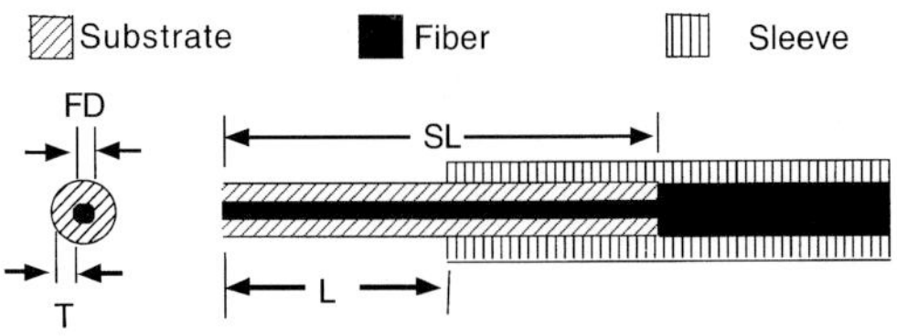

Fig. 12.2. SPME apparatus for dilution analysis. *FD* Fiber diameter; *L* substrate exposed to the vapor phase outside the sleeve; *SL* total fiber length; *T* substrate thickness

greater extraction of some compounds from the sample. If the sample is large, the amount extracted is insignificant in relation to the equilibrium state and thus equilibration is rapid. Separating the gas phase (headspace) from the solid or liquid sample and then extracting the headspace reduces SPME system equilibration time by simplifying the system to only the SPME and the gas phase equilibrating. For steady-state systems such as sampling the RAS effluent, the SPME fiber comes to equilibrium with the flow without a reduction in the flow concentration.

12.2.2 SPME CharmAnalysis

For a dilution analysis using SPME, varying fiber thickness, exposure length, and multiple fibers can be combined to produce the dilution series (Deibler et al. 1999). Figure 12.2 depicts the SPME apparatus. Each fiber thickness can be fully exposed, two thirds exposed, and one third exposed to the effluent from the RAS to achieve a dilution step. Exposure length (L in Fig. 12.2) can be controlled by creating two additional notches in the SPME holder or by placing an O-ring at a marked location on the injection assembly. It is important to measure the exact amount of fiber that is exposed at the two intermediate settings. Also, the actual fiber thickness may be different from the nominal thickness for commercial fibers. After piercing the GC injector septum with the manual SPME assembly, the appropriate amount of fiber corresponding to the amount exposed during sample adsorption is desorbed onto the column. Seeping of compounds past the stainless-steel shield was found to be insignificant for most compounds. Three fibers can be exposed to the headspace and desorbed in the GC while cryofocusing the compounds with a cold liquid such as liquid nitrogen.

12.2.3 Quantification of SPME

Because SPME selectively absorbs compounds, it must be calibrated for quantification. This can be achieved using the apparatus in Fig. 12.3. Each analyte is prepared at 0.01% in a solvent by GC/MS for purity. The standard compounds are then prepared at 0.1% in a nonaqueous, low-volatility solvent (i.e., propylene glycol). Solutions must be prepared such that compound and impurity peaks do not overlap. The standard solution (45 ml) is stirred with a magnetic stirrer in a 100-ml round-bottom flask with a 100-ml addition funnel topped with a screw-cap with septum adapter (Fig. 12.3). After at least 30 min of stirring, the addition funnel

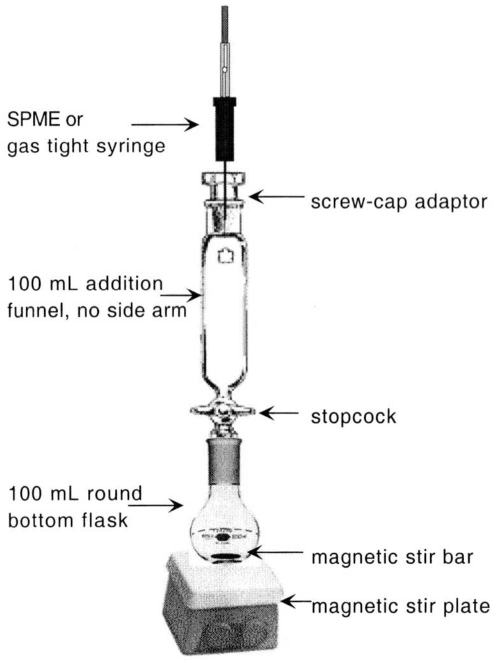

Fig. 12.3. Apparatus for quantification of headspace SPME

should be closed from the solution isolating the headspace. The headspace is sampled with a 5-ml gas-tight syringe and cryogenically focused on the GC/MS or by the SPME fiber being calibrated and desorbed in the injection port for GC/MS analysis. The solvent injection of the standard amount is used to determine the quantity of compound per unit of peak area ($C_{solvent}/V_{solvent} \times$ peak $area_{solvent}$). The gas-tight syringe extraction of the headspace gives the absolute quantity per peak area of the compound in the headspace (V_{vapor}/peak $area_{vapor}$). The selectivity of SPME is thus accounted for in quantification using Eq. (1). The conversion constant (k_{SPME}) characterizes the specific fiber and compound and may be used for calculating any SPME headspace extraction of that compound by multiplying the measured peak area by the conversion constant.

$$K_{SPME} = \frac{C_{solvent} \times V_{solvent}}{\text{peak area}_{solvent}} \times \frac{\text{peak area}_{vapor}}{V_{vapor}} \times \frac{1}{\text{peak area}_{SPME}}$$

where

k_{SPME}	Conversion constant for SPME fiber and specific compound
$C_{solvent}$	Concentration of compound injected in solvent injection
$V_{solvent}$	Volume of solvent injected
V_{vapor}	Volume of gas in gas-tight syringe injection
peak $area_{solvent}$	GC peak area of aroma compound produced from solvent injection
peak $area_{vapor}$	GC peak area produced from gas-tight syringe injection
peak $area_{SPME}$	GC peak area of aroma compound produced from SPME injection

12.3 Example of SPME Dilution Analysis

12.3.1 Methods

12.3.1.1 SPME Extraction

SPME fibers with 95-, 30-, and 7-µm thickness of polydimethylsiloxane (PDMS) coating and the manual holder were obtained from Supelco Co. (Bellefonte, PA). PDMS was selected to demonstrate this method since it was commercially available in three thicknesses. Before initial use, the fibers were conditioned for 1 h at 250 °C for the 95- and 30-µm fibers and for 2 h at 320 °C for the 7-µm fiber. Before each extraction the fiber was held at 225 °C for 5 min and allowed to come to room temperature for 10 min. The plunger depth was set at 3 cm to allow for maximum desorption into the GC by injecting into the hottest part of the injection port. It was found to be important that the holder be tightly assembled to obtain consistent results. When sampling, the fiber was injected 1 cm above the solution surface. The fiber was in the injection port to desorb volatiles for 2 min.

12.3.1.2 GC Parameters

An HP5890 gas chromatograph (GC) with a fused silica capillary column DB1 liquid phase (0.33 µm) with column dimensions of 25 m◊0.20 mm was used. Carrier gas was helium at 25 cm/s linear velocity. The temperature program was an initial 3 min at 40 °C then a ramp rate of 4 °C/min to 200 °C followed by 25 °C/min to 250 °C and held at 250 °C for 15 min. The injection purge on the GC was off for the initial 1 min.

12.3.1.3 Optimization of Exposure Time

The 95-µm fiber was exposed to the standard solution for 1, 5, 7, 10, 15, 20, 25, 60, and 90 min. The shortest time where the 95% of maximum product was extracted for all compounds was determined to be the optimum exposure time for the PDMS fiber.

12.3.1.4 Dilution Analysis

Each fiber was exposed to the headspace of the standard solution for the appropriate time. Each fiber was fully exposed (1.0 cm), approximately two thirds exposed (0.58 cm), and approximately one third exposed (0.28 cm) to the headspace standard solution. Exposure length was controlled by creating two additional notches in the SPME holder. After piercing the injector septum with the manual SPME instrument, the appropriate amount of fiber corresponding to the amount exposed during sample adsorption was desorbed onto the column for 1.5 min, followed with a full exposure desorption in another injection port for 5 min. Relative concentration was calculated based on the area of the corresponding GC peak. Each measurement was conducted in triplicate.

12.3.1.5 CharmAnalysis of Coffee

CharmAnalysis was conducted on an HP5890 GC modified by DATU, Inc. (Geneva, NY) (Acree et al. 1984; Acree 1997). GC/O conditions were the same as the GC parameters described above. Commercially brewed coffee (50 ml) was held in a 100-ml glass bottle with a Teflon-faced butyl septum lid at 28 °C. The coffee was stirred with a magnetic stirring bar and allowed to equilibrate for 1 h prior to extraction (15 min). To calculate relative retention indices, the SPME was exposed to the headspace of an ethanol solution of ethyl esters from C4–C18 using identical sampling and measurement procedures except detection was by FID.

12.3.2 Results of Example

1,8-Cineol was the slowest to achieve 95% of maximum absorption and thus the limiting compound, requiring 10 min to achieve absorption. The optimum exposure time was thus concluded to be 10 min for this analysis.

Actual fiber dimensions were obtained from Supelco. All fibers were 10 mm long. The thickness were 95, 30, and 7 µm with nominal coating volumes of 0.612, 0.132, and 0.0121 mm^3 respectively. However, measurement of the 7-µm fiber with an Olympus microscope at 100 times magnification showed an actual thickness of 13 µm giving a fiber volume of 0.054 mm^3. When fractions of the fiber were exposed (100, 58, or 28%) the resulting volumes of the 95-, 30-, and 7-µm fibers were: 0.612, 0.349, 0.171; 0.132, 0.075, 0.037; and 0.054, 0.036, 0.018 mm^3 respectively. For each fiber thickness the GC peak areas were reproducible at less than 5% standard deviation. Each compound produced linear plots of peak area versus exposed fiber volume ($R^2 > 0.90$). Slopes of each line are indicative of the quantity of analyte adsorbed to the SPME indicative of the amount of analyte in the gas phase and its affinity for the fiber. Since there were no deviations from the linear trend in any of the compounds tested, it was assumed that this method can be applied generally

Table 12.2. CharmAnalysis results of SPME (solid phase microextraction) extracted coffee headspace compared to CharmAnalysis results using serial solvent extraction with freon 113 and ethyl acetate (Deibler et al. 1998) and stable isotope dilution assay producing odor-active values (OAV) (Semmelroch and Grosch 1996). For comparison all data were converted to OSV [OSV = $(\text{potency}/\text{potency max})^{0.5} \times 100$] (Acree 1997)

Compound	SPME	Charm	OAV	CAS #
2,4,5-Trimethylthiazole	100	33	0	13623-11-5
3-Methoxy-2-isobutyl pyrazine	54	42	32	24683-00-9
Sotolon	34	100	37	28664-35-9
Abhexon	31	25	0.2	698-10-2
Vanillin	29	64	0.7	121-33-5
Furaneol	26	57	49	3658-77-3
4-Vinyl guaiacol	24	70	20	7786-61-0
4-Ethyl guaiacol	15	19	0.2	2785-89-9
2-Isopropyl-3-methoxy pyrazine	15	10	0	25773-40-4
2-Furfurylthiol	14	90	100	98-02-2

to volatiles. An exposure of the fiber fully sheathed (zero exposure) to the experimental conditions resulted in less than 5% of the 1,8-cineol and ethanol, and no other compounds could be detected, thus analyte movement along the protected fiber did not significantly affect the results. There was no noticeable difference in peak shapes with the different size fibers.

Table 12.2 lists the most odor-potent compounds detected in brewed coffee headspace using GC/O-SPME, their OSVs (odor spectrum values) and the OSVs from previously published experiments (Semmelroch and Grosch 1996; Deibler et al. 1998). An OSV is an expression of odor potency normalized to the most potent odorant detected and transformed as the square root and allows for comparison of results between separate studies (Acree 1997). All the compounds have been previously found in brewed coffee, thus verifying the method is at least as good as the other methods in identifying the presence of odor-active volatiles. Since GC/O-SPME allows for the detection of compounds that would coelute with solvents, this method could replace solvent extraction for GC/O.

12.4 Conclusions

GC/O dilution analysis of the headspace above a food system can be achieved with SPME by varying the fiber thickness, exposure length, and number of fibers. Using the model mouth generates a headspace that corresponds to the retronasal breath of humans during eating. The combination of SPME, a mouth simulator, GC/O, and appropriate standardization is a powerful tool for food aroma analysis.

References

Acree TE (1997) GC/olfactometry GC with a sense of smell. Anal Chem 69:170A–175A

Acree TE, Barnard J, Cunningham DG (1984) A procedure for the sensory analysis of gas chromatographic effluents. Food Chem 14:273–286

Aurthur CL, Potter D, Bucholz K, Pawliszyn JB (1992) Solid phase microextraction for the direct analysis of water: theory and practice. LCGC 10(9):656–661

Chaintreau A, Grade A, Munoz-Box R (1995) Determination of partition coefficients and quantitation of headspace volatile compounds. Anal Chem 67:3300–3304

Deibler KD, Acree TE, Lavin EH (1998) Aroma analysis of two coffee brew methods by gas chromatography olfactometry. In: Contis ET, Mussinan CJ, Parliment TH, Shahidi F, Spanier AM (eds) Food flavors: formulation, analysis and packaging influences. Elsevier, Amsterdam, pp 69–78

Deibler KD, Lavin LH, Acree TE (1999) Solid phase microextraction for dilution analysis based gas chromatography/olfactometry. J Agric Food Chem 4:1616–1618

Deibler KD, Acree TE, Lavin EH, Taylor AJ, Linforth RST (2000) Flavor release measurements with retronasal aroma simulator. In: Rothe M (ed) Proc 6th Wartburg Aroma Symp, Eisenach

DeRoos KB, Wolswinkel K (1994) Non-equilibrium partition model for predicting flavour release in the mouth. In: Maarse H, van den Heij DG (eds) Trends in flavour research. Elsevier, Amsterdam

Elmore JS, Langley KR (1996) Novel vessel for the measurement of dynamic flavor release in real time from liquid foods. J Agric Food Chem 44:3560–3563

Grosch W (1993) Detection of potent odorants in foods by aroma extract dilution analysis. Trends Food Sci Technol 4:68–73

Lee WE III (1986) A suggested instrumental technique for studying dynamic flavor release from food products. J Food Sci 51:249–250

Pawliszyn J (1997) Solid phase microextraction: theory and practice. Wiley, New York

Roberts DD, Acree TE (1995) Simulation of retronasal aroma using a modified headspace technique: investigating the effects of saliva, temperature, shearing, and oil on flavor release. J Agric Food Chem 43:2179–2186

Roberts DD, Acree TE (1996a) Effects of heating and cream addition on fresh raspberry aroma using a retronasal aroma simulator and gas chromatography olfactometry. J Agric Food Chem 44:3919–3925

Roberts DD, Acree TE (1996b) Retronasal flavor release in oil and water model systems with an evaluation of volatility predictors. ACS Symp (Flavor-Food Interactions) Ser 633:179–187

Roberts DD, Lavin EH, Acree TE (1994) Simulation and analysis of retronasal aroma. In: Proc 4th Wartburg Aroma Symp, Eisenach

Semmelroch P, Grosch W (1996) Studies on character impact odorants of coffee brews. J Agric Food Chem 44:537–543

Springett MB, Rozier V, Bakker J (1999) Use of fiber interface direct mass spectrometry for the determination of volatile flavor release from model food systems. J Agric Food Chem 47(3): 1123–1131

vanRuth SM, Roozen JP, Cozijnsen JL (1994) Comparison of dynamic headspace mouth model systems for flavour release from rehydrated bell pepper cuttings. In: Maarse H, van den Heij DG (eds) Trends in flavour research. Elsevier, Amsterdam, pp 59–64

Zhang Z, Pawliszyn J (1993) Headspace solid-phase microextraction. Anal Chem 85:1843–1852

Zhang Z, Pawliszyn J (1995) Quantitative extraction using an internally cooled solid phase microextraction device. Anal Chem 67:34–43

13 RNA Gel Blot Analysis to Determine Gene Expression of Floral Scents

J. Boatright and N. Dudareva

13.1 Introduction

To attract pollinators to flowers, flowering plants use diverse biosynthetic pathways to produce a broad spectrum of low molecular weight (100- to 200-Da) floral volatile compounds (>700 described structures exist in 60 families of plants) (Knudsen et al. 1993; Dobson 1994). In most plant species floral volatiles are emitted primarily by the petals, although other parts of the flower may also contribute to the total scent output. Often, a specific floral component is emitted from several parts of the flower, although not necessarily at the same amount or rate. However, there are also examples of emission of specific compounds from only a subset of the floral organs.

The chemical composition of flower fragrances varies widely among species in terms of the number, identity, and relative amounts of constituent volatile compounds (Knudsen and Tollsten 1993; Knudsen et al. 1993). Closely related plant species, which rely on different insects for pollination, produce different odors. Characteristic floral odors are usually correlated with the type of pollinators. Species pollinated by bees and flies tend to have scents that are described as "sweet," whereas those pollinated by beetles have "musty," "spicy," or "fruity" odors (Dobson 1994).

Plant scents are dominated by monoterpenoid, sesquiterpenoid, phenylpropanoid, and benzenoid compounds. Fatty acid derivatives and a range of other chemicals, especially those including nitrogen or sulfur, are also sometimes floral bouquet constituents. All of these compounds are regularly defined as "secondary metabolites" (Knudsen and Tollsten 1993; Knudsen et al. 1993). Until recently, floral scent research remained largely descriptive and failed to identify specific enzymes or genes involved in the biosynthesis of floral volatiles. As a result, most of the biochemical pathways and the enzymes involved in the synthesis of these compounds have not yet been elucidated. Although several hundred compounds have been identified from various floral scents and the chemical structures of most are known today, only a few studies have focused on the biosynthesis of these compounds in plants. Many of the volatiles found in floral scent are also synthesized in vegeta-

Department of Horticulture and Landscape Architecture, Purdue University, West Lafayette, Indiana 47907, USA tive tissue under specific conditions (mostly for defense purposes) (McGarvey and Groteau 1995; Shulaev et al. 1997), and of these, some information concerning their biosynthesis is available (Bohlmann et al. 1998).

Terpenes, especially monoterpenes such as linalool, limonene, myrcene, and *trans*-β-ocimene, but also some sesquiterpenes, such as farnesene, nerolidol, and caryophyllene, are very common constituents of floral scent. In work done mostly with vegetative tissue, it was found that monoterpenes are synthesized in the plastidic compartment. In this cellular compartment, isopentenyl pyrophosphate (IPP) is derived from the mevalonate-independent "Rohmer" pathway (Lichtenthaler et al. 1997). IPP can be isomerized to dimethylallyl diphosphate (DMAPP), and one molecule of IPP is condensed with one molecule of DMAPP in a reaction catalyzed by the enzyme geranyl pyrophosphate synthase (GPPS) to form GPP, the universal precursors of all the monoterpenes. Similar work with vegetative tissue has revealed that in the cytosol, IPP is derived from the mevalonic acid pathway (McCaskill and Croteau 1995), and two molecules of IPP and one molecule of DMAPP are condensed in a reaction catalyzed by the enzyme farnesyl pyrophosphate synthase (FPPS) to form FPP, the universal precursors of all the sesquiterpenes (McGarvey and Groteau 1995).

In the last few years, genes encoding the enzymes responsible for the synthesis of many monoterpenes and sesquiterpenes have been identified and characterized (Bohlmann et al. 1998). However, to date only the enzyme that catalyzes the formation of the acyclic monoterpene linalool has been characterized in floral tissue (Pichersky et al. 1994, 1995).

The phenylpropanoids, derived from phenylalanine, constitute a large class of secondary metabolites in plants. Many are intermediates in the synthesis of structural cell components (e.g., lignin), pigments (e.g., anthocyanins), and defense compounds. These are not usually volatile. However, several phenylpropanoids whose carboxyl group at C9 is reduced (to either the aldehyde, alcohol, or alkane/alkene) and/or that contain alkyl additions to the hydroxyl groups of the benzyl ring or to the carboxyl group (i.e., ethers and esters) are volatiles. Recent work with *C. breweri* flowers has resulted in the identification and characterization of three enzymes that catalyze the formation of floral volatiles from this group: (iso) methyleugenol, benzylacetate, and methylsalicylate. The enzymes are, respectively, *S*-adenosyl-L-methionine:(iso)eugenol *O*-methyltransferase (IEMT), acetyl-CoA:benzylalcohol acetyltransferase (BEAT), and *S*-adenosyl-L-methionine: salicylic acid carboxyl methyltransferase (SAMT) (Wang et al. 1997; Dudareva et al. 1998a,b; Wang and Pichersky 1998; Ross et al. 1999). In addition, the enzyme *S*-adenosyl-L-methionine:benzoic acid carboxyl methyltransferase (BAMT), which catalyzes the formation of methylbenzoate, has been identified and characterized in snapdragon flowers (Dudareva et al. 2000; Murfitt et al. 2000). cDNAs encoding all these enzymes have also been characterized.

Although production of scent volatile compounds appears to be very widespread in the plant kingdom, our knowledge of their biosynthesis and regulation still remains quite limited and is based (to our knowledge) on the analysis of two model systems – *Clarkia* and snapdragon. In both *C. breweri* and snapdragon, flowers produce scent compounds de novo in the tissues from which they are emitted, and that their emission levels, corresponding enzyme activities, and mRNA levels are all spatially and temporally correlated. A positive correlation between amounts of emitted scent compounds, corresponding enzyme activity, and mRNA indicates that the levels of activity of enzymes involved in scent production, and, indirectly, scent emission, are regulated mainly by transcription of the corresponding genes at the site of emission (Dudareva et al. 1999). Expression of these genes is relatively uniform, being highest in petals, and is restricted to surfaces of floral tissues (epidermal cells) (Dudareva et al. 1996; Dudareva and Pichersky 2000). In addition, in snapdragon flowers we found that the earlier biochemical steps in the pathway are also involved in the regulation of volatile production and emission by controlling the supply of the substrate for the reaction (Dudareva et al. 2000). These results suggest that common regulatory mechanisms are involved in floral scent production in different plant species.

Molecular characterization of any gene usually includes a thorough analysis of the temporal and spatial distribution of RNA expression. A number of procedures are widely used to determine the abundance of a particular mRNA in a total or poly(A)$^+$ mRNA sample, including RNA blot analysis, ribonuclease protection assays, and reverse transcription polymerase chain reaction (RT-PCR). RNA gel blot analysis remains the standard procedure to detect the tissue and developmental patterns of a particular gene expression, if a probe for the gene of interest is available. It is also the easiest method of determining transcript size and the presence of alternatively spliced or multiple transcripts generated from a single locus. RNA gel blot analysis is the only technique that not only can determine the molecular weight (size) of mRNA of interest, but also measures steady-state levels of mRNA. Steady-state levels of mRNA transcripts reflect the combined effects of two parallel processes: the rate of transcription and the rate of degradation (stability) of the RNA in question. In addition, this technique permits the quantification of developmental changes in gene expression as well as the quantification of effects of various stimuli in the mature plant. Because RNA gel blot analysis measures steady-state levels of selected mRNAs, it allows comparison between different gene transcripts. RNA gel blot analysis has been used to study tissue specificity and developmental regulation of several floral scent genes, such as LIS (Dudareva et al. 1996), IEMT (Wang et al. 1997), BEAT (Dudareva et al. 1998a), and BAMT (Dudareva et al. 2000).

13.2 RNA Gel Blot Analysis

In RNA gel blot analysis, the RNA is first electrophoretically size-separated on an agarose gel under denaturing conditions ensuring that molecules migrate according to their size and are unaffected by secondary structures. The separated individual RNA species are then transferred to a membrane and immobilized. Specific mRNAs are then detected by hybridization to labeled DNA or RNA probes and the quantification of transcripts can be obtained by using a molecular imager system.

Hybridization is often carried out in formamide solution, which permits incubation at a relatively low temperature, reducing degradation of the membrane-bound RNA. The amount of cellular RNA needed to detect a specific mRNA species depends on its prevalence in the mRNA population. High abundant transcripts can be detected with 7–10 μg of total RNA. For rare transcripts, more total RNA (~30 μg) or poly(A)$^+$ mRNA of about 1–5 μg are used. Although RNA gel blot analysis is a very powerful and sensitive technique, it can not detect different tissues (type of cells) within an organ. In this chapter we describe the isolation of RNA from floral tissues, RNA gel electrophoresis, blotting of the electrophoretically separated RNA, preparation and labeling of a DNA probe, and hybridization.

13.2.1 RNA Isolation

The preparation of uncontaminated, intact population of full-length RNA is an absolute prerequisite for the accurate quantitative and qualitative analysis of gene expression. RNA molecules are very sensitive to degradation not only by ribonucleases (RNases), but also by certain physicochemical conditions such as high temperature and alkaline pH. RNases are ubiquitous and highly stable enzymes, which generally require no cofactors to function. They are widely found as contaminants in tissues, from which RNA is extracted, as well as on human skin and in different reagents. Therefore, only minute amounts of RNase in an RNA preparation are sufficient to destroy RNA and will create a real problem. To avoid contamination problems, the following precautions should be taken when working with RNA:

1. To prevent RNase contamination from fingers, wear double gloves (latex or vinyl) at all times when handling RNA and all solutions and equipment for RNA work. Change gloves frequently and keep tubes and solutions closed.
2. Keep separate stocks of chemicals and disposable plasticware for RNA work.
3. To eliminate RNases, glassware should be oven baked at 300 °C for at least 4 h.
4. All solutions should be treated with 0.1% (v/v) of ribonuclease inhibitor, diethylpyrocarbonate (DEPC) at 37 °C overnight followed by autoclaving to remove traces of DEPC. Tris solutions cannot be treated with DEPC and should be made from RNAse-free stocks (from unopened bottle of Tris) using DEPC-treated sterile water.
5. During the procedure, work quickly and keep everything on ice where possible to avoid degradation of RNA by endogenous or residual RNases.

Plant tissue is well known for being a difficult source for isolation of high-quality RNA with good yield. This difficulty is primarily due to the presence of naturally occurring polysaccharides and/or polyphenols that are released during cell disruption. These products typically bind to RNAs during extraction with the formation of either a viscous, insoluble jelly leading to large losses of nucleic acids, or of brownish adducts completely devoid of biological activity. The large number of publications of RNA isolation procedures reflects these difficulties and also demonstrates that the conditions required for successful isolation of RNA can differ significantly between species (Schneiderbauer et al. 1991; Lewinsohn et al. 1994; Schultz et al. 1994).

The study of the expression of genes regulated at the transcriptional level often requires RNA extraction from large number of samples. In addition, for genes involved in floral scent production very often a limited quantity of tissue is available and RNA has to be isolated from small quantities of starting material (stigma or style, for example). For these reasons, we recommend the use of commercially available RNeasy Plant Mini Kit (Qiagen Inc., Chatsworth, CA) which allows multiple, simultaneous processing of a wide variety of samples (maximum of 100 mg of plant starting material) in less than 30 min. In this procedure a specialized high-salt buffer allows up to 100 µg of RNA longer than 200 nucleotides to bind to a silica-gel-based membrane and contaminants are efficiently washed away. High-quality RNA is then eluted in 30–50 µl of water and stored at –80 °C. This procedure can be applied on both fresh and frozen plant tissues. In the latter case, tissue should be flash frozen immediately after harvesting in liquid nitrogen and stored at –80 °C. It is very important that tissue is pulverized in liquid nitrogen and is not allowed to thaw before adding a guanidinium isothiocyanate-containing buffer, which immediately inactivates RNases. It is also critical that frozen tissue is not allowed to thaw during handling such as weighing. We successfully used this kit for RNA isolation from different floral tissues (sepals, petals, stigma, style, and stamens) and leaves of *Clarkia breweri* and *Clarkia conncina*, which are rich in polysaccharides and phenolic compounds, and also from snapdragon, *Antirrhinum majus*, tobacco *Nicotiana suaveolens*, and *Petunia er Mitchell.*

Isolated RNA may be quantified by UV spectrophotometry at 260 and 280 nm. To quantify, dilute RNA by adding 5 µl of sample to 500 µl of water (1/100 dilution). In a pure RNA solution, which has an A_{260}:A_{280} ratio of 1.8 to 2.0, an A_{260} reading of 1.0 corresponds to an RNA concentration of 40 µg/ml. In general, deviations from the expected ratio may be caused by contaminants including DNA, phenol (when used in the procedure), polysaccharides, and proteins. In our RNA samples an A_{260}:A_{280} ratio was 1.8 to 2.0 and we were able to isolate depending on tissue 20 to 90 µg of RNA from 100 mg of starting material. The RNA preparations were free of DNA, a common problem with many other protocols, and there was no apparent degradation of RNA as judged by the clarity and intactness of ribosomal RNA bands (Figs. 13.1 and 13.2).

13.2.2 RNA Fractionation by Agarose-6 M Urea Gel Electrophoresis

Once total RNA has been isolated, individual transcripts are separated by agarose gel electrophoresis under denaturing conditions. Several denaturants can be used, including formaldehyde (Lehrach et al. 1977), glyoxal and dimethyl sulfoxide (DMSO) (McMaster and Carmichael 1977), urea (Locker 1979), or highly toxic methylmercuric hydroxide (Bailey and Davidson 1976). Most RNAs, being single-stranded, are able to form secondary structures by intramolecular base pairing and

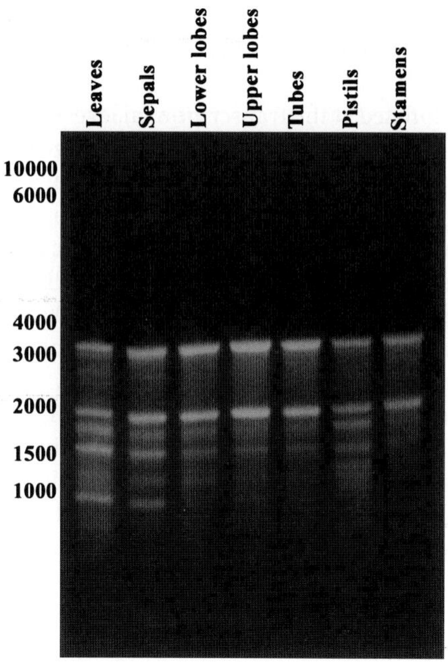

Fig. 13.1. An example of electrophoretic analysis of RNA isolated from young leaves, sepals, pistil, stamens, upper and lower petal lobes, and tubes of 6-day-old snapdragon flower. Seven micrograms of total RNA was electrophoresed on a 1.3% agarose-6 M urea gel and stained with ethidium bromide. *Numbers* on the *left* show positions of RNA molecular weight markers run in an adjacent lane

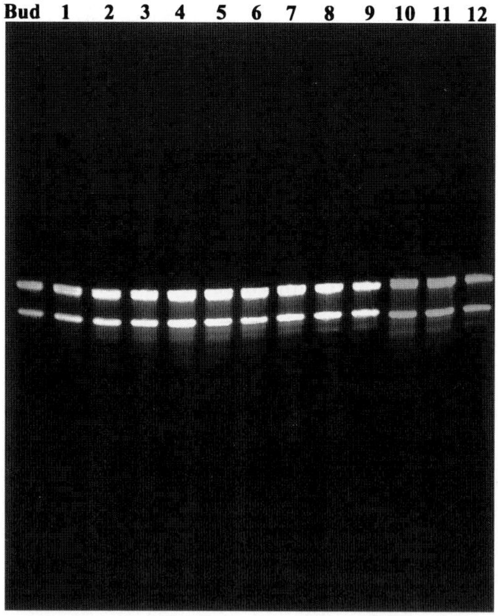

Fig. 13.2. An example of electrophoretic analysis of RNA from upper and lower petal lobes of snapdragon flower at different stages of development, from mature flower buds to 12 days after anthesis. Three micrograms of total RNA was electrophoresed on a 1.3% agarose-6 M urea gel and stained with ethidium bromide

must therefore be electrophoresed under denaturing conditions. Denaturation of the RNA is important not only for accurate assessment of RNA size, but also for efficient transfer of RNA from agarose gels. Denaturation is achieved by adding either formaldehyde or urea to the gel and loading buffers or by treating the RNA with glyoxal and DMSO prior to loading. Comparisons of the RNA gel electrophoresis under four different denaturing conditions have been done by Lehrach and coworkers (1977). We routinely run agarose-urea gels and describe this protocol here. Agarose-6 M urea is a medium that provides highly reproducible gel electrophoresis with high resolution of RNA (the addition of urea to agarose gels causes a marked increase in the sharpness of RNA bands; Figs. 13.1 and 13.2). In addition, RNA can be transferred from agarose-urea gels immediately, without pretreatment of the gels. This eliminates the diffusion of RNA bands that occurs during washes used to remove formaldehyde or methyl mercury (Alwine et al. 1977).

13.2.2.1 Preparation of Vertical Agarose-6 M Urea Gel

1. Set up the agarose gel apparatus. We recommend using a vertical gel electrophoresis unit from Sigma. It is recommended that an apparatus be dedicated exclusively to RNA use, in order to avoid contamination with RNases. Seal the bottom of the vertical gel with 4% acrylamide gel in 1 × Locker buffer.
2. For a 16.5 × 22 cm (w × h), 1.3% agarose-6 M urea gel, mix 1.95 g of agarose, 15 ml of 10 × Locker buffer (0.4 M Tris-HCl, pH 7.4, 0.36 M NaH_2PO_4, and 10 mM EDTA), and adjust volume to 150 ml with sterile RNase-free water.
3. Heat until all components are completely dissolved and mix well. Try to avoid boiling.
4 Pour the gel while hot and let the gel cool to room temperature. Place the gel at 4 °C and allow to solidify for several hours. We usually make a gel the day before electrophoresis and let it solidify overnight at 4 °C in a horizontal position.

13.2.2.2 Gel Electrophoresis

1. Carefully remove comb, place the gel in the apparatus and fill buffer tanks with 1 × Locker buffer. Rinse wells with running buffer before loading samples.
2. Standardize RNA samples by absorbance measurements at 260 nm before loading gel. In our experiments, 7 μg of total RNA was loaded per lane in blots to determine tissue-specific expression and 3 μg per lane in blots to determine variations in gene expression in petals or other floral organs over the life span of the flower. As an RNA molecular-weight marker we recommend using 5 μg of 0.2- to 10-kb RNA markers from Sigma. The markers are usually loaded into the outside lanes of the gel so that they can be cut from the gel after

electrophoresis and staining with ethidium bromide. Molecular-weight markers do not need to be run every experiment, since rRNA molecules present in total RNA appear as sharp bands that can be used as internal markers. Denature the RNA samples and RNA markers at 65 °C for 10 min and chill on ice.

3. Spin briefly to collect sample at the bottom of the tube, add an equal volume of gel-loading buffer, containing 10% (v/w) sucrose, 10 M urea, 5 mM EDTA, and 0.25% bromophenol blue to RNA samples, and immediately load into gel.
4. Run the gel at 5 V/cm for 3–5 h at 4 °C until the dye front reaches two-thirds of the gel length and stain with ethidium bromide (1 µl of EtBr, 10 mg/ml per 100 ml of water) for 20 min in an RNase-free glass dish, and photograph by ultraviolet illumination. Following electrophoresis and staining of total RNA samples, 28 S and 18 S ribosomal RNA (rRNA) bands should be clearly visible. In leaf tissue, bands of chloroplast rRNA may also be apparent (Figs. 13.1 and 13.2). Trim away any unused areas of the gel with a razor blade and cut off the bottom left-hand corner of the gel for orientation purposes.

13.2.3 Transfer RNA from Gel to Membrane

1. Place a piece of Plexiglas to form a support (platform) that is longer and wider than the gel inside a large baking dish or a tray filled with 10 × SSC (1.5 M NaCl and 0.15 M sodium citrate, pH 7.0). Cut two pieces of Whatman 3MM paper to the same width, but longer length, than the Plexiglas. Place them on the Plexiglas such that the ends reach into the 10 × SSC and wet them completely in 10 × SSC. Remove all air bubbles from the Plexiglas surface by rolling a plastic pipette back and forth.
2. Place the gel on Whatman 3MM paper and remove air bubbles between the gel and paper as before. Place four strips of Parafilm around the gel to prevent liquid flow directly from the reservoir to paper towels placed on the top of the gel rather than passing through it. This type of short-circuiting is a major reason for inefficient transfer of RNA from the gel to the filter.
3. Cut a piece of Hybond-N^+ membrane (Amersham) to the size of the gel and wet it by placing it on the surface of the water for 5 min.
4. Place the wetted membrane on the surface of the gel and remove air bubbles as before. Do not move the filter once it has been applied to the surface of the gel.
5. Wet two pieces of 3MM paper (cut to exactly the same size as the gel) in 10 × SSC and place them on the top of the membrane. Remove all air bubbles as before.
6. Place paper towels (unfolded and cut to the size of the gel) on the top of the 3MM paper to a height of about 6–8 cm. Put a glass plate and a weight of about 500 g on top of the paper stack.
7. Cover the apparatus with Saran wrap to prevent evaporation. Allow the transfer to proceed overnight.

8. After capillary transfer, mark the membrane using pencil, for orientation purposes, and remove it carefully. Rinse it in 5 × SSC briefly and allow it to air-dry. Fix RNA to the membrane by UV cross-linking: wrap the dry membrane in UV-transparent plastic wrap (polyvinylidene, Saran Wrap), place RNA-side-down on a UV transilluminator (254-nm wavelength), and irradiate for 2 min. The membrane must be completely dry before UV cross-linking. UV transilluminators must be repeatedly calibrated as the output of the lights changes with age. A UV light box (e.g., Stratagene Stratalinker) can be used instead of a UV transilluminator (follow manufacturer's instructions). UV cross-linking is recommended for a nylon membrane because it leads to covalent attachment and enables the membrane to be reprobed several times. However, irradiation may increase background hybridization and extensive cross-linking can severely reduce hybridization efficiency (Belin 1994). The optimal exposure time can be found in the manufacturer's support literature for the membrane. Membranes can be stored dry between sheets of Whatman 3MM filter paper for several months at room temperature.

13.2.4 Hybridization

Prehybridization and hybridization are performed in a commercial hybridization oven or a water bath at the appropriate temperature. Hybridization temperature is dependent on salt concentration, formamide concentration, GC-content, and degree of homology between probe and target RNA. Usually, temperature is in the range of 37–42 °C. In our experiments we used 50% formamide with a homologous probe and performed hybridization at 37 °C. RNA gel blot analysis is exceptionally flexible in the type of probe used during hybridization. High specific activity random-primed or nick-translated DNA probes, in vitro transcribed RNA probes, and oligonucleotide probes can be used successfully. Specific activity of the probe should be $>10^8$ dpm/μg. We usually use DNA fragments (500–1,500 bp in size) labeled by random priming (Multiprime DNA labeling system, Amersham Pharmacia Biotech, NJ). DNA fragments after restriction digestion and separation on agarose gel are extracted and purified from a gel by QIAquick gel extraction kit (Qiagen Inc., CA). Using this kit, DNA can be purified from agarose, proteins, salts, and ethidium bromide in less than 20 min. Following the labeling reaction, unincorporated radioactive nucleotides were removed by Sephadex G-50 column chromatography.

1. Hybridize in 5 × SSPE (1 × SSPE is 0.15 M NaCl, 10 mM sodium phosphate, pH 7.4, 0.1 mM EDTA), 50% formamide, 5 × Denhardt's solution (1 × Denhardt's solution is 0.02% Ficoll, 0.02% PVP, 0.02% BSA), and 0.5% SDS. Heat the hybridization solution to 37 °C.
2. Place the blot into the container or tube and pour the prewarmed prehybridization solution over the blot.

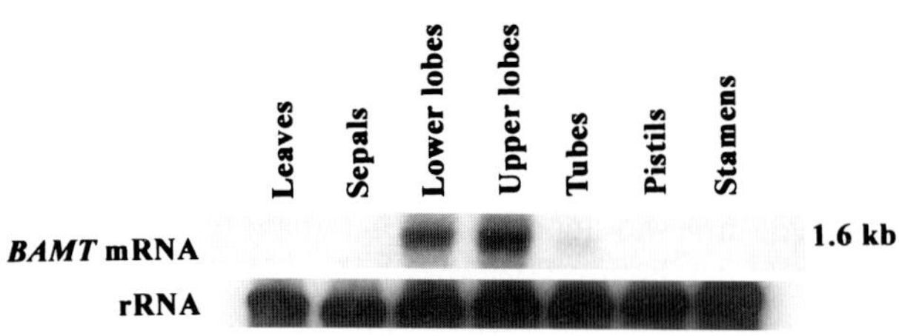

Fig. 13.3. Tissue specificity of *S*-adenosyl-L-methionine:benzoic acid carboxyl methyltransferase (BAMT) gene expression. Electrophoretically separated RNA from floral tissues and leaves shown on Fig. 13.1were blotted to Hybond-N+ and hybridized with a BAMT probe (*above*). Length of the BAMT mRNA was estimated at 1.6 kb by using RNA molecular markers in an adjacent lane. Autoradiography was carried out for 24 h. The blot was rehybridized with an 18 S rDNA probe (*below*) to standardize samples

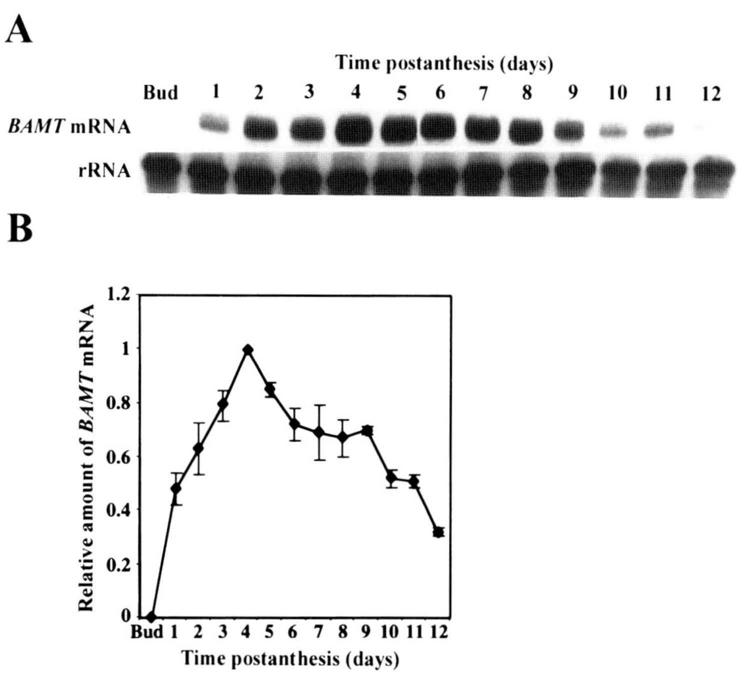

Fig. 13.4. Analysis of developmental changes in steady-state BAMT mRNA levels in upper and lower lobes of snapdragon petals. **A** Electrophoretically separated RNA from petals at different stages of flower development shown in Fig. 13.2 were blotted to Hybond-N+ and hybridized with a BAMT probe (*above*). Autoradiography was carried out for 24 h. The blots were rehybridized with an 18 S rDNA probe (*below*) to standardize samples. **B** Plot of variations in levels of BAMT mRNA in upper and lower petal lobes throughout the life span of the flower. Relative amounts of BAMT transcripts were obtained by scanning RNA blots with a Phosphor Imager (Storm 860 Phosphor Imager, Molecular Dynamics, Sunnyvale, CA). Each *point* is the average of five different experiments (including the one shown in A), and values were corrected by standardizing for 18 S rRNA levels. Standard error values are indicated by *vertical bars*

3. After 1–2 h discard the prehybridization solution and add fresh hybridization solution with denatured (5 min at 100 °C) probe in it. Hybridize overnight at 37°C.
4. After overnight hybridization, wash the blot once with 5 × SSPE, 0.5% SDS at 37 °C for 30 min (low stringency), and twice with 2 × SSPE at 65 °C for 30 min (high stringency). Carry out the washes under gentle shaking.
5. Air-dry the membrane and expose it to Kodak X-OMAT X-ray film at −80 °C with an intensifying screen. Exposure time can be varied according to the abundance of the transcript to be analyzed and, to a lesser degree, the specific activity of the labeled probe.
6. For normalization of the specific transcript among samples, the internal control, which does not change in response to experimental treatment, should be analyzed by subsequent probing of the same blot. Housekeeping genes, such as α-tubulin, actin, or ribosomal rRNA are common examples used for this purpose. In our experiments blots were rehybridized with the 18 S rDNA as a hybridization probe.

Typical data from RNA gel blot analysis are shown in Figs. 13.3 and 13.4A. For quantitative analysis, the Phosphor Imager analyzer, available now from different companies (e.g., Molecular Dynamics, Bio-Rad, or Fuji), can be used for direct counting of the specific radioactive bands. We used both the Molecular Dynamics and the molecular imager system model GS-363 from Bio-Rad at different times (Dudareva et al. 1996, 1998a, 2000; Wang et al. 1997). In our experiments, each band was quantified with a molecular imager and normalized using an internal rRNA control transcript, to calculate the relative amount of mRNA of interest in each sample (Fig. 13.4B). In addition, all gels contained a standard RNA sample for equalizing signals among gels (Dudareva et al. 1996).

Acknowledgments. This work is supported by National Science Foundation grant IBN-9904910 and by grants from the Fred C. Gloeckner Foundation, Inc.

References

Alwine JC, Kemp DJ, Stark GR (1977) Method for detection of specific RNAs in agarose gels by transfer to diazobenzyloxymethyl-paper and hybridization with DNA probes. Proc Natl Acad Sci USA 74:5350–5354

Bailey JM, Davidson N (1976) Methylmercury as a reversible denaturing agent for agarose gel electrophoresis. Anal Biochem 70:75–85

Belin D (1994) The use of riboprobes for the analysis of gene expression. In: Harwood AJ (ed) Protocols for gene analysis. Methods in molecular biology, vol 31. Humana Press, Totowa, pp 257–272

Bohlmann J, Meyer-Gauen G, Croteau R (1998) Plant terpenoid synthases: molecular biology and phylogenetic analysis. Proc Natl Acad Sci USA 95:4126–4133

Dobson HEM (1994) Floral volatiles in insect biology. In: Bernays E (ed) Insect–plant interactions, vol 5. CRC Press, Boca Raton, pp 47–81

Dudareva N, Pichersky E (2000) Biochemical and molecular genetic aspects of floral scents. Plant Physiol 122:627–633

Dudareva N, Cseke L, Blanc VM, Pichersky E (1996) Evolution of floral scent in *Clarkia*: novel patterns of S-linalool synthase gene expression in the *C. breweri* flower. Plant Cell 8:1137–1148

Dudareva N, D'Auria JC, Nam KH, Raguso RA, Pichersky E (1998a). Acetyl CoA: benzylalcohol acetyltransferase – an enzyme involved in floral scent production in *Clarkia breweri*. Plant J 14:297–304

Dudareva N, Raguso RA, Wang J, Ross JR, Pichersky E (1998b) Floral scent production in *Clarkia breweri*. III. Enzymatic synthesis and emission of benzenoid esters. Plant Physiol 116:599–604

Dudareva N, Piechulla B, Pichersky E (1999) Biogenesis of floral scent. Hortic Rev 24:31–54

Dudareva N, Murfitt LM, Mann CJ, Gorenstein N, Kolosova N, Kish CM, Bonham C, Wood K (2000) Developmental regulation of methyl benzoate biosynthesis and emission in snapdragon flowers. Plant Cell 12:949–961

Knudsen JT, Tollsten L (1993) Trends in floral scent chemistry in pollination syndromes: floral scent composition in moth-pollinated taxa. Bot J Linn Soc 113:263–284

Knudsen JT, Tollsten L, Bergstrom G (1993) Floral scents – a checklist of volatile compounds isolated by head-space techniques. Phytochemistry 33:253–280

Lehrach H, Diamond D, Wozney JM, Boedtker H (1977) RNA molecular weight determinations by gel electrophoresis under denaturing conditions, a critical reexamination. Biochemistry 16:4743–4751

Lewinsohn E, Steele CL, Croteau R (1994) Simple isolation of functional RNA from woody stems of gymnosperms. Plant Mol Biol Rep 12:20–25

Lichtenthaler HK, Rohmer M, Schwender J (1997) Two independent biochemical pathways for isopentenyldiphosphate and isoprenoid biosynthesis in higher plants. Physiol Plant 101:643–652

Locker J (1979) Analytical and preparative electrophoresis of RNA in agarose-urea. Anal Biochem 98:358–367

McCaskill D, Croteau R (1995) Isoprenoid synthesis in peppermint (*Mentha* × *piperita*): development of a model system for measuring flux of intermediates through the mevalonic acid pathway in plants. Biochem Soc Trans 23:290S

McGarvey DJ, Croteau R (1995) Terpenoid metabolism. Plant Cell 7:1015–1026

McMaster GK, Carmichael GG (1977) Analysis of single- and double-stranded nucleic acids on polyacrylamide and agarose gels by using glyoxal and acridine orange. Proc Natl Acad Sci USA 74:4835–4838

Murfitt LM, Kolosova N, Mann CJ, Dudareva N (2000) Purification and characterization of S-adenosyl-L-methionine:benzoic acid carboxyl methyltransferase, the enzyme responsible for biosynthesis of the volatile ester methyl benzoate in flowers of *Antirrhinum majus*. Arch Biochem Biophys 382:145–151

Pichersky E, Raguso RA, Lewinsohn E, Croteau R (1994) Floral scent production in *Clarkia* (Onagraceae). I. Localization and developmental modulation of monoterpene emission and linalool synthase activity. Plant Physiol 106:1533–1540

Pichersky E, Lewinsohn E, Croteau R (1995) Purification and characterization of S-linalool synthase, an enzyme involved in the production of floral scent in *Clarkia breweri*. Arch Biochem Biophys 316:803–807

Ross JR, Nam KH, D'Auria JC, Pichersky E (1999) *S*-adenosyl-L-methionine: salicylic acid carboxyl methyltransferase, an enzyme involved in floral scent production and plant defense, represents a new class of plant methyltransferases. Arch Biochem Biophys 367:9–16

Schneiderbauer A, Sandermann H Jr, Ernst D (1991) Isolation of functional RNA from plant tissues rich in phenolic compounds. Anal Biochem 197:91–95

Schultz D, Craig R, Cox-Foster DL, Mumma RO, Medford JI (1994) RNA isolation from recalcitrant plant tissue. Plant Mol Biol Rep 12:310–316

Shulaev V, Silverman P, Raskin I (1997) Airborne signalling by methyl salicylate in plant pathogen resistance. Nature 385:718–721

Wang J, Pichersky E (1998) Characterization of *S*-adenosyl-L-methionine:(iso)eugenol *O*-methyltransferase involved in floral scent production in *Clarkia breweri*. Arch Biochem Biophys 349:153–160

Wang J, Dudareva N, Bhakta S, Raguso RA, Pichersky E (1997) Floral scent production in *Clarkia breweri* (Onagraceae). II. Localization and developmental modulation of the enzyme *S*-adenosyl-L-methionine:(iso)eugenol *O*-methyltransferase and phenylpropanoid emission. Plant Physiol 114:213–221

Index